T0139864

Machine Learning: Foundations, Methodologies, and Applications

Series Editors

Kay Chen Tan, Department of Computing, Hong Kong Polytechnic University, Hong Kong, China

Dacheng Tao, University of Technology, Sydney, Australia

Books published in this series focus on the theory and computational foundations, advanced methodologies and practical applications of machine learning, ideally combining mathematically rigorous treatments of a contemporary topics in machine learning with specific illustrations in relevant algorithm designs and demonstrations in real-world applications. The intended readership includes research students and researchers in computer science, computer engineering, electrical engineering, data science, and related areas seeking a convenient medium to track the progresses made in the foundations, methodologies, and applications of machine learning.

Topics considered include all areas of machine learning, including but not limited to:

- Decision tree
- Artificial neural networks
- Kernel learning
- Bayesian learning
- Ensemble methods
- Dimension reduction and metric learning
- Reinforcement learning
- Meta learning and learning to learn
- Imitation learning
- Computational learning theory
- Probabilistic graphical models
- Transfer learning
- Multi-view and multi-task learning
- Graph neural networks
- Generative adversarial networks
- Federated learning

This series includes monographs, introductory and advanced textbooks, and state-of-the-art collections. Furthermore, it supports Open Access publication mode.

More information about this series at https://www.springer.com/series/16715

Fangfang Zhang · Su Nguyen · Yi Mei ·
Mengjie Zhang

Genetic Programming for Production Scheduling

An Evolutionary Learning Approach

 Springer

Fangfang Zhang
School of Engineering and Computer
Science
Victoria University of Wellington
Wellington, New Zealand

Yi Mei
School of Engineering and Computer
Science
Victoria University of Wellington
Wellington, New Zealand

Su Nguyen
La Trobe Business School
La Trobe University
Bundoora, VIC, Australia

Mengjie Zhang
School of Engineering and Computer
Science
Victoria University of Wellington
Wellington, New Zealand

ISSN 2730-9908   ISSN 2730-9916 (electronic)
Machine Learning: Foundations, Methodologies, and Applications
ISBN 978-981-16-4861-8   ISBN 978-981-16-4859-5 (eBook)
https://doi.org/10.1007/978-981-16-4859-5

This Springer imprint is published by the registered company Springer Nature Singapore Pte Ltd.
The registered company address is: 152 Beach Road, #21-01/04 Gateway East, Singapore 189721,
Singapore

Foreword

Scheduling, i.e., the assignment of resources to tasks and their sequencing, is an important challenge in many areas, including manufacturing, health care, construction, and even when scheduling processes within a computer. Given its wide ranging importance, it is not surprising that scheduling is one of the oldest and most researched topics in Operational Research. Yet despite the progress made over the past decades, it remains a challenging and interesting research area.

Since many scheduling problems are NP-hard and thus probably never solvable to optimality in polynomial time, many people have turned to meta-heuristics like evolutionary algorithms to tackle the hardest problems.

The book **Genetic Programming for Production Scheduling—An Evolutionary Learning Approach** looks at scheduling from the perspective of hyper-heuristics: Rather than searching directly for the best *solution* to a particular problem instance, hyper-heuristics search for a *heuristic* that can then be applied to construct solutions to many different problem instances.

Construction heuristics have been known in scheduling for a long time, most notably in the form of dispatching rules. Whenever multiple tasks compete for the same resource, the dispatching rule decides which task is to be prioritised. Examples include simple rules such as "first in first out" or "shortest processing time first", but the scientific literature is full of sometimes quite complex dispatching rules.

Because dispatching rules are based on just partial (usually local) information about the problem instance, they are very quick to compute and can quickly construct solutions even to very large problems. Furthermore, they are easy to understand and to implement, and thus are popular in practice.

What is new in the approach followed in this book is that these dispatching rules and construction heuristics are to be generated *automatically* via Genetic Programming, a variant of evolutionary algorithms that is able to generate arbitrarily complex functions, rather than optimising a fixed set of prespecified parameters.

Being able to generate dispatching rules automatically means it now becomes possible to generate rules that are specifically designed for a particular problem class—for example, for a specific configuration of a shop floor, with a specific product mix. As has been shown in numerous studies in recent years, this is an extremely powerful idea. Automatically generated rules are frequently shown to outperform manually designed rules from the literature by a large margin.

The biggest benefit, however, probably comes when applying this idea to scheduling problems that are dynamic and stochastic—as is common in practice, with new jobs arriving over time, machines breaking down, or stochastic processing times. In such environments, perhaps the only disadvantage of dispatching rules, namely that they cannot guarantee to find the optimal solution, becomes less relevant. When the problem is changing over time, even seemingly optimal decisions now may turn out to be poor choices later, once the problem has changed and new information is revealed. In fact, there is growing evidence that greedy heuristics perform better in dynamic stochastic environments than using a rolling horizon approach with each stage solved to optimality. More importantly, dispatching rules naturally cope with the dynamic and stochastic events by postponing decisions to the very last minute, namely when several tasks compete for the same resource, thus always taking into account the latest information. And because they can be computed so quickly, they work even in very volatile environments.

Without any doubt, the idea of rule-based scheduling with rules automatically generated via Genetic Programming is extremely powerful, especially in dynamic and stochastic environments. It also fits very well to the latest Industry 4.0 movement based on an Internet of Things and decentralised decision-making.

The book **Genetic Programming for Production Scheduling—An Evolutionary Learning Approach** is the first comprehensive book on the topic, written by scientists at the forefront of research. It covers the basics, as well as advanced topics such as the use of surrogates, the handling of multiple objectives, or the latest trend, multitask approaches. It also devotes four chapters to dynamic production scheduling problems.

Overall, it is an invaluable resource for everyone interested in this area.

Prof. Juergen Branke
Professor of Operational Research &
Systems
Warwick Business School, The
University of Warwick
Coventry, UK

Preface

Production scheduling is an important optimisation problem that reflects the practical and challenging issues in real-world scheduling applications such as order picking in warehouses, the manufacturing industry, and grid/cloud computing. Most real-world production scheduling problems are dynamic, with new jobs arriving over time or stochastic processing times. The traditional optimisation algorithms cannot handle such dynamic problems effectively and efficiently. Scheduling heuristics have been successfully used to handle dynamic scheduling heuristics, e.g., routing rule to allocate an operation to one of its candidate machines, and sequencing rule to decide which operation to be processed next when a machine is idle. In recent years, hyper-heuristic approaches such as genetic programming have been widely used to learn scheduling heuristics for dynamic scheduling.

This book introduces an evolutionary learning approach, especially genetic programming for production scheduling. This book is divided into six parts with 16 chapters. Part I provides the background knowledge of this book including production scheduling, machine learning, and genetic programming. This lays the foundation for readers to read the following chapters. Part II proposes genetic programming-based algorithms for learning schedule construction heuristics, schedule improvement heuristics, and augment operations research in static production scheduling problems. Part III develops genetic programming algorithms with different machine learning techniques, i.e., surrogate, feature selection, and specialised genetic operators, for dynamic production scheduling problems. Part IV studies genetic programming in multi-objective production scheduling with a single decision or multiple decisions. Multiple decisions are handled with a cooperative coevolutionary strategy or a multi-tree representation in genetic programming. Part V develops multitask genetic programming algorithms for production scheduling problems including multitask genetic programming in the hyper-heuristic domain, adaptive multitask genetic programming and surrogate-assisted multitask genetic programming. In Part VI, this book gives the conclusions and prospectus about the studies in this book.

This book benefits scientists, engineers, researchers, postgraduates, and undergraduates in the areas of machine learning, artificial intelligence, evolutionary computation, operations research, and industrial engineering. This book is expected to promote the research of evolutionary learning in various production scheduling problems by providing a systematic review of state-of-the-art methods.

Wellington, New Zealand   Fangfang Zhang
Melbourne, Australia   Su Nguyen
Wellington, New Zealand   Yi Mei
Wellington, New Zealand   Mengjie Zhang
June 2021

Acknowledgements

We would like to express our appreciation to our colleagues and friends who encouraged and motivated us to write this book. We would like to thank Prof. Kay Chen Tan at The Hong Kong Polytechnic University for his advice and support throughout the writing of the book.

We are grateful to our friends and colleagues from the Evolutionary Computation Research Group (ECRG) at Victoria University of Wellington (VUW), New Zealand. Thanks go to the discussions and knowledge sharing, which helped and supported us to finish this book. We would like to thank the School of Engineering and Computer Science and Victoria University of Wellington for the resources provided to complete this book.

This work is supported mainly by the Marsden Fund of New Zealand Government under Contract VUW1509 and Contract VUW1614; in part by the Science for Technological Innovation Challenge (SfTI) fund under Grant E3603/2903; and in part by the MBIE SSIF Fund under Contract RTVU1914. The work of Fangfang Zhang is also supported by the China Scholarship Council/Victoria University Scholarship.

Last but not least, we wish to thank our families for their great support and understanding. Thanks go to our friends for their friendship and support. You all have always been the source of love that helps us complete this book.

Contents

Part VI   Conclusions and Prospects

Acronyms

ATC	Apparent Tardiness Cost
CCGP	Genetic Programming with Cooperative Coevolution
CDRs	Construct Composite Dispatching Rules
CP	Constraint Programming
CR	Critical Ratio
DDARs	Due-date Assignment Rules
DFJSS	Dynamic Flexible Job Shop Scheduling
DJSS	Dynamic Job Shop Scheduling
DMOCC	Diversified Multi-objective Cooperative Coevolution
DRs	Dispatching Rules
EMO	Evolutionary Multi-objective Optimisation
FIFO	First In First Out
GCP	Genetic-Based Constraint Programming
GD	Generational Distance
GP	Genetic Programming
HV	Hypervolume
HVR	Hypervolume Ratio
IDR	Iterative Dispatching Rule
IGD	Inverted Generational Distance
IMR	Iterative Multiple Regression
JSS	Job Shop Scheduling
KNN	K-Nearest Neighbour
LB	Lower Bound
LSRW	Large-Step Random Walk
MFEA	Multifactorial Evolutionary Algorithm
MO-DJSS	Multi-objective Dynamic Job Shop Scheduling
MOGP	Multi-objective Genetic Programming
MTGP	Genetic Programming with Multi-tree Representation
ND	Non-dominated
NDLH	Number of Dominating Learned Heuristics
NSGA-II	Non-dominated Sorting Genetic Algorithm II

OBJW	Objective-wise
PC	Phenotypic Characterisation
REPW	Replication-Wise
SB	Shifting Bottleneck
SFJSS	Static Flexible Job Shop Scheduling
SPEA2	Strength Pareto Evolutionary Algorithm 2
SPT	Shortest Processing Time
VNS	Variable Neighbourhood Search

List of Figures

List of Tables

Part I
Introduction

Chapter 1
Introduction

1.1 Production Scheduling

Scheduling [81, 187] is a decision-making process to optimise the use of the resources, which is commonly used in many applications, especially in the manufacturing and services industries [137, 186]. The resources can be machines in a job shop, gates at an airport, the staff at the hospital, and so on. In other words, scheduling is the activity of planning the starting times of particular tasks. The scheduling objectives can be different, such as minimising the total cost to get more benefits or minimising the total operating time to deliver a particular service to customers on time. Production scheduling is the allocation of available production resources (i.e., machines) to perform a number of tasks (i.e., jobs) over time to meet the desired performance criteria such as customer satisfaction and production efficiency [81]. Due to the limited production resources, jobs usually have to wait in the shop queue significantly longer than their actual processing times. Production scheduling has been one of the most popular research topics in operations research, management science, and artificial intelligence for production efficiency improvement.

Production scheduling has been widely used in real-world applications such as order picking in a warehouse [99], the manufacturing industry [74, 212], and grid/cloud computing [157]. A good production schedule can help improve the economic benefits for a production enterprise by improving the production delivery speed to customers, and customer loyalty [246]. An overview of production planning and control activities in a production enterprise is shown in Fig. 1.1. For a production enterprise, the work is to purchase materials from suppliers to produce products for customers. The goal is to maximise the profit from the difference between the selling price and the cost. Figure 1.1 shows that production scheduling is the key in executing the orders from customers along with the supports from demand planning, production planning, and material requirements planning. Production scheduling also has a big impact on different kinds of planning. For example, efficient production schedul-

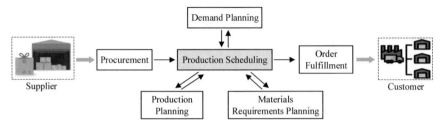

Fig. 1.1 An overview of the production planning and control activities in a production enterprise

ing can support material requirements planning, since the products can be produced quickly. In this book, we focus on optimising production scheduling to improve production efficiency.

There are three common models of production scheduling [24], i.e., flow shop, open shop, and job shop. A job in *flow shop scheduling* consists of an ordered list of operations. The number of operations of each job equals the number of machines. All operations have the same processing order through the machines. This means that the processing order for each operation in flow shop scheduling is the same, i.e., the ith operation of the job must be executed on the ith machine. Flow shop scheduling can occur in a manufacturing site with 100% standardisation operated in an assembly production line. *Open shop scheduling* is the same as flow shop scheduling except that the order of processing operations comprising one job may be arbitrary. This means that there are no ordering constraints on operations. A job in *job shop scheduling* (JSS) is composed of an ordered list of operations. Each job can have different processing orders through the machines. A job shop is complex because of the different production processes for jobs. JSS is an important combinatorial optimisation problem, and it is an NP-hard problem. A job shop can occur in a business with 100% customisation with a typical batch size of 1, which implies that every finished product is unique. JSS provides a flexible form of manufacturing to reflect real-world applications, and thus is the focus of this book.

In JSS, a job shop contains a number of jobs that need to be processed by a set of machines. The goal of JSS is to optimise the machine resources to achieve production efficiency, such as minimising the makespan [130] to reduce the total processing time, and tardiness [185] to reduce the production delays. Depending on whether the information of jobs is known in advance or not, JSS can be classified as *static (classical) JSS* and *dynamic JSS* [163]. Depending on whether a job can be processed on more than one machine, JSS can be categorised into *flexible JSS* and *non-flexible JSS* [236].

Static Job Shop Scheduling: In the typical version of JSS [141], a number of jobs need to be processed on a set of machines. Each job has a sequence of operations. The operations of a job need to be executed in a predefined order, and each operation can *only* be processed at a specified machine. In addition, *information about jobs is available* when making a schedule. The flowtime of a job is the duration from the job arrival at the shop floor to its completion time. An example of the flowtime of

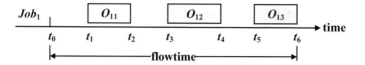

Fig. 1.2 An example of the flowtime of a job (Job_1)

Job_1 with three operations (i.e., O_{11}, O_{12}, and O_{13}) is shown in Fig. 1.2. We can see that Job_1 arrives at time t_0, and the starting processing time of its first operation O_{11} is t_1. The last operation O_{13} of Job_1 is finished at time t_6. Therefore, the flowtime of Job_1 is $t_6 - t_0$.

Dynamic Job Shop Scheduling: Static JSS implies that the information of jobs is known when we make a schedule for the production in a job shop. However, in practice, the *job shop environments is usually dynamic*, e.g., new jobs can arrive in real time [67, 234, 237], and the machine may break down unexpectedly [174, 224]. Dynamic JSS (DJSS) [152, 163] is used for considering the scheduling situations with dynamic events. This book considers the dynamic job arrivals, since it is the most common dynamic factor in real-world scheduling applications [94]. For example, it is common for a company to receive orders from customers over time, and it is not easy to predict the information of an order before it arrives. This implies that in DJSS, information about jobs is unknown until they arrive at the job shop floor.

Flexible Job Shop Scheduling: Flexible JSS [29, 235] is an extension of JSS in which *each operation can be processed by a set of candidate machines* rather than a specific machine. Each operation can be processed on one of its candidate machines, and its processing time depends on the machine that processes it. Two decisions need to be made simultaneously in flexible JSS. One is *machine assignment* (i.e., routing decision) for allocating a ready operation to a machine, and the other is *operation sequencing* (i.e., sequencing decision) for choosing an operation to be processed next when a machine is idle and its queue is not empty. Given a number of jobs and a set of machines, flexible JSS aims to determine which machine to process each job and how the jobs are sequenced in their allocated machines.

Dynamic Flexible Job Shop Scheduling: Dynamic flexible JSS (DFJSS) [239] needs to do machine assignment and operation sequencing simultaneously under a dynamic environment with unpredicted events such as dynamic job arrivals [244] and machine breakdown unexpectedly [224]. DFJSS is more challenging than static JSS, dynamic JSS, and flexible JSS, since DFJSS involves more than one decision and dynamic environments compared with other variations of JSS.

The details about the availability of job information and the required decisions of different types of JSS are shown in Table 1.1. Table 1.1 shows that DFJSS is the most complex one among the four types of JSS with routing and sequencing decisions, and unknown job information.

In general, DJSS and flexible JSS are variants of the classical JSS. Furthermore, DFJSS is a combination of DJSS and flexible JSS. The studies of production schedul-

Table 1.1 The availability of job information and decision requirements in different types of job shop scheduling

	Problem	Static JSS	Dynamic JSS	Flexible JSS	Dynamic flexible JSS
Job information	Known	✓		✓	
	Unknown		✓		✓
Decision	Routing			✓	✓
	Sequencing	✓	✓	✓	✓

Fig. 1.3 The types of investigated job shop scheduling problems in this book

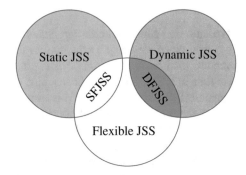

ing problems in this book can be classified into two main streams, which can be found in Fig. 1.3. The first one focuses on static JSS problems (i.e., highlighted in blue) where information of all jobs is available. The second one is DJSS problems (i.e., highlighted in orange with a special case as DFJSS which is highlighted in purple) where information of jobs is not available before their arrival. It is noted that SFJSS in Fig. 1.3 indicates static flexible JSS, which is not the focus of this book.

1.2 Machine Learning

Machine learning [149] aims at learning generalisable models from available data to improve the performance of the investigated tasks. Machine learning is a hot topic of artificial intelligence [199], which focuses on the learning aspect of artificial intelligence, developing algorithms that best represent a set of data. **Generalisation** is a term used to describe a model's performance on unseen test data, which is central to the success of a trained model. A good model is expected to have a good generalisation ability.

1.2.1 Training Set and Test Set

In machine learning, we try to build a model to predict the output of the unseen data. To measure the generalisation of the algorithms, the data are normally divided into two subsets, i.e., **training set** and **test set**. The training set is a subset of the data to derive a model. The test data is unseen to the trained models, and is used to measure the performance of the trained models. Normally, data are randomly split into training and test data with a split percentage such as 0.7 (i.e., 70% data are training data and 30% data are test data) [60].

In static JSS, we classify the data as **training data** and **test data**. Simulation [57] is a promising technique to investigate the complex real-world problems such as health care [122] and manufacturing [165]. In DJSS including DFJSS, the studies are mainly based on simulation [18] (i.e., the data are from the simulation), since we need to mimic the dynamic environments for evaluating the quality of schedules. Different simulations in DJSS are designed by assigning different seeds for the simulations, and the training simulations are distinguished from the test simulations. In DJSS, we classify the simulation data as **training instances** and **test instances**.

1.2.2 Types of Machine Learning Tasks

Machine learning contains a set of methods to learn models from data. According to the characteristics of available data for machine learning techniques to learn from, there are four main types of learning tasks in machine learning [9]: (1) supervised learning, (2) unsupervised learning, (3) semi-supervised learning, and (4) reinforcement learning.

Supervised learning is the machine learning task of learning a mapping function from the inputs to the outputs. In supervised learning problems, the target outputs of the available data for a problem are known (i.e., labelled data) for machine learning techniques to learning from. The most commonly investigated supervised learning tasks are regression [69] and classification [111].

Unsupervised learning is the machine learning task of learning to discover patterns where the information of the data cannot supervise the model. In unsupervised learning problems, the target outputs for the available data are unknown (i.e., unlabelled data), and the models need to discover information by themselves. The commonly investigated unsupervised learning tasks are anomaly detection [38] and clustering [126, 226].

Semi-supervised learning is a machine learning task that consists of a small amount of labelled data and a large amount of unlabelled data [253]. Semi-supervised learning is between supervised and unsupervised learning and is particularly useful for datasets that contain both labelled and unlabelled data. Semi-supervised learning is often used for medical images, where a physician might label a small subset of images and use them for training a model.

Reinforcement learning is the technique of training an algorithm for a specific task where no single answer is correct, but an overall outcome is desired [208]. In reinforcement learning problems (e.g., game), the learner interacts with the environment and receives an immediate reward for each action. Reinforcement learning differs from supervised learning in not needing labelled data instances to be presented, and the agent learns from interactions with the environment.

1.2.3 Machine Learning Paradigms

According to the ways to learn models/solutions, the paradigms in machine learning can be divided into five categories, i.e., case-based learning [2], inductive learning [95], analytic learning [148], connectionist learning [188], and genetic learning [252].

1.2.3.1 Case-Based Learning

Case-based learning, also known as instance-based learning and memory-based learning such as K-nearest neighbour [179], uses specific cases or experiences and relies on flexible matching methods to retrieve similar cases to predict the output of unlabelled cases. This approach is normally used for solving supervised learning tasks, and its performance depends on the case representations.

1.2.3.2 Inductive Learning

Inductive learning [144], also known as discovery learning, is a process where the learner discovers rules such as decision tree [167] by observing the examples. Inductive learning methods typically start with the specifics and are used to acquire general knowledge from examples. The goal of inductive learning is to induce a generalised rule from a set of observed instances.

1.2.3.3 Analytic Learning

Analytic learning systems [148] represent knowledge as rules in the logic form. A common technique is to represent knowledge as Horn clauses, then to phrase problems as theorems and to search for proofs. Analytic learning allows the learner to process information, break it into its component parts, and generate hypotheses using critical and logical thinking skills.

1.2.3.4 Connectionist Learning

Connectionist learning [188] is a method of modelling cognition as the interaction of neuro-like units. A popularly used connectionist learning is the artificial neural network [222] derived from the research of human brain behaviour. The network is generally constructed from nodes, links, weights, biases, and transfer functions. The network weights are automatically updated through network training by the learning algorithm, which can be selected or developed according to the architecture of the network.

1.2.3.5 Genetic/Evolutionary Learning

Genetic learning normally refers to evolutionary computation approaches [10], which is a kind of search algorithm based on the mechanism of natural selection and natural genetics. Genetic operators such as selection, crossover, and mutation are used to improve the quality of solutions generation by generation during the evolutionary process. Fixed-length vector-based representation is normally used to represent the candidate solutions, such as the representation of genetic algorithms [52] and particle swarm optimisation [108]. There are also other types of representations to present the individuals, such as variable-length tree-based representation in genetic programming [118].

In this book, we focus on the last machine learning paradigm, i.e., genetic/evolutionary learning, particularly genetic programming for production scheduling.

1.3 Evolutionary Learning and Genetic Programming

Evolutionary learning applies evolutionary computation to address optimisation problems in machine learning [252]. Evolutionary computation [10] is a computational intelligence technique inspired by natural evolution based on population. Evolutionary computation consists of a family of algorithms. The success of evolutionary computation relies on the improvement of individuals generation by generation. There are two main categories in EC, which are *evolutionary algorithms* such as genetic algorithms [52], genetic programming [190], evolution strategies [22], and evolutionary programming [72], and *swarm intelligence* such as particle swarm optimisation [109] and ant colony optimisation [64]. Evolutionary algorithms, especially genetic programming, are the focus in this book.

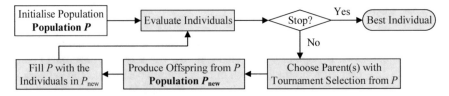

Fig. 1.4 The flowchart of a typical evolutionary computation algorithm

0.73	0.21	0.43	0.98	0.66

Fig. 1.5 An example of vector-based program

1.3.1 Evolutionary Computation

Figure 1.4 shows the flowchart of a typical evolutionary algorithm. An evolutionary algorithm starts by generating a population with many individuals that represent solutions to the problem. The initial population could be created randomly. The individuals are evaluated with a fitness function, and the output of the function shows how well the individuals solve the investigated problem. Then, genetic operators, such as crossover, mutation, and reproduction, are applied to the selected parents to generate new individuals. Individuals with higher quality have higher chances to be selected as the parent(s) to produce offspring. This process continues until the termination criterion (e.g., reaching a certain number of generations) is met. The best individual is selected as the output of the evolutionary algorithm.

Representation is an important characteristic to distinguish different evolutionary algorithms. One popular representation is the fixed-length vector-based (i.e., binary or real-valued) representation, which uses a vector to represent an individual. Typical vector-based evolutionary algorithms include genetic algorithms and particle swarm optimisation. An example individual with the vector-based representation is shown in Fig. 1.5.

1.3.2 Genetic Programming

Genetic programming (GP) [116] is one of the most popular evolutionary algorithms, which can automatically generate computer programs to solve problems. The main distinguishing characteristic of GP from other evolutionary algorithms is its variable-length tree-based representation. As GP is the focused learning technique in this book, we provide its description in detail. In the rest of this section, the main concepts of GP are presented to show how GP works.

Fig. 1.6 An example of
tree-based genetic
programming program

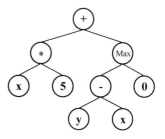

1.3.2.1 Representation

Representation [68] is a way of representing individuals in evolutionary algorithms. A typical GP uses the tree structure to represent individuals. Specifically, GP generates tree-based individuals based on a terminal set (for leaf nodes) and a function set (for non-leaf nodes). Figure 1.6 shows an example of a GP program $5x + \text{Max}(y - x, 0)$. In this program, the terminals consist of the variables $\{x, y\}$ and two constants $\{5, 0\}$, and the functions are composed of $\{\times, +, -, \text{Max}\}$, where $\times$ is indicated by $*$. The program is the combination of the components in the terminal set and the function set. Except for tree-based GP, there are also some other popular representations such as linear GP [17, 168], graph-based GP [189], grammar-guided GP [117], and Cartesian GP [145].

It is noted that the selection of the terminals and functions is critical for GP to succeed. The terminal set and the function set should be selected to satisfy the requirements of *sufficiency* and *closure* [190]. Sufficiency means that there must be some combinations of terminals and functions that can solve the problem, while closure means that any function can accept any input value returned by any function and terminal.

1.3.2.2 Initialisation

Initialisation [68] is the first step of GP to generate a population with a number of individuals randomly. For GP, *full* and *grow* [116] methods are commonly used to initialise population. A maximum depth can be determined for each GP individual to restrict the size of one program. For the full method, the terminals can only be sampled at the maximum depth of the trees. On the other hand, in the grow method, the terminals can be sampled at any position of the tree to early prune some branches. As a result, the full method always generates full trees at a given depth, while the grow method can generate unbalanced trees that are much smaller than the full trees. To improve the diversity of the initial population, these two methods are commonly combined to initialise the population, which is known as *ramped half-and-half*. Specifically, this hybrid method generates half of the population by the full method, and the other half by the grow method. Examples of a generated individual

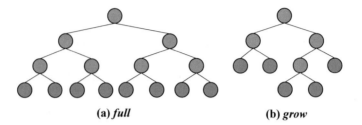

(a) *full*   (b) *grow*

Fig. 1.7 An example of genetic programming programs generated by *full* method and *grow* method

by the full method and the grow method are shown in Fig. 1.7a, b, where the functions
are marked in orange and the terminals are highlighted in grey.

1.3.2.3   Evaluation

Evaluation is an important step to measure the quality of the individuals in the GP
population according to the fitness function. The fitness function plays a significant
role in GP to guide the search to find good programs. Individuals with bad quality
are eliminated generation by generation. In general, the fitness function is defined
based on the objectives of the problem. For example, fitness can be defined as the
classification accuracy in the classification problem, and the total flowtime in the JSS
problem.

1.3.2.4   Selection

Selection is the stage for evolutionary algorithms to decide which individual to choose
from a population (i.e., parent) for later breeding. There are a number of methods
for selection such as roulette wheel selection [134] and tournament selection [114].
For roulette wheel selection, the probability of choosing an individual for breeding
offspring to the next generation is proportional to its fitness. The better the fitness
is, the higher the chance for that individual to be chosen. Tournament selection also
chooses parent(s) based on fitness. It randomly selects a number of individuals first,
and then selects the one with the best fitness as the parent. Tournament selection is
widely used in GP [92, 159].

1.3.2.5   Evolution

Evolution is the main process of GP to generate offspring for the next generation.
There are three genetic operators for GP, which are *crossover*, *mutation*, and *reproduction*. These operators aim at generating a new population by inheriting good
materials from the parent population.

- The crossover operator produces offspring by exchanging genetic materials between individuals. For example, two individuals are selected as the parents with the selection method. First, a subtree will be selected randomly from each parent. Then, these two subtrees from the parents are swapped to produce two new individuals, and they will be put into the new population.
- The mutation operator produces offspring by replacing partial genetic material with newly generated genetic material. One individual is selected as the parent with the selection method. First, a subtree of the parent will be selected randomly. Then, the chosen subtree will be replaced by a newly generated subtree.
- The reproduction operator produces offspring by copying the selected individuals into the next generation directly. An individual will be firstly chosen by the selection method, then the selected individual is copied into the new population directly. It is noted that the elitism mechanism (i.e., a special case of reproduction) picks up the top individuals from the current population. The selected individuals are inserted into the population of the next generation. This aims to ensure that the best individuals will not be lost when generating new populations.

1.4 Framework of Genetic Programming for Production Scheduling

GP for production scheduling has been very successful in recent years. With flexible representations, GP can represent and evolve effective scheduling heuristics to deal with a wide range of scheduling problems. In addition, since GP does not rely on any specific assumptions, it can be easily extended to deal with different production scheduling problems.

Figure 1.8 shows a unified framework for automated heuristic design with GP for production scheduling. In this book, **the production scheduling task** to be solved can be one task or multiple tasks. In other words, some studies will focus on solving a single production scheduling task at a time, while some studies will work on solving multiple production scheduling tasks simultaneously. If more than one scheduling production task are solved simultaneously, there will be multiple corresponding **evaluation models**.

Which **scheduling heuristic(s) need(s) to be learned** depends on the type of the investigated production scheduling. If it is a flexible production scheduling task, two scheduling heuristics (i.e., the routing rule for machine assignment and the sequencing rule for operation sequencing) need to be learned simultaneously. If it is a non-flexible production scheduling task, only the sequencing rule needs to be learned for operation sequencing. Accordingly, two terminal sets and two function sets are defined to learn the routing rule and the sequencing rule with the variable-length and tree-based representation for flexible production scheduling, respectively. However, one terminal set and one function set are set to learn the sequencing rule in non-flexible production scheduling.

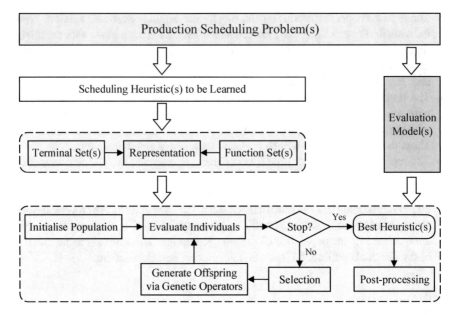

Fig. 1.8 A unified framework of genetic programming for production scheduling

In the lower part of Fig. 1.8, the evolutionary process of GP is presented. GP starts with population initialisation, and the output of the GP algorithm is the learned best heuristic(s). It is noted that if multiple production scheduling problems are solved simultaneously, there will be multiple independent heuristics corresponding to each problem. **Post-processing** is an important step to gain a better knowledge of the learned heuristics. The following are some post-processing techniques to analyse scheduling heuristics commonly used in the recent literature:

- Simplification of the learned scheduling heuristics;
- Size analyses of the learned scheduling heuristics;
- Feature importance analyses within the learned scheduling heuristics;
- Analyse the fragments of the learned scheduling heuristics.

1.5   Interpretable Machine Learning

Interpretability is the degree to which a human can understand the cause of a decision [146] or the degree to which a human can consistently predict the model's result [110]. A machine learning model with higher interpretability is easier for one to understand why the model makes the decisions. Interpretable machine learning can greatly promote the adoption of machine learning techniques for production scheduling in real-world applications. This is because the elements in production scheduling are highly related to each other, and a decision of the former can greatly affect the choice of the latter. The machine learning techniques are easier to be accepted by the schedulers in production scheduling if they can understand how the machine learning techniques work.

Interpretable machine learning has a great potential for improving the efficiency of production scheduling. First, an interpretable production scheduling process makes the human operators feel confident with the decisions, which leads to a better social acceptance [146]. Second, an interpretable production scheduling process makes it easier to maintain the schedules and find the limitations of the decisions, which can bring positive financial effects. Last but not least, if the orders are delayed, the enterprise can explain to the customers well, which can help keep a good relationship with the customers rather than losing them. In other words, the explanation is social, representing a conversation or interaction, thus relative to the explainer's beliefs about the explainee's beliefs.

However, in general, a major disadvantage of using machine learning is that the insights about the data and the task to be solved are hidden in the increasingly complex models such as neural networks [207]. The most likely explanation is not always the best explanation for a person. Importantly, using statistical generalisations to explain why events occur is unsatisfying, unless accompanied by an underlying causal explanation for the generalisation itself. In this book, we focus on using GP to learn scheduling heuristics for production scheduling. The learned scheduling heuristics with tree structures have natural visualisation with their symbolic nodes and edges.

1.6 Terminology

To avoid confusion due to ambiguity, below are the definitions of the terms commonly used in this book:

- A **problem** is a high-level proposition we aim at solving, such as JSS.
- A **dataset** is the available data from which the machine learning algorithms learn models. The term dataset is used in *static* JSS.
- A **simulation** is the process to represent the environment of a problem, which is used in *dynamic* JSS.
- An **instance** is a specific simulation with a fixed random seed. The term instance is used in *dynamic* JSS.
- A **scenario** represents a specific problem to be solved with the instances generated by the same problem configuration, e.g., the same objective and utilisation level. An instance is an example of a scenario.
- A **task** is a specific problem to be solved which is represented by a scenario. Solving the problems in different scenarios simultaneously is a multitask learning problem.

1.7   Organisation of the Book

The book is divided into six parts according to the types of the investigated JSS problems and the studied machine learning techniques.

- In Part I, there are two chapters. In this chapter, we give an introduction of production scheduling, (interpretable) machine learning, and evolutionary learning and GP. A unified framework of GP for production scheduling is provided. In addition, we describe a number of commonly used terms in this book, and the organisation of this book. In Chap. 2, the definition of JSS is given. Characteristics of different approaches, including exact methods, heuristics, and hyper-heuristics, hyper-heuristics in evolutionary learning, a brief introduction of GP-based hyper-heuristics for production scheduling, and evaluation criteria are also presented in this chapter.
- In Part II, we present different ways including construction heuristics, improvement heuristics, and augment operations research algorithms that GP can be employed to solve static production scheduling problems and their connections with conventional operations research methods.
- In Part III, we introduce how to design GP algorithms for dynamic production scheduling problems and advanced techniques to enhance GP's performance, including surrogate modelling, feature selection, and specialised genetic operators.
- In Part IV, we introduce how to learn heuristics for multi-objective production scheduling problems, and present an advanced multi-objective approach with the cooperative coevolution technique. In addition, how to use multi-objective optimisation for DFJSS is introduced in this part.
- In Part V, we introduce how to use multitask learning in the hyper-heuristics domain with production scheduling. We also present how assisted task selection strategy and surrogate techniques can benefit multitask optimisation with GP for learning heuristics in production scheduling.
- In Part VI, we give the conclusions and prospectus of this book.

This book targets researchers who are not familiar with GP and/or production scheduling as well as experienced researchers who are interested in special research topics. Readers with a background in machine learning, operations research, and industrial engineering can enjoy this book and explore the interfaces between operations research/optimisation and machine learning. Research students can start with Part I to gain fundamental understandings of the subjects and incrementally explore different applications and algorithms in the latter parts. Operations Research or Industrial Engineering researchers who are familiar with combinatorial optimisation and production scheduling will find Part I a good transition to refresh their knowledge and get used to machine learning and GP's terminologies before exploring hybrid Operations–Research–GP methods in Part II, and other applications in dynamic production environments in Parts III–V. Machine learning researchers can familiarise themselves with production scheduling applications from this book and understand the basic settings for applying machine learning algorithms to production scheduling from Parts I–III. Operations Research, Industrial Engineering, and Machine Learning practitioners will find Parts I–III most useful to understand how machine learning can be applied to solve production scheduling problems. Advanced applications of machine learning presented in Parts III–IV will be of great interest for many machine learning researchers. Researchers interested in multitask learning, especially in combinatorial optimisation, will find Part V quite useful.

Chapter 2
Preliminaries

2.1 Job Shop Scheduling

In JSS, n jobs $\mathcal{J} = \{J_1, J_2, \ldots, J_n\}$ need to be processed by m machines $\mathcal{M} = \{M_1, M_2, \ldots, M_m\}$. Each job has a sequence of operations $O_j = (O_{j1}, O_{j2}, \ldots, O_{jl_j})$. The job is completed upon the completion of the last operation. In DJSS and DFJSS, dynamic events are necessary to be taken into account when constructing schedules. This book focuses on one type of dynamic event, i.e., stochastic and continuous job arrivals. The details of a job are unknown until it reaches the shop floor. In DJSS, each operation O_{ji} can be handled by a single predefined machine, and its proceeding time is a fixed value. Only the sequencing decision needs to be made in DJSS. In DFJSS, each operation O_{ji} can only be done by one of its optional machines $M(O_{ji}) \subseteq \pi(O_{ji})$, and the time required to process it $\delta(O_{ji}, M(O_{ji}))$ is determined by the chosen machine. In DFJSS, routing and sequencing decisions should be determined simultaneously.

The parameters, variables, and constraints used in this book are described as follows.

Parameters:

- $\mathcal{J}$: the set of jobs in the job shop.
- n: the number of jobs in the job shop.
- $\mathcal{M}$: the set of machines in the job shop.
- m: the number of machines in the job shop.
- j: the index of the job.
- i: the index of the operation.
- l_j: the number of operations for job J_j, $l_j \leq m$.
- $\mathcal{O}_j = (O_{j1}, O_{j2}, \ldots, O_{jl_j})$: the set of operations of job J_j.
- w_j: the weight of job J_j.
- d_j: the due date of job J_j.
- $M(O_{ji})$: the machine that processes operation O_{ji}.

© The Author(s), under exclusive license to Springer Nature Singapore Pte Ltd. 2021
F. Zhang et al., *Genetic Programming for Production Scheduling*, Machine Learning:
Foundations, Methodologies, and Applications,
https://doi.org/10.1007/978-981-16-4859-5_2

- $\delta(O_{ji}, M(O_{ji}))$: the processing time of operation O_{ji}.
- $\pi(O_{ji})$: the set of optional machines of O_{ji}, $\pi(O_{ji}) \subseteq M$. This parameter is used in flexible JSS which will be described later.

Variables:

- C_j: the completion time of job J_j.
- $r(O_{ji})$: the release/ready time of operation O_{ji}. That is, the time that ith operation of job J_j is allowed to be processed. For the first operation of each job, it is set to the arrival time of the job. Otherwise, it is set to the completion time of its preceding operation.
- $r(j)$: the release time of job J_j. It is the release time of the first operation of a job.

Constraints:

- The $(i + 1)$th operation of job J_j (i.e., $O_{j(i+1)}$) can only be processed after its preceding operation O_{ji} has been processed (i.e., precedence constraint).
- At most, each machine can handle one operation at a time.
- The scheduling is non-preemptive, which means that once an operation is started, it cannot be interrupted or paused until it is finished.
- Each operation O_{ji} can be processed on one of the corresponding set of machines $\pi(O_{ji}) \subseteq M$ with processing time $\delta(O_{ji}, M(O_{ji}))$.

Objectives:

- Max-flowtime: $\max_{j=1}^{n}\{C_j - r(j)\}$.
- Mean-flowtime: $\frac{\sum_{j=1}^{n}\{C_j - r(j)\}}{n}$.
- Mean-weighted-flowtime: $\frac{\sum_{j=1}^{n} w_j * \{C_j - r(j)\}}{n}$.
- Max-tardiness: $\max_{j=1}^{n} \max\{0, C_j - d_j\}$.
- Mean-tardiness: $\frac{\sum_{j=1}^{n} \max\{0, C_j - d_j\}}{n}$.
- Mean-weighted-tardiness: $\frac{\sum_{j=1}^{n} w_j * \max\{0, C_j - d_j\}}{n}$.

For the sake of convenience, Fmax, Fmean, WFmean, Tmax, Tmean, and WTmean are used to describe max-flowtime, mean-flowtime, mean-weighted-flowtime, max-tardiness, mean-tardiness, and mean-weighted-tardiness, respectively.

2.2 Exact, Heuristic, and Hyper-heuristic Approaches

Over the years, a number of approaches, including exact approaches, heuristic approaches, and hyper-heuristic approaches, have been adapted for solving the JSS problems. Different types of these approaches are discussed as follows.

JSS has a number of challenges. For example, in DJSS, production scheduling environments are dynamic (e.g., job arrivals, job cancellations, and machine breakdown), which causes computational difficulties for most optimisation techniques.

The complexities of the production systems caused by heterogeneous production processes (e.g., the machine-dependent processing in flexible JSS) also makes scheduling tasks hard. Moreover, production scheduling has to take into account multiple conflicting objectives to ensure that the obtained schedules are approved and applicable.

Exact Approaches: Many techniques that search for optimal solutions, which are known as exact approaches such as dynamic programming [39], branch-and-bound [124], and integer linear programming [202], have been investigated for static JSS [36, 37, 39]. However, many JSS problems are NP-hard [203], and exact approaches are only limited to solve *small-scale static JSS* problems. Exact approaches are too time-consuming when the problems are getting large. It is also hard for exact approaches to handle dynamic problems where a lot of real-time decisions need to be made.

Heuristic Approaches: A heuristic approach is designed for solving a problem more quickly when the exact approaches are too slow, or for finding an approximate solution when the exact approaches fail to find the optimal solution [177]. A heuristic approach aims to produce a good enough solution for the problem at hand in a reasonable time. There are two different types of heuristics, i.e., *improvement heuristic* and *construction heuristic*, depending on how they build solutions. An improvement heuristic starts with a complete initial solution, and improves the solution iteratively until a stopping criterion is met [217]. A construction heuristic starts with an empty solution and incrementally updates the partial solutions until a complete solution is obtained [112].

Improvement heuristics such as bee colony algorithm [51, 129], tabu search [166], simulated annealing [218], and genetic algorithms [46, 181] have been proposed to find "good enough" solutions for solving JSS problems [3, 79, 178, 214] in a reasonable time. However, they cannot find solutions efficiently to handle the dynamic events in the DJSS problems, since they face rescheduling problems which are time-consuming. Construction heuristics such as dispatching rules have been popularly used for DJSS due to its efficiency to make real-time decisions (i.e., dispatching rules are used as priority functions to prioritise jobs and machines at the decision points). Three attractive characteristics of the dispatching rules are efficiency, reactiveness, and interpretability. Comprehensive comparison among a large number of dispatching rules can be found in [201]. It is noted that a scheduling heuristic in DFJSS is made of two rules, a routing rule for machine assignment and a sequencing rule for operation sequencing [236, 239]. Scheduling heuristics make decisions according to the *priority values* of machines or operations *only at the decision points*. This leads to two major reasons for the success of scheduling heuristics in DFJSS [228, 239]. One is its ability to efficiently address large-scale problems. The other is its efficiency to make real-time decisions in the face of changing circumstances. However, the scheduling heuristics, such as SPT (i.e., shortest processing time) and some composite rules [101], are often manually designed by experts [50, 63, 101, 102, 220]. The designing process is time-consuming, and the designed rules are typically too specific to be applied to different scenarios [25, 78, 106].

Hyper-heuristic Approaches: Specific heuristic approaches do not always perform well when applied to other problem domains without significant modification. This is a primary motivation for the development of general-purpose problem-independent heuristic search approaches, called hyper-heuristic approaches [66]. A hyper-heuristic [30, 31] is a heuristic search approach to selecting and generating heuristics to solve hard computational search problems. The term hyper-heuristics was first used to describe heuristics to choose heuristics in combinatorial optimisation [53]. There are two ways to classify hyper-heuristics. In terms of the nature of the search space of the heuristics, there are two kinds of hyper-heuristic approaches [33]. One is *heuristic selection*, which aims to select from existing heuristics for different situations. The other is *heuristic generation*, which generates new high-level heuristics from existing low-level heuristics. In terms of the sources of feedback information, the hyper-heuristics approaches can be grouped into *online learning hyper-heuristics* and *offline learning hyper-heuristics*. For the online learning hyper-heuristics, the learning algorithm aims at solving an instance of a problem [31]. For the offline learning hyper-heuristics, the learning algorithm learns knowledge in the form of rules or programs from a set of training instances. It expects the learned rules or programs to have good generalisation to unseen instances.

Overall, hyper-heuristic approaches have advantages for solving dynamic JSS problems compared with exact approaches and heuristic approaches. The fundamental difference between the heuristic and hyper-heuristic approaches is that heuristic approaches work on solution space, while hyper-heuristic approaches work on heuristic space.

2.3   Hyper-heuristics in Evolutionary Learning

Evolutionary algorithms have been widely used in hyper-heuristic approaches [183]. As mentioned in Chap. 1, there are two types of hyper-heuristic approaches, i.e., hyper-heuristic selection and hyper-heuristic generation. The hyper-heuristics types applied for evolutionary algorithms naturally depend on the representations of the evolutionary algorithms. The evolutionary algorithms with fixed-length vector-based representations such as genetic algorithms [120] and particle swarm optimisation [113] have been successfully used as hyper-heuristic selection approaches to learning an order of low-level heuristics to apply to different situations for different problems, e.g., training scheduling [88] and school timetabling [196]. The evolutionary algorithms with variable-length tree-based representations such as GP hyper-heuristic approaches [32] have been widely used to learn high-level heuristics for different problems, e.g., 2-dimensional strip packing [34], JSS [236], and arc routing [7, 221].

2.4 Scheduling Heuristics for Job Shop Scheduling

Only the *ready operations* are permitted to be allocated to machines due to the precedence constraint. There are two kinds of operations that will become ready operations. One is the first operation of a job. The other is the subsequent operation whose preceding operation has been finished. As scheduling heuristics, routing rules and sequencing rules work together to do machine assignment and operation sequencing in DFJSS. It is noted that we can consider that there is a fixed routing rule for allocating each operation to the corresponding machine in DJSS, since the mapping of operations to machines is predefined. Taking the general characteristic into consideration, this section takes DFJSS as an example to show how scheduling heuristics are used for making schedules in JSS.

Both routing and sequencing rules are numerical priority functions, which are used to prioritise the machines or operations in different decision situations [241]. When a new operation becomes ready, the routing rule will be applied to prioritise its candidate machines, and the operation will be assigned to the machine with the highest priority (e.g., has the least workload under the least-work-in-queue rule). When a machine becomes idle, the sequencing rule will be triggered to prioritise the operations in its queue, and the operation with the highest priority (e.g., the one with the shortest processing time if the shortest-processing-time rule is used) will be chosen to be processed next.

Figure 2.1 shows an example of how to use scheduling heuristics to make decisions in DFJSS. There are three machines, each with several operations waiting in its queue. The operation O_{81} is being processed on *Machine 3*. A routing (sequencing) *decision situation* includes a temporal job shop state, the given operation (machine), and the

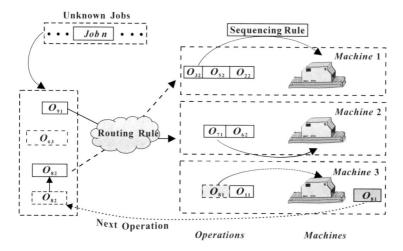

Fig. 2.1 An example of decision-making processes of dynamic flexible job shop scheduling with scheduling heuristics (i.e., routing rule and sequencing rule)

candidate machines (operations). The routing decision and sequencing decision are made with the routing and sequencing decision situation, respectively.

Routing Decision: If there is a ready operation (*a routing decision situation is encountered*), the operation will be assigned to the machine with the highest priority according to the *routing rule*. For example, when a new job (J_9) arrives at the job shop, its first operation O_{91} is assigned to *Machine 2*, which has the highest priority value among the three machines according to the routing rule. In addition, when O_{81} is finished, its next operation (O_{82}) becomes a ready operation and is allocated to *Machine 1* by the routing rule.

Sequencing Decision: If a machine (e.g., *Machine 1*) becomes idle and its queue is not empty (*a sequencing decision situation is encountered*), the *sequencing rule* will be used to calculate the priority value of each operation in its queue. The operation with the highest priority is then chosen as the next operation to be processed (e.g., in this case, O_{32} is selected to be processed on *Machine 1*).

2.5   Genetic Programming for Production Scheduling Heuristics

GP, as one of the popular evolutionary computation methods, has been successfully used to evolve scheduling heuristics for JSS, especially in DJSS [225]. In this book, we use GP as a hyper-heuristic approach, i.e., heuristic generation, and an offline learning hyper-heuristic approach. The goal is to generate effective heuristics that have good generalisation for solving JSS problems.

2.5.1   Advantages of Genetic Programming for Production Scheduling

Compared with other machine learning techniques such as decision tree and artificial neural networks, GP has a number of key advantages.

- GP has flexible representations which can represent various scheduling heuristics.
- GP has promising search mechanisms which can operate in the heuristic search space to find optimal or near-optimal scheduling heuristics.
- GP can simultaneously explore both the structure and corresponding parameters of a scheduling heuristic.
- The heuristics learned by GP can be partially interpretable, which makes it easier for people to accept them.
- The learned heuristics of GP, as priority functions, are very efficient to construct schedules, which enhances its applicability in practice.

2.5.2 Overall Process of Genetic Programming for Job Shop Scheduling

The decision difference among classical static JSS, DJSS, and DFJSS is that classical static JSS and DJSS only need to make the sequencing decision, while DFJSS needs to make the sequencing and routing decisions simultaneously. This section takes DFJSS (i.e., a relatively complex type of JSS problem) as an example to show how GP is used for JSS. The pseudo-code of GP to learn heuristics for DFJSS is shown in Algorithm 2.1. The input of the proposed algorithm is a task that is expected to be solved, and the output is the learned heuristic h^* with a routing heuristic r^* and sequencing heuristic s^*. GP starts with a randomly initialised population (line 1). It is noted that each GP individual contains two trees [239] in Algorithm 2.1. The first tree represents the routing rule, and the second tree represents the sequencing rule. The fitness of heuristics is evaluated based on the objective functions (from line 7 to line 11). Specifically, a simulation is run with the heuristic h_i to get a schedule $Schedule_i$ (line 9). The fitness of the heuristic h_i is assigned by calculating the objective value of its obtained schedule $Schedule_i$ (line 10). A new population is generated by recombining the heuristics (crossover), mutating the heuristics (mutation), or copying

---

Algorithm 2.1 Framework of genetic programming to learn routing and sequencing heuristics for dynamic flexible job shop scheduling

---

Input: A DFJSS scheduling task
Output: The learned scheduling heuristics h^* with routing rule r^* and sequencing rule s^*
1: **Initialisation**: Randomly initialise the population
2: set $r^* \leftarrow null$
3: set $s^* \leftarrow null$
4: set $h^* \leftarrow r^* \cup s^*$ and $fitness(h^*) \leftarrow +\infty$
5: $gen \leftarrow 0$
6: **while** $gen < maxGen$ **do**
7:     //**Evaluation**: Evaluate the individuals in the population
8:     **for** i = 1 to $popsize$ **do**
9:         Run a DFJSS simulation with h_i to get the schedule $Schedule_i$
10:         $fitness_{h_i} \leftarrow Obj(Schedule_i)$
11:     **end for**
12:     **for** i = 1 to $popsize$ **do**
13:         **if** $fitness_{h_i} < fitness_{h^*}$ **then**
14:             $h^* \leftarrow h_i$
15:         **end if**
16:     **end for**
17:     **if** $gen < maxGen - 1$ **then**
18:         **Parent selection**: Select parents to generate offspring
19:         **Evolution**: Generate a new population by crossover, mutation, and reproduction
20:     **end if**
21:     $gen \leftarrow gen + 1$
22: **end while**
23: **return** routing rule r^* and sequencing rule s^*

---

the heuristics with good fitness directly (reproduction and elitism) (line 19) to the next generation.

2.5.3 Extracting High-Level Heuristic from Low-Level Heuristics

Figure 2.2 shows how to generate high-level heuristics from low-level heuristics with GP. The circles with different colours indicate three different low-level heuristics. The boxes with light grey background mean the corresponding heuristics are evaluated. The low-level heuristics are set as the terminal set of GP and combined with the function set to form a GP individual. The quality of the individuals in the population is improved generation by generation with genetic operators (i.e., mutation, crossover, and reproduction). An offspring is a function/rule that can be used as a heuristic to optimise problems by assigning priority values for machines and operations in production scheduling. Taking a generated offspring as an example, its corresponding heuristic is obtained based on the GP individual. The heuristic is then executed on training instances to get its fitness value. If the algorithm fulfils the stopping condition, the current best learned heuristic will be treated as the final obtained high-level heuristic. Otherwise, the search continues.

Fig. 2.2 An example of generating high-level heuristics from simple low-level heuristics with genetic programming

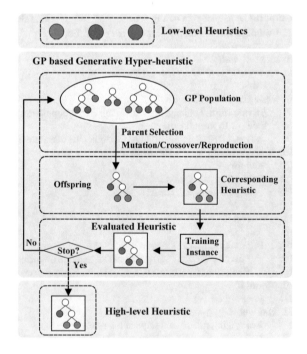

2.6 Evaluations of Genetic Programming Hyper-heuristics

Continue to take DFJSS as an example; Fig. 2.3 shows the overall process of GP for DFJSS in this book. In the training phase, GP is used as a hyper-heuristic approach to learning heuristics based on a set of training instances. The outputs of the training process are heuristics (i.e., routing and sequencing rules) rather than solutions (schedules). It is noted that only sequencing rules are learned in DJSS. The learned heuristics in the training phase are tested on a number of unseen instances in the test phase. Specifically, to calculate the test performance of a trained heuristic, it is applied to each test instance to construct a schedule. The test performance is then defined as the average objective value (e.g., mean-flowtime) of the constructed schedules on the test instances.

In this book, there are mainly four measures for the performance of the algorithms. Different measures are chosen based on the characteristics of the proposed algorithms.

Quality of Learned Best Scheduling Heuristics: The objectives considered in this book are mentioned in Sect. 2.1. We consider them separately as a single objective in different scenarios or we consider a number of them simultaneously as a multi-objective optimisation.

The utilisation level is a commonly used parameter [28, 182] to generate a number of job shop scenarios in order to access the effectiveness of the algorithms. It is noted that a higher degree of utilisation level will result in a more difficult scheduling task. This book focuses on using the objective function and the utilisation level to construct multiple test scenarios because our preliminary studies have shown that the performance of the learned scheduling heuristics is influenced significantly by these two factors.

Training Efficiency: Training time is a common way to measure the efficiency of algorithms. If an algorithm can achieve comparable scheduling heuristics with a shorter training time, the algorithm is more efficient.

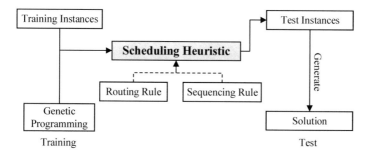

Fig. 2.3 The overall process of genetic programming for dynamic flexible job shop scheduling

Sizes of Learned Scheduling Heuristics: The rule size is defined as the number of nodes in this book, and the rule with a smaller size is preferred. There are a number of advantages of evolving smaller rules. First, smaller rules tend to be more interpretable by decision-makers, which is particularly important for the floor operators of the job shop. Second, smaller rules are easier to implement in real-world applications, which are more efficient to make real-time decisions with dynamic events compared with larger rules.

Number of Unique Features of Learned Scheduling Heuristics: The number of unique features is the number of different features needed to construct the rules. The number of unique features in the scheduling heuristics is one indicator of the complexity of learned rules [236]. With fewer unique features, the scheduling heuristics can be interpreted more easily.

It is noted that except for the above measures, other measures (i.e., the measures are defined specifically in each chapter) will also be designed based on the characteristics of the investigated algorithm.

2.7   Chapter Summary

In this part (Chap. 1 and this chapter), we first introduce the concepts of production scheduling, machine learning, evolutionary learning, and GP. The basics of JSS and the corresponding approaches are also discussed. The advantages of GP for solving the JSS problems are highlighted. In addition, how to use GP for JSS is described in detail. The contents in this part help prepare the reading of this book by giving details of related concepts and techniques, and guiding the research in the following parts. The following four parts will focus on the studies with various JSS problems and different machine learning techniques.

Part II
Genetic Programming for Static Production Scheduling Problems

Most studies in the production scheduling literature focus on static scheduling problems in which all information of jobs and machines is available before scheduling decisions are made. One of the reasons for the popularity of static scheduling problems is that they are easily formulated as mathematical optimisation problems which can be solved by off-the-shelf optimisation solvers or specialised algorithms. Moreover, solutions from static scheduling problems are useful for the production environments where change frequency and uncertainty are low. While simpler than dynamic scheduling problems, solving static scheduling problems is still very challenging due to a large number of variables and complex real-world constraints. Many static production scheduling problems such as job shop scheduling have been proved to be NP-hard and optimal solutions can only be found for very small instances. Therefore, many new methods, e.g., meta-heuristics, heuristics, and math-heuristics, have been proposed in the literature to efficiently find good or near-optimal solutions for these problems.

In this part, we will introduce different ways in which genetic programming (GP) can be employed to deal with static production scheduling problems. Different from other methods above which explore the solution search space to find a good solution for each problem instance, GP is used as a hyper-heuristic which explores the heuristic search space to find a competitive and efficient heuristic for solving a set of problem instances. Here, GP, similar to machine learning techniques, has two phases: (1) training and (2) test. In training, GP tries to optimise heuristics in its population to cope better with a set of training instances. In test, the best heuristics evolved by GP are applied to a set of unseen instances to evaluate their effectiveness, efficiency, and generalisation.

Part II is organised into three chapters:

- Chapter 3 introduces how GP is used to design scheduling construction heuristics, the most fundamental components of most scheduling algorithms when solving static production scheduling problems. This chapter also presents and compares different representations of heuristics in GP.

- Chapter 4 further explores how GP can evolve improvement heuristics based on the representations developed for construction heuristics. This chapter also discusses how ideas from optimisation approaches can be used to enhance the performance of evolved improvement heuristics.

- Chapter 5 shows how GP can be integrated with generic optimisation solvers to improve their efficiency while eliminating the need for meta-algorithms. Different from other chapters, combining GP and optimisation solvers can help the users search near-optimal solutions efficiently and systematically, and potentially prove the optimality of found solutions.

Chapter 3
Learning Schedule Construction Heuristics

Schedule construction heuristics are the most fundamental element of scheduling algorithms, from simple dispatching rules to sophisticated optimisation algorithms. In production scheduling, schedule construction is a procedure in which jobs and operations are gradually inserted into the schedule. The output of a schedule construction heuristic is a complete and (usually) feasible schedule for a problem instance. For some special production scheduling problems such as minimising makespan for 2-machine flow shops [185], it is possible to quickly create optimal schedules with schedule construction heuristics. For more complex problems, schedule construction heuristics can be used within optimisation algorithms (e.g., genetic algorithms and particle swarm optimisation) to generate schedules, usually by transforming (or decoding) solutions represented (or encoded) by those algorithms to feasible schedules. It has been shown in the literature that the quality of schedule construction heuristics can improve the performance of optimisation algorithms by narrowing down their search space or producing good initial solutions.

3.1 Challenges and Motivations

Schedule construction heuristics have been studied carefully in the early operation research literature, and many heuristics have been designed by researchers and practitioners for a number of specific production environments (e.g., job shops and flow shops). Those schedule construction heuristics are extensively adopted by researchers in the 1990s and 2000s to design optimisation algorithms. However, since those available heuristics are designed for specific environments, there is no guarantee that they can be successfully applied to less related or new environments. For example, conventional schedule construction heuristics do not consider energy efficiency or the availability of big data obtained by smart devices. The growth of emerging tech-

© The Author(s), under exclusive license to Springer Nature Singapore Pte Ltd. 2021
F. Zhang et al., *Genetic Programming for Production Scheduling*, Machine Learning: Foundations, Methodologies, and Applications,
https://doi.org/10.1007/978-981-16-4859-5_3

nologies related to digital factories and Industry 4.0 presents new opportunities and challenges for scheduling, which have not yet been fully addressed in the literature.

Designing schedule construction heuristics is a time-consuming process. The pressure of quick responses to the new and changing environments makes the manual design process less efficient while adopting available heuristics is not effective. To address this challenge, hyper-heuristics and machine learning techniques have been used to automate the design of scheduling heuristics [21]. GP has been extensively investigated in the last decade for automated design of scheduling construction heuristics. The variable and flexible representation of GP is one of the main reasons for the popularity of GP in automated heuristic design. From arithmetic representations to user-friendly grammar-based representation, GP presents an unlimited number of ways in which schedule construction heuristics can be represented and learned. The choice of representations is usually driven by practical requirements such as interpretability and effectiveness. In this chapter, we will explore three common and natural representations of GP for schedule construction heuristics and their applications in static production scheduling problems. Specially, we try to shed light on the following questions:

- What are the computational advantages and disadvantages of different representations?
- What are the practical implications of each representation?
- How do learned heuristics perform as compared to available heuristics in the literature?

In Sect. 3.2, details of three GP representations for scheduling heuristics are presented. Readers who are not familiar with GP are recommended to read this section carefully as these representations (and the schedule construction procedure) will further explore and extend in the later chapters. Section 3.3 shows how to set up experiments and compare the performance of GP with different algorithms. Detailed analyses of evolved heuristics, including their interpretability, will also be provided in this chapter.

3.2 Algorithm Design and Details

This section will show how schedule heuristics can be used to construct a complete schedule and their representations in GP.

3.2.1 Meta-algorithm for Schedule Construction

Constructing a schedule for JSS is more complex than other production environments such as flow shops or single machines because of the precedence constraints. Figure 3.1 shows a generic procedure or *meta-algorithm* to construct an active schedule, a non-delay schedule, or a hybrid of both active and non-delay schedules with

```
1: Ω ← {o₁₁, o₂₁, ..., oₙ₁}
2: repeat
3:     let t(Ω) = min_{σ∈Ω}{max{r(σ), U_{M(σ)}} + δ(σ, M(σ))}   ▷ U_{M(σ)} is ready time of
       machine M(σ)
4:     let σ* be the operation that minimum is achieved, M* = M(σ*), and Ω* = {σ ∈
       Ω|M(σ) = M*}
5:     let S(m*) = max{min_{σ∈Ω*}{r(σ)}, U_{M*}}
6:     let Ω' = {σ ∈ Ω*|r(σ) ≤ S(M*) + α(t(Ω) − S(M*))}
7:     apply a rule H on Ω' to find the next operation σ' to be scheduled on M*
8:     remove σ' from Ω' and include next(σ) into Ω if next(σ) ≠ null ▷ next(σ) is the
       next operation from the same job
9: until all operations have been scheduled
```

Fig. 3.1 Meta-algorithm for schedule construction

a dispatching rule H. In this procedure, Ω contains all the operations that are ready to be scheduled. The procedure will first determine the next operation with the earliest completion time and its corresponding machine M^* (ties are broken arbitrarily). The non-delay factor $\alpha \in [0, 1]$ controls the look-ahead ability of the algorithm and determines the set Ω' of jobs which should be considered to be processed next on machine M^* by the dispatching rule H. If $\alpha = 0$, the procedure can only construct non-delay schedules, and only the operations that have already joined the queue are considered for scheduling. On the other hand, if $\alpha = 1$, the procedure will consider all the potential jobs that are ready to join the queue of machine M^* before the earliest completion time of M^*. In general, α determines the interval of time that a machine is allowed to wait even when there are operations in the queue. As a result, there are fewer operations to be considered in each step of the algorithm when α is small than when α is large. The purpose of the dispatching rule DR is for the next operation in Ω' to be processed next on the machine M^*. The most common form of DR is priority functions which are used to assigned priorities to operations in Ω', and the one with the highest priority will be selected. In this chapter, all schedule construction heuristics will use the meta-algorithm in Fig. 3.1 with two components: (1) a dispatching rule DR and (2) a non-delay factor α to generate schedules. The performance of a heuristic is determined by the dispatching rule DR and the non-delay factor α that it employs. In the next section, we will show that GP can represent and learned heuristics H with these two key components.

3.2.2 Representations of Scheduling Construction Heuristics

Three representations of schedule construction heuristics are considered. The first representation (R_1) provides a way to incorporate machine attributes into the GP program along with simple dispatching rules and the hybrid scheduling strategy (between non-delay and active scheduling). The second representation (R_2) is the traditional arithmetic representation like that employed in [213]. The purpose of this representation is to generate composite dispatching rules (referring to rules that employ multiple jobs and machine attributes). The last representation (R_3) is a combination of R_1 and R_2, in which different composite dispatching rules exist and are logically applied to JSS based on the machine and system attributes.

3.2.2.1   Decision-Tree-Like Representation (R_1)

The key idea of this representation is to provide heuristics with the ability to apply different simple dispatching rules based on machine attributes. In this case, dispatching rules are represented in a decision-tree form. To make the heuristics more interpretable, the proposed grammar in Fig. 3.2 is used when building the GP programs and performing the genetic operators (e.g., dispatching nodes must contain two arguments, which are a value of the non-delay factor α and a single dispatching rule $\mathcal{DR}$). Two example heuristics based on this grammar are shown in Fig. 3.3 (it is noted that the numbers in this figure are only for demonstration purpose). In Fig. 3.3a, the heuristic will use the SPT rule and the non-delay factor $\alpha = 0.084$ to construct schedules. The heuristic in Fig. 3.3b is a bit more sophisticated. The heuristic firstly checks the workload ratio WR (the ratio of the total processing times of jobs in the queue to the total processing times of all jobs that have to be processed at the machine) of the considered machine M^*; if the workload ratio is less than or equal to 20%, dispatching rule SPT is applied with $\alpha = 0.221$; otherwise, dispatching rule FIFO is applied with $\alpha = 0.078$. This heuristic can be considered as a variant of FIFO/SPT, in which the workload of the machine is used as the switch instead of the waiting times of jobs in the queue. Different from other applications [104, 180] where a single non-delay factor is involved, the proposed GP system using this representation allows different values of non-delay factors to be employed based on the status of the shop.

For this representation, we will consider six attributes which indicate the status of machines in the shop. Let Λ be the set of operations that are planned to visit the considered machine M^*, and K and I are the sets of all operations that have and have not

```
Start ::= <action>
<action> ::= <if> | <dispatch>
<if> ::= if <attributetype> <op> <threshold>
then <action> else <action>
<op> ::= ≤ | >
<attributetype> ::= WR | MP | DJ | CMI | CWR | BWR
<threshold> ::= 10%|20%|30%|...|90%|100%
<dispatch> ::= assign <nondelayfactor> assign <rule>
<nondelayfactor> ::= uniform[0,1]
<rule> ::= FIFO | SPT | LPT | LSO | LRM | MWKR | SWKR | MOPR | EDD | MS | WSPT
```

Fig. 3.2   Grammar for the proposed GP system with R_1

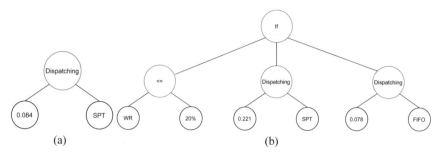

Fig. 3.3 Example program trees based on representation R_1

yet been processed by M^*, respectively ($\Lambda = K \cup I$). In the shop, we call a machine *critical* if it has the greatest total remaining processing time $\sum_{\sigma \in I} \delta(\sigma, M(\sigma))$, and a machine is called *bottleneck* if it has the largest workload $\sum_{\sigma \in \Omega'} \delta(\sigma, M(\sigma))$ in Ω'. The following definitions of the machine and system attributes are used in R_1:

- *Workload ratio*, $\text{WR} = \frac{\sum_{\sigma \in \Omega'} \delta(\sigma, M(\sigma))}{\sum_{\sigma \in I} \delta(\sigma, M(\sigma))}$: it indicates the workload in Ω' compared to the total remaining workload that M^* has to process (including the operations in the queue and operations that have not yet visited M^*).
- *Machine progress*, $\text{MP} = \frac{\sum_{\sigma \in K} \delta(\sigma, M(\sigma))}{\sum_{\sigma \in \Lambda} \delta(\sigma, M(\sigma))}$: it indicates the progress of M^*, calculated as the ratio of the total processing time that M^* has processed to the total processing time of all operations in Ω' that have to visit M^*.
- *Deviation of jobs*, $\text{DJ} = \frac{\min_{\sigma \in \Omega'}\{\delta(\sigma, M(\sigma))\}}{\max_{\sigma \in \Omega'}\{\delta(\sigma, M(\sigma))\}}$: it is a simple ratio of minimum processing time to the maximum processing time of operations in Ω'.
- *Critical machine idleness*, CMI: it is the workload ratio WR of the critical machine.
- *Critical workload ratio*, $\text{CWR} = \frac{\sum_{\sigma \in \Omega^c} \delta(\sigma, M(\sigma))}{\sum_{\sigma \in \Omega'} \delta(\sigma, M(\sigma))}$: it is the ratio of the workload of operations in Ω^c to the workload in Ω' where $\Omega^c \subset \Omega'$ is the set of operations belonging to the jobs that have operations that still need to be processed at the critical machine after being processed at M^*.
- *Bottleneck workload ratio*, $\text{BWR} = \frac{\sum_{\sigma \in \Omega^b} \delta(\sigma, M(\sigma))}{\sum_{\sigma \in \Omega'} \delta(\sigma, M(\sigma))}$: it is the ratio of the workload of operations in Ω^b to the workload in Ω' where $\Omega^b \subset \Omega'$ is the set of operations belonging to the jobs that have operations that still need to be processed at the bottleneck machine after being processed at M^*.

While the first three attributes provide the local status at M^*, the last three attributes indicate the status of the shop with a special focus on the critical and bottleneck machines. The machine and system attributes here appear in the scheduling literature in different forms. There are also other attributes in the literature, but they are mainly designed for special manufacturing environments which are not useful (or applicable) for this study. The key difference between our attributes and the attributes used in other studies is that our attributes have been scaled from 0 to 1. The scaled (normalised) attribute values aim to enhance the generality of the learned rules and also make the learned rules easier to understand. For example, two scheduling instances can have very different processing times. If the machine progress is important for our scheduling decisions, it is very difficult to design a general heuristic for two instances with the unnormalised value $\sum_{\sigma \in K} \delta(\sigma, M(\sigma))$ (total processing time that a machine has processed). Also, heuristics with normalised attributes are more interpretable (e.g., "50% of workload has been done" rather than "100 processing hours have been done").

For representation R_1, 11 simple dispatching rules are considered as the candidate rules. The aim of these rules is to determine which operation σ in Ω' will be processed next.

- *FIFO*: operations are sequenced *first in first out*.
- *SPT*: select the operation with the *shortest processing time*.
- *LPT*: select the operation with the *longest processing time*.
- *LSO*: select the operation belonging to the job that has the *longest subsequent operation*.
- *LRM*: select the operation belonging to the job that has the *longest remaining processing time* (excluding the operation under consideration).
- *MWKR*: select the operation belonging to the job that has the *most work remaining*.
- *SWKR*: select the operation belonging to the job that has the *smallest work remaining* .
- *MOPR*: select the operation belonging to the job that has the *largest number of operations remaining*.
- *EDD*: select the operation belonging to the job that has the *earliest due date*.
- *MS*: select the operation belonging to the job that has the *minimum slack*.
- *WSPT*: select the operation that has the maximum *weighted shortest processing time*.

The function set for this representation contains *If, Dispatch*, $\leq, >$ to help construct the logic of the heuristic as demonstrated by the examples in Fig. 3.3. *Dispatch* represents the combination of a single dispatching rule and its corresponding non-delay factor. The non-delay factor is treated as ephemeral random constants (ERC) in GP [115]. The values of the non-delay factor will initially be a random number from 0 to 1. Meanwhile, attribute type, attribute threshold, and dispatching rule terminals are randomly chosen from their candidate values as described in the previous section with equal probabilities.

3.2.2.2 Arithmetic Representation (R_2)

For this representation, the focus is to design composite dispatching rules that include different pieces of information from jobs and machines. In this case, only dispatching rules $\mathcal{DR}$ of heuristics are learned, and these rules are represented as priority functions that can be used to calculate the priorities for operations in Ω' and the operation with the highest priority will be scheduled to be processed next on the machine M^*. Different from R_1 heuristics that become more effective by logical choices of single dispatching rules, R_2 heuristics create their sophisticated behaviour by

Table 3.1 Terminal set for the R_2 representation

Notation	Description
RJ	Operation ready time
RO	Number of remaining operations of job
RT	Work remaining of job
PR	Operation processing time
W	Weight
DD	Due date
RM	Machine ready time
#	Constant Uniform [0, 1]

Fig. 3.4 Example GP trees based on representation R_2

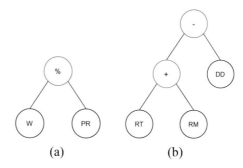

(a)   (b)

arithmetically combining various terms into the priority functions. The advantage of this representation is that more information can be directly considered to determine the priorities of operations when sequencing decisions need to be made.

In our GP system, the priority function set consists of $+$, $-$, $*$, % (protected division), If, min, max, and abs. The terminal set contains popular terms that are used in existing dispatching rules. The descriptions of the terminals used for calculating the priority of operation σ are shown in Table 3.1. Figure 3.4 shows two simple examples when WSPT and MS are represented by R_2 heuristics. The non-delay scheduling strategy will be used along with this representation like common applications of composite dispatching rules.

3.2.2.3   Mixed Representation (R_3)

This representation tries to combine the advantages of R_1 and R_2 to create sophisticated heuristics. Within R_3, the incorporation of both system/machine status and composite dispatching rules is considered. The representation R_3 inherits the grammar in R_1, and the composite dispatching rules will be used to calculate the priorities of operations for sequencing decisions besides the use of simple dispatching rules. An

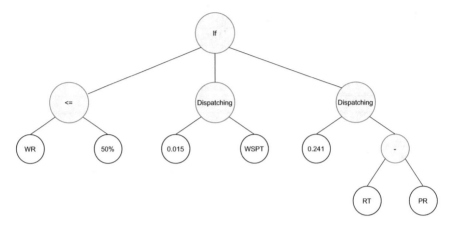

Fig. 3.5 An example GP tree based on representation R_3

example of an R_3 heuristic is shown in Fig. 3.5. The function set for R_3 representation is the combination of function sets of R_1 and R_2.

3.2.3 Fitness Evaluation

The focus of this study is to learn effective heuristics for JSS to minimise the makespan or total weighted tardiness. In order to estimate the effectiveness of a learned heuristic, it will be applied to solve a set of instances in the training set and the resulting objective values from all instances are recorded. Since the objective values obtained by a heuristic $\mathcal{H}$ for each instance are very different, we will measure the quality of an obtained schedule by the relative deviation of its objective value from its corresponding reference objective value as shown in Eq. (3.1).

$$dev(\mathcal{H}, I_n) = \frac{Obj(\mathcal{H}, I_n) - Ref(I_n)}{Ref(I_n)} \tag{3.1}$$

In this equation, $Obj(\mathcal{H}, I_n)$ is the objective value obtained when applying heuristic $\mathcal{H}$ to instance I_n, and $Ref(I_n)$ is the reference objective value for instance I_n. The fitness of $\mathcal{H}$ on the training set is calculated by either Eq. (3.2) or (3.3); $dev_{average}(\mathcal{H})$ and $dev_{max}(\mathcal{H})$ will correspondingly measure the average performance and worst-case performance of $\mathcal{H}$ across T instances in the dataset.

$$dev_{average}(\mathcal{H}) = \frac{\sum_{I_n \in \{I_1,...,I_T\}} dev(\mathcal{H}, I_n)}{T} \tag{3.2}$$

$$dev_{max}(\mathcal{H}) = \max_{I_n \in \{I_1,...,I_T\}} \{dev(\mathcal{H}, I_n)\} \tag{3.3}$$

The objective of the GP system is to minimise these fitness functions. In the case of $Jm||C_{max}$, the reference objective value is the lower bound obtained by other approaches (refer to [172] for a list of lower bound values obtained for popular benchmark instances). Since the lower bounds are used in this case, the fitness values for the GP programs are always non-negative. If the fitness value is close to zero, it indicates that the learned heuristics can provide near-optimal solutions. For $Jm||\sum w_j T_j$, because the lower bound values are not available for all instances, we will use objective values obtained by EDD as the reference objective values for all instances in the dataset since it is a widely used dispatching rule for due-date-related problems. Because EDD is just a simple rule, it can be dominated by more sophisticated rules. For that reason, the fitness value of the GP programs for $Jm||\sum w_j T_j$ can be negative, which means that the learned heuristics perform better than EDD.

3.2.4 Proposed Genetic Programming Algorithm

Figure 3.6 shows the GP algorithm used in this study to evolve heuristics for JSS. The GP system first sets up the training set D and randomly initialises the population. At a generation, each heuristic (or individual) $\mathcal{H}_i$ will be applied to solve all instances in the training set D to find the relative deviation $dev(\mathcal{H}_i, I_n)$ for each instance. Then, the fitness value of each heuristic is calculated by using $dev_{average}(\mathcal{H}_i)$. If the evaluated heuristic is better (has smaller fitness value) than the best heuristic $\mathcal{H}^*$,

---

Inputs: training instances $D \leftarrow \{I_1, I_2, \ldots, I_T\}$
Output: the best learned rule $\mathcal{H}^*$
1: randomly initialise the population $P \leftarrow \{\mathcal{H}_1, \ldots, \mathcal{H}_S\}$
2: set $\mathcal{H}^* \leftarrow null$ and $fitness(\mathcal{H}^*) = +\infty$
3: $generation \leftarrow 0$
4: **while** $generation \leq maxGeneration$ **do**
5:     **for all** $\mathcal{H}_i \in P$ **do**
6:         **for all** $I_n \in D$ **do**
7:             $dev(\mathcal{H}_i, I_n) \leftarrow$ solve I_n with $\mathcal{H}_i$
8:         **end for**
9:         evaluate $fitness(\mathcal{H}_i) \leftarrow dev_{average}(\mathcal{H}_i)$
10:         **if** $fitness(\mathcal{H}_i) < fitness(\mathcal{H}^*)$ **then**
11:             $\mathcal{H}^* \leftarrow \mathcal{H}_i$
12:             $fitness(\mathcal{H}^*) \leftarrow fitness(\mathcal{H}_i)$
13:         **end if**
14:     **end for**
15:     $P \leftarrow$ apply reproduction, crossover, mutation to P
16:     $generation \leftarrow generation + 1$
17: **end while**
18: **return** $\mathcal{H}^*$

Fig. 3.6 GP algorithm to evolve schedule construction heuristics for JSS

it will be assigned to the best heuristic $\mathcal{H}^*$ and the best fitness value $fitness(\mathcal{H}^*)$ is also updated. After all individuals in the population are evaluated, the GP system will apply genetic operators such as reproduction (elitism), crossover, and mutation to the programs in the current population to generate new individuals for the next generation. More details of the genetic operators used in this study will be provided in the next section. When the maximum number of generations is reached, the GP algorithm will stop and return the best-found heuristic $\mathcal{H}^*$, which will be applied to the test set to evaluate the performance of the GP system.

3.3 Empirical Study

This section discusses the configuration of the GP system and the datasets used for training and test.

3.3.1 Parameter Settings

The GP system for learning heuristics is developed based on the ECJ20 library [139] (a java-based evolutionary computation research system). The parameter settings of the GP system used in the rest of this chapter are shown in Table 3.2. The parameters in this table are commonly used for GP [17, 115] and also show good results from our pilot experiments. The population size of 1000 is used to ensure that there is enough diversity in the population. The initial GP population is created using the ramped-half-and-half method [115]. Since the R_1 and R_3 heuristics are created based on the grammar in Fig. 3.2, we use strongly typed GP to ensure that the GP nodes will provide proper return types as determined by the grammar. In this case, the crossover and mutation operators of GP are only allowed if they do not violate the grammar. For crossover, the GP system uses the subtree crossover [115], which creates new individuals for the next generation by randomly recombining subtrees from two selected parents. Meanwhile, mutation is performed by subtree mutation [115], which randomly selects a node of a chosen individual and replaces the subtree rooted at that node with a new randomly generated subtree. The combinations of three levels of crossover rates and mutation rates will be tested in our experiments to examine the influence of these genetic operators on the performance of GP. When generating random initial programs or applying crossover/mutation, the maximum depth of eight is used to restrict the program from becoming too large. Greater maximum depths can also be used here to extend the search space of GP; however, we choose this maximum depth to reduce the computational times of the GP system and make the learned heuristics easier to analyse. Tournament selection with the tournament size of seven is used to select individuals for genetic operations [115].

Table 3.2 GP parameters

Parameters	Values
Population size	1000
Crossover probability	95%, 90%, 85%
Mutation probability	0%, 5%, 10%
Reproduction probability	5%
Number of generations	50
Maximum depth for GP trees	8
Selection	Tournament selection with size 7
Function set (R_1)	*If, Dispatch, $\leq$, $>$*
Terminal set (R_1)	*Attribute type, attribute threshold, non-delay factor, dispatching rule*
Function set (R_2)	$+, -, *, \%,$ If, min, max, abs
Terminal set (R_2)	*As shown in Table*
Function set (R_3)	*Function set (R_1) and Function set (R_2)*
Terminal set (R_3)	*Terminal set (R_1) and Terminal set (R_2)*
Fitness	$dev_{\text{average}}(\mathcal{H})$

3.3.2 Datasets

There are many datasets in the JSS literature that are generated by different scheduling researchers [4, 59, 125, 209] to measure the performance of different heuristics and optimisation methods. The instances in these datasets are still very useful because they include a wide range of instances with different levels of difficulty. Moreover, lower bounds for these instances are available and can be used to calculate the fitness of the learned heuristics as described in Sect. 3.2.3. Descriptions of the datasets used for the experiments are shown in Table 3.3 (n is the number of jobs and m is the number of machines). In our experiment, we combine these datasets and distribute them into the training set and test set used by the proposed GP system. The training set and the test set are created to include halves of the instances of each individual

Table 3.3 Benchmark datasets for static job shop scheduling

Dataset	Notation	No. of instances	Size ($n \times m$)	Reference
LA	la01–la40	40	10×5 to 15×15	[125]
ORB	orb01–orb10	10	10×10	[5]
TA	ta01–ta80	80	15×15 to 100×20	[209]
DMU	dmu01–dmu80	80	20×15 to 50×20	[59]

dataset in Table 3.3. In particular, the training set will contain {la01, la03, ..., la39}, {orb01, ..., orb09},{ ta01, ..., ta79}, {dmu01, ..., dmu79}. The other (even index) instances will be included in the test set. This allows a fair distribution of problems with different instance sizes into both the training set and the test set. There are 105 instances in each of the training sets and test sets. For the case of $Jm||\sum w_j T_j$, the due dates for jobs in each instance will be generated (following Baker [11]) by a due-date assignment rule:

$$due\_date = release\_time + h \times total\_processing\_time \qquad (3.4)$$

The parameter h is used to indicate the tightness of due dates. We choose $h = 1.3$ for all instances in the training set and test set because it is the common value used in previous research [119, 249]. For the weights of jobs, we employ the $4 : 2 : 1$ rule which has been used in [119, 249]. This rule is inspired by Pinedo and Singer [184] when their research showed that 20% of the customers are very important, 60% are of average importance, and the remaining 20% are of less importance. For that reason, in $Jm||\sum w_j T_j$, the weight of 4 is assigned to the first 20% of jobs, the next 60% are assigned a weight of 2, and the last 20% of jobs are assigned a weight of 1.

3.3.3 Performance of Learned Heuristics

The proposed GP systems, with different settings, will be applied to evolve new schedule construction heuristics. This section shows the results obtained from the GP system with three representations, three levels of crossover/mutation rates, two fitness functions, and two objectives. In total, we need to run $3 \times 3 \times 2 \times 2 = 36$ experiments. For each experiment, 30 independent runs are performed with different random seeds. Tables 3.4 and 3.6 show the means and standard deviations of fitness values obtained from all experiments on the training set and test set. The upper part and lower part of each table show the statistics of $dev_{average}(\mathcal{H})$ and $dev_{max}(\mathcal{H})$ when they are used as the fitness function for the GP system. The triple $\langle c, m, r \rangle$ indicates the GP parameters used in a specific experiment. For example, $\langle 85, 10, 5 \rangle$ represents the experiment where the crossover rate is 85%, the mutation rate is 10% and the reproduction rate is 5%. All statistical tests discussed in this section are the standard Wilcoxon tests, and they are considered significant if the obtained p-value is less than 0.05.

3.3.3.1 Makespan

As shown in Table 3.4, when $dev_{average}(\mathcal{H})$ is used as the fitness function, the learned heuristics based on R_1 show a performance close to those obtained by R_2 and R_3 on the training set. It is also noted that the R_1 heuristics learned with higher mutation rates are significantly better than those learned without lower mutation (all p-values

Table 3.4 Performance of learned heuristics for $Jm||C_{max}$ on training set and test set (*mean ± standard deviation*)

R_1		Training	Test
$dev_{average}(\mathcal{H})$	$\langle 95, 0, 5 \rangle$	0.188 ± 0.008	0.197 ± 0.007
	$\langle 90, 5, 5 \rangle$	0.181 ± 0.003	0.192 ± 0.005
	$\langle 85, 10, 5 \rangle$	0.180 ± 0.003	0.191 ± 0.006
$dev_{max}(\mathcal{H})$	$\langle 95, 0, 5 \rangle$	0.378 ± 0.011	0.460 ± 0.032
	$\langle 90, 5, 5 \rangle$	0.372 ± 0.007	0.445 ± 0.025
	$\langle 85, 10, 5 \rangle$	0.370 ± 0.008	0.457 ± 0.039
R_2		Training	Test
$dev_{average}(\mathcal{H})$	$\langle 95, 0, 5 \rangle$	0.181 ± 0.003	0.187 ± 0.004
	$\langle 90, 5, 5 \rangle$	0.181 ± 0.003	0.188 ± 0.005
	$\langle 85, 10, 5 \rangle$	0.180 ± 0.003	0.188 ± 0.004
$dev_{max}(\mathcal{H})$	$\langle 95, 0, 5 \rangle$	0.381 ± 0.009	0.474 ± 0.050
	$\langle 90, 5, 5 \rangle$	0.386 ± 0.012	0.461 ± 0.032
	$\langle 85, 10, 5 \rangle$	0.385 ± 0.009	0.472 ± 0.038
R_3		Training	Test
$dev_{average}(\mathcal{H})$	$\langle 95, 0, 5 \rangle$	0.183 ± 0.005	0.186 ± 0.005
	$\langle 90, 5, 5 \rangle$	0.181 ± 0.005	0.184 ± 0.006
	$\langle 85, 10, 5 \rangle$	0.179 ± 0.005	0.184 ± 0.004
$dev_{max}(\mathcal{H})$	$\langle 95, 0, 5 \rangle$	0.384 ± 0.010	0.457 ± 0.034
	$\langle 90, 5, 5 \rangle$	0.376 ± 0.011	0.457 ± 0.033
	$\langle 85, 10, 5 \rangle$	0.376 ± 0.010	0.452 ± 0.028

<0.0163) on the training set. Since R_1 heuristics contain many ERCs, higher mutation rates seem quite useful to improve the performance of the GP system with the R_1 representation. However, the performances of R_1 heuristics are quite poor on the test set. This indicates the overfitting issue of R_1 heuristics when learning with the training set. The reason for this problem comes from the fact that the candidate rules used in R_1 are too simple, and therefore the heuristics have to depend strongly on the machine and system statuses to provide better sequencing decisions on the training instances. However, the overuse of the machine and system attributes makes R_1 heuristics less effective when dealing with unseen instances in the test set.

Learned R_2 heuristics show a more consistent performance on both the training set and test set. Different from R_1, the statistical tests indicate that the choice of GP parameters does not have a significant influence (all p-values >0.1511) when R_2 is used as the representation of the dispatching rule. These results indicate that mutation is not really useful in this case and the crossover operator is sufficient for the GP system to evolve good individuals. Since R_2 provides only one way to sequence operations, the effectiveness is obtained by good combinations of different terms. Hence, they are also less affected when working with unseen instances like R_1 rules.

Taking the advantages of R_1 and R_2, the learned R_3 heuristics show very promising performance. Different from R_1, the incorporation of the machine and system attributes into the R_3 heuristics is supported by better composite dispatching rules, and therefore the R_3 heuristics do not need to depend heavily on the use of machine and system attributes to be effective. The performance on the test set shows that the learned R_3 heuristics also have good generalisation qualities like that of the R_2 heuristics. Mutation does not affect the GP system with R_3 as strongly as the GP system with R_1. The significant difference is only observed between the experiment with no mutation and the experiment with the mutation rate of 10% (p-value <0.0042). Although there is no obvious difference in the performance of learned heuristics from different configurations on the training set, R_2 heuristics and R_3 heuristics (learned with non-zero mutation rate) are significantly better than R_1 heuristics on the test set (all p-values <0.0421 over $(3 + 2) \times 3 = 15$ statistical tests). Also, the learned R_3 heuristics from the GP system with non-zero mutation rates are significantly better than R_1 and R_2 heuristics on the test set (all p-values <0.017 over $2 \times (3 + 3) = 12$ statistical tests).

When $dev_{\max}(\mathcal{H})$ is used as the fitness function, the learned R_1 heuristics do not show overfitting issues compared to other representations as shown in the case when $dev_{\text{average}}(\mathcal{H})$ is used. This is because the learned heuristics tend to focus on hard instances that caused high relative deviations and integrate essential features into the learned heuristics to avoid worst-case scenarios. In this case, R_1 and R_3 heuristics from the GP system with non-zero mutation rates are significantly better than R_2 heuristics in the training set (all p-values <0.0213 over $(2 + 2) \times 3 = 12$ statistical tests). This suggests that learned heuristics that employ the machine and system attributes are more effective to improve worst-case scenarios for $Jm||C_{\max}$. However, no obvious difference is recorded from the performance of learned heuristics on the test set. This indicates that the GP system with $dev_{\max}(\mathcal{H})$ as the fitness function does not provide good generalisation for its learned heuristics.

Table 3.5 shows the performance of the learned heuristics with heuristics from different combinations of four benchmark dispatching rules for $Jm||C_{\max}$ and non-delay factors. These four benchmark rules are selected since they are known to reduce the flowtimes (completion time C_j in static JSS) of jobs. The values in this table are the statistics of relative deviations using Eq. (3.1) obtained when applying the learned rules to the instances in the training set and test set. The values of Mean and Max can be calculated by using Eqs. (3.2) and (3.3) to measure the average and worst-case performance of a dispatching rule on a given set of instances. The values of Min, like Max, indicate the performance of the learned heuristics in extreme cases, but they are used to show the best-case performance instead. From each representation and each fitness function, two learned heuristics that show the best fitness in the training stage are used here for comparison. Heuristic $\mathcal{H}_{R_x}^{yvz}$ is the vth best rule that was learned by using the representation R_x, fitness function z to minimise objective function y for the JSS. For example, $\mathcal{H}_{R_2}^{c2a}$ is the second best heuristic learned with the representation R_2, $dev_{\text{average}}(\mathcal{H})$ as the fitness function (m will be used to indicate $dev_{\max}(\mathcal{H})$), and $C_{\max}$ as the objective function (t will be used to indicate the total weighted tardiness).

Table 3.5 Relative deviations obtained by learned heuristics and other heuristics for $Jm||C_{max}$

Heuristics		Training set			Test set		
		Min	Mean	Max	Min	Mean	Max
FIFO	Active	0.012	0.325	0.691	0.000	0.325	0.654
	Non-delay	0.012	0.325	0.691	0.000	0.325	0.654
SPT	Active	0.322	0.694	1.252	0.316	0.711	1.091
	Non-delay	0.029	0.292	0.576	0.092	0.312	0.664
LRM	Active	0.020	0.321	0.745	0.000	0.319	0.723
	Non-delay	0.000	0.224	0.556	0.000	0.225	0.529
MWKR	Active	0.047	0.323	0.736	0.000	0.328	0.713
	Non-delay	0.000	0.253	0.590	0.000	0.254	0.584
$dev_{\text{average}}(\mathcal{H})$	$\mathcal{H}_{R_1}^{c1a}$	0.000	0.173	0.448	0.000	0.192	0.525
	$\mathcal{H}_{R_1}^{c2a}$	0.000	0.174	0.428	0.000	0.183	0.479
	$\mathcal{H}_{R_2}^{c1a}$	0.000	0.174	0.479	0.000	0.190	0.442
	$\mathcal{H}_{R_2}^{c2a}$	0.000	0.175	0.457	0.000	0.191	0.446
	$\mathcal{H}_{R_3}^{c1a}$	0.000	0.171	0.490	0.000	0.185	0.572
	$\mathcal{H}_{R_3}^{c2a}$	0.000	0.172	0.447	0.000	0.177	0.428
$dev_{\text{max}}(\mathcal{H})$	$\mathcal{H}_{R_1}^{c1m}$	0.000	0.194	0.356	0.000	0.193	0.416
	$\mathcal{H}_{R_1}^{c2m}$	0.000	0.190	0.361	0.000	0.189	0.412
	$\mathcal{H}_{R_2}^{c1m}$	0.023	0.220	0.362	0.000	0.219	0.482
	$\mathcal{H}_{R_2}^{c2m}$	0.000	0.187	0.367	0.000	0.198	0.420
	$\mathcal{H}_{R_3}^{c1m}$	0.000	0.185	0.360	0.000	0.184	0.457
	$\mathcal{H}_{R_3}^{c2m}$	0.000	0.179	0.361	0.000	0.185	0.478

The results show that the heuristic that combines LRM and non-delay scheduling strategy is better than other hand-crafted heuristics. The performance of the learned heuristics is better than all of the hand-crafted heuristics on the training set and test set. The results from the learned heuristics show that the GP system can find the heuristics that minimise the fitness function. On the training set, heuristics learned by $dev_{\text{average}}(\mathcal{H})$ show better average relative deviations than those learned with $dev_{\text{max}}(\mathcal{H})$ as the fitness function. On the other hand, the heuristics learned by $dev_{\text{max}}(\mathcal{H})$ show better worst-case relative deviations than those learned with $dev_{\text{average}}(\mathcal{H})$ as the fitness function. However, not all learned heuristics can produce good results when dealing with unseen instances. Heuristics $\mathcal{H}_{R_1}^{c2a}$ and $\mathcal{H}_{R_3}^{c2a}$ are the heuristics that provide the best performance on the test set even though they are not the best rules on the training set. The two R_1 heuristics show the best worst-case performance on the training set and test set among heuristics learned with $dev_{\text{max}}(\mathcal{H})$. These results again suggest that the incorporation of machine and system attributes plays an important role in improving the worst-case performance. Generally, it seems difficult to develop heuristics that produce good average and worst-case performance. One explanation is that the heuristics learned with $dev_{\text{max}}(\mathcal{H})$ only aim at a group of hard instances (which cause high relative deviations) and cannot capture the general useful features. Meanwhile, the rules learned with $dev_{average}(\mathcal{H})$ focus on the over-all performance, therefore, they may ignore some extreme cases. Learned heuristic $\mathcal{H}_{R_3}^{c2a}$ is one rare case when the best average performance and very good worst-case performance are achieved.

3.3.3.2 Total Weighted Tardiness

According to Table 3.6, the experiments with R_3 as the representation can produce heuristics with better fitness than experiments with R_1 or R_2 as the representation. When $dev_{average}(\mathcal{H})$ is used as fitness, the statistical tests show that learned R_3 heuristics from the GP system with non-zero mutation rates are significantly better than other learned R_1 and R_2 heuristics on both training set and test set (all p-values <0.0007 over $2 \times (3 + 3) \times 2 = 24$ statistical tests). These results again confirm the effectiveness of the R_3 representation. Also, the R_1 heuristics from the GP system with non-zero mutation rates are significantly better than R_2 heuristics in this case (all p-values $<2 \times 10^{-16}$ over $2 \times 3 \times 2 = 12$ statistical tests). This suggests that the machine attributes and non-delay factors are important when dealing with

Table 3.6 Performance of learned heuristics for $Jm||\sum w_j T_j$ on training set and test set (*mean* $\pm$ *standard deviation*). Negative values mean the learned heuristics are better than EDD

R_1		Training	Test
$dev_{average}(\mathcal{H})$	$\langle 95, 0, 5 \rangle$	-0.227 ± 0.004	-0.221 ± 0.005
	$\langle 90, 5, 5 \rangle$	-0.235 ± 0.006	-0.223 ± 0.003
	$\langle 85, 10, 5 \rangle$	-0.237 ± 0.006	-0.222 ± 0.003
$dev_{max}(\mathcal{H})$	$\langle 95, 0, 5 \rangle$	0.002 ± 0.016	0.108 ± 0.139
	$\langle 90, 5, 5 \rangle$	-0.001 ± 0.017	0.061 ± 0.085
	$\langle 85, 10, 5 \rangle$	-0.001 ± 0.008	0.018 ± 0.049
R_2		Training	Test
$dev_{average}(\mathcal{H})$	$\langle 95, 0, 5 \rangle$	-0.216 ± 0.003	-0.203 ± 0.005
	$\langle 90, 5, 5 \rangle$	-0.216 ± 0.004	-0.205 ± 0.006
	$\langle 85, 10, 5 \rangle$	-0.216 ± 0.004	-0.204 ± 0.006
$dev_{max}(\mathcal{H})$	$\langle 95, 0, 5 \rangle$	0.000 ± 0.000	0.000 ± 0.000
	$\langle 90, 5, 5 \rangle$	0.000 ± 0.001	0.007 ± 0.036
	$\langle 85, 10, 5 \rangle$	0.000 ± 0.000	0.000 ± 0.000
R_3		Training	Test
$dev_{average}(\mathcal{H})$	$\langle 95, 0, 5 \rangle$	-0.240 ± 0.015	-0.233 ± 0.015
	$\langle 90, 5, 5 \rangle$	-0.245 ± 0.012	-0.233 ± 0.014
	$\langle 85, 10, 5 \rangle$	-0.247 ± 0.013	-0.235 ± 0.013
$dev_{max}(\mathcal{H})$	$\langle 95, 0, 5 \rangle$	0.006 ± 0.030	0.167 ± 0.148
	$\langle 90, 5, 5 \rangle$	-0.009 ± 0.026	0.132 ± 0.115
	$\langle 85, 10, 5 \rangle$	-0.017 ± 0.033	0.143 ± 0.125

Table 3.7 Relative deviations obtained by learned heuristics and other hand-crafted heuristics for $Jm||\sum w_j T_j$

Heuristics		Training set			Test set		
		Min	Mean	Max	Min	Mean	Max
FIFO	Active	−0.318	0.455	1.600	−0.159	0.442	1.196
	Non-delay	−0.318	0.455	1.600	−0.159	0.442	1.196
LRM	Active	0.193	0.832	2.129	−0.125	0.836	1.813
	Non-delay	−0.189	0.507	1.571	−0.236	0.494	1.425
MS	Active	0.106	0.601	1.519	0.035	0.607	1.321
	Non-delay	−0.040	0.387	1.205	−0.133	0.363	0.822
WSPT	Active	−0.133	0.148	0.874	−0.154	0.160	0.946
	Non-delay	−0.394	−0.169	0.168	−0.459	−0.161	0.253
W(CR+SPT)	Active	−0.133	0.140	0.670	−0.234	0.147	0.858
	Non-delay	−0.398	−0.173	0.177	−0.459	−0.165	0.435
W(S/RPT+SPT)	Active	−0.110	0.149	0.867	−0.154	0.161	0.853
	Non-delay	−0.394	−0.168	0.168	−0.459	−0.161	0.253
COVERT	Active	−0.171	0.146	0.722	−0.235	0.147	0.853
	Non-delay	−0.394	−0.173	0.177	−0.459	−0.160	0.253
ATC	Active	−0.258	0.145	0.757	−0.260	0.140	0.795
	Non-delay	−0.394	−0.168	0.168	−0.459	−0.163	0.253
$dev_{average}(\mathcal{H})$	$\mathcal{H}_{R_1}^{t1a}$	−0.586	−0.247	0.020	−0.560	−0.220	0.150
	$\mathcal{H}_{R_1}^{t2a}$	−0.531	−0.244	0.003	−0.507	−0.224	0.018
	$\mathcal{H}_{R_2}^{t1a}$	−0.467	−0.223	−0.003	−0.517	−0.199	0.326
	$\mathcal{H}_{R_2}^{t2a}$	−0.529	−0.223	0.151	−0.523	−0.206	0.116
	$\mathcal{H}_{R_3}^{t1a}$	−0.506	−0.265	−0.033	−0.616	−0.246	−0.002
	$\mathcal{H}_{R_3}^{t2a}$	−0.581	−0.265	−0.022	−0.555	−0.253	0.031
$dev_{max}(\mathcal{H})$	$\mathcal{H}_{R_1}^{t1m}$	−0.469	−0.212	−0.043	−0.492	−0.211	0.129
	$\mathcal{H}_{R_1}^{t2m}$	−0.493	−0.206	−0.041	−0.559	−0.203	0.121
	$\mathcal{H}_{R_2}^{t1m}$	−0.456	−0.170	−0.005	−0.490	−0.177	0.198
	$\mathcal{H}_{R_2}^{t2m}$	0.000	0.000	0.000	0.000	0.000	0.000
	$\mathcal{H}_{R_3}^{t1m}$	−0.487	−0.221	−0.098	−0.448	−0.202	0.086
	$\mathcal{H}_{R_3}^{t2m}$	−0.497	−0.215	−0.072	−0.514	−0.212	0.065

$Jm||\sum w_j T_j$. When $dev_{max}(\mathcal{H})$ is used as the fitness function, the learned heuristics cannot easily dominate EDD because the obtained $dev_{average}(\mathcal{H})$ are not significantly smaller than zero. This observation shows that $Jm||\sum w_j T_j$ is a very hard objective.

The comparison of the learned heuristics and other hand-crafted heuristics is shown in Table 3.7. It is easy to see that non-due-date-related rules such as FIFO and LRM have a very poor performance on $Jm||\sum w_j T_j$ even though LRM achieves good performance on $Jm||C_{max}$. MS and WSPT show better performance than FIFO and LRM because information about the due date and the weight of a job is considered in these rules. While MS is still much worse than EDD, WSPT shows good performance even though it still cannot totally beat EDD. Sophisticated due-date-related rules W(CR+SPT), W(S/RPT+SPT), COVERT, and ATC (see [151] for a detailed description of these rules) which have not been included in the list of candidate dispatching rules are also presented here. The expected waiting time in COVERT and ATC is calculated based on the standard method in which the expected waiting time $W = b \times PR$, where PR is the operation processing time. For each method, two parameters need to be specified which are k (look-ahead parameter) and b. 25 combinations of $b \in \{0.5, 1.0, 1.5, 2.0, 2.5\}$ and $k \in \{1.5, 2.0, 2.5, 3.0, 3.5\}$ are examined on the training set, and the combination that gives the best average performance is selected for comparison ($k = 2.5$, $b = 2.5$ for COVERT and $k = 1.5$, $b = 0.5$ for ATC). The performance of these two rules is quite good since they are customised to deal with weighted tardiness problems. However, in the worst case, they still cannot provide better schedules than those obtained by EDD.

On both the training set and the test set, all of these GP learned heuristics show better average performance than the hand-crafted heuristics (except for the R_2 heuristics learned with $dev_{\max}(\mathcal{H})$ as the fitness function). Similar to what has already been stated, R_1 and R_3 heuristics are also much better than R_2 heuristics. However, it is still not easy to find a heuristic that totally dominates EDD. Among all the learned rules, the learned R_3 heuristics are the most promising ones. The two best learned R_3 heuristics obtained very good average relative deviations for both training set and test set compared to those obtained by R_1 and R_2 heuristics. Heuristic $\mathcal{H}_{R_3}^{t1a}$ is also the only learned heuristic that totally dominates EDD on all training and test instances. Meanwhile, $\mathcal{H}_{R_3}^{t2a}$ produces a very good average performance, even better than $\mathcal{H}_{R_3}^{t1a}$ on the test set and it is just slightly worse than $\mathcal{H}_{R_3}^{t1a}$ in the worst case. The impressive results of R_3 heuristics again confirm the need for integrating machine and system attributes with sophisticated dispatching rules to generate generalised and effective heuristics.

In the case where $dev_{\max}(\mathcal{H})$ is used as the fitness function, the learned heuristics show a very poor generalisation quality. Their worst-case performance on the test set is sometimes worse than those obtained by heuristics using $dev_{\text{average}}(\mathcal{H})$ as the fitness function. Some promising heuristics can be found when R_3 is used as the representation. However, these heuristics are still much worse than their counterparts $\mathcal{H}_{R_3}^{t1a}$ and $\mathcal{H}_{R_3}^{t2a}$. This suggests that $dev_{\text{average}}(\mathcal{H})$ is better than $dev_{\max}(\mathcal{H})$ as the fitness function.

3.3.4 Further Analyses

In this section, we will further investigate the learned heuristics to figure out how they can produce a good performance.

3.3.4.1 Interpretability of the Learned Heuristics

In the previous experiment, there are 1080 heuristics learned with different GP parameters and representations for $Jm||C_{max}$ and $Jm||\sum w_j T_j$. The performance of the two best-found heuristics for each representation and each objective function of JSS were shown in Tables 3.5 and 3.7. As an example, we pick one learned heuristic from each representation among these heuristics based on their overall performance on the training set and test set for further analysis (other learned heuristics have a similar pattern).

Learned Heuristics for $Jm||C_{max}$: Here, $\mathcal{H}_{R_1}^{c1a}$, $\mathcal{H}_{R_2}^{c1a}$, and $\mathcal{H}_{R_3}^{c2a}$ are chosen for analysis. The detailed representations of these heuristics are shown in Fig. 3.7. For $\mathcal{H}_{R_1}^{c1a}$ (Fig. 3.7a), the first observation is that even though the heuristics look complicated, it is just a combination of four simple dispatching rules (LRM, SPT, LPT, and WSPT) and three machine attributes (CMI, CWR, and DJ). Since WSPT is less relevant to $Jm||C_{max}$ and LPT is not very effective in this case, they only appear once in the heuristic. The root condition of $\mathcal{H}_{R_1}^{c1a}$ checks the critical machine idleness, and it is noted that LRM is the main dispatching rule when the idleness of the critical machine is greater than 10%. When CMI is small, rules that favour small processing time operations like SPT and WSPT occur more in this case. This heuristic suggests that when the critical machine seems to be idle, the considered machine should focus on completing operations with small processing times in order to feed more work to the critical machine and keep it busy; otherwise, LRM should be used to prevent certain jobs from being completed so late and increase the makespan.

Different from $\mathcal{H}_{R_1}^{c1a}$, $\mathcal{H}_{R_2}^{c1a}$ (Fig. 3.7b) is just a pure mathematical function (i.e., priority function to prioritise operations). In order to make it easy for analysis, we will simplify the whole function by eliminating terms which appear to be less relevant. The simplification step is as follows:

$$
\begin{aligned}
\mathcal{H}_{R_2}^{c1a} &= \frac{(PR + RM)W}{PR} - \frac{(RJ + RM)PR}{RT} + RT + \frac{DD}{(PR + RM)} RT \frac{W}{(PR^2)} + \frac{RT}{PR}(RJ + RM) \\
&\approx RT + \frac{RT}{PR} \frac{DD}{(PR + RM)} \frac{W}{(PR)} + \frac{RT}{PR}(RJ + RM) \\
&\approx RT + k \times \frac{RT}{PR}
\end{aligned}
$$

Since the first and second terms of $\mathcal{H}_{R_2}^{c1a}$ do not make much sense in this case, we just drop them from the priority function. The rest of the function can be grouped into two parts. The first part has RT, like rule MWKR, and the second part contains $\frac{RT}{PR}$, which is a combination of SPT and MWKR. When considering other terms in the second part as a constant k, we have the approximation of $\mathcal{H}_{R_2}^{c1a}$ as a linear combination of MWKR and SPT/MWKR. This rule is actually not new. If we omit RT in the approximation function, the rest is known as the shortest processing time by total work (SPTtwk) rule in the literature [49].

Heuristic $\mathcal{H}_{R_3}^{c2a}$ (see Fig. 3.7c) is the most interesting heuristic in this case because both arithmetic rules and simple dispatching rules are employed. However, with the

```
(IF (> CMI 10%)
 (IF (> CWR 20%)
  (IF (> CWR 80%) (DISPATCH 0.131 LRM)
   (IF (≤ DJ 30%) (DISPATCH 0.198 SPT) (DISPATCH 0.102 LRM)))
   (IF (> CWR 10%) (DISPATCH 0.102 LRM) (DISPATCH 0.131 LRM)))
 (IF (> CWR 10%)
  (IF (> CWR 80%) (DISPATCH 0.014 WSPT)
   (IF (≤ DJ 30%) (DISPATCH 0.198 SPT) (DISPATCH 0.131 LRM)))
  (IF (> CWR 80%) (DISPATCH 0.830 LPT)
   (IF (≤ DJ 20%) (DISPATCH 0.198 SPT) (DISPATCH 0.102 LRM)))))
```

(a) Learned heuristic $\mathcal{H}_{R_1}^{c1a}$

```
(+ (+ (-
        (*(+ PR RM) (/ W PR))
        (/(+ RJ RM) (/ RT PR)))
     RT)
  (-(*
        (/ DD(+ PR RM))
        (*(/ RT PR) (/ W PR)))
  ((-1*) (*(/ RT PR) (+ RJ RM)))))
```

(b) Learned heuristic $\mathcal{H}_{R_2}^{c1a}$

```
(IF (> CWR 90%)
 (DISPATCH 0.069 (/ (* (+ RJ 0.594) (+ RT PR)) (+ W PR)))
 (IF (≤ MP 100%) (DISPATCH 0.128 (/ (- RT PR) (+ W PR)))
  (IF (≤ CWR 100%) (IF (≤ MP 100%)
   (DISPATCH 0.166 WSPT) (DISPATCH 0.282 LPT))
   (IF(≤ MP 100%)(DISPATCH 0.044 LRM) (DISPATCH 0.736 FIFO)))))
```

(c) Learned heuristic $\mathcal{H}_{R_3}^{c2a}$

Fig. 3.7 Selected learned heuristics for $Jm||C_{max}$

condition that MP has to be less than 100% at the second level, we can totally eliminate the subtrees that contain the simple dispatching rules. After some simplification steps, it is also noted that the arithmetic rules are also variants of SPTtwk. With the support of the machine attribute, the obtained arithmetic rules become much easier to analyse. The simplified version of $\mathcal{H}_{R_3}^{c2a}$ is as follows:

$$\mathcal{H}_{R_3}^{c2a} = \begin{cases} \langle \frac{(RJ+0.594845)(RT+PR)}{W+PR}, \alpha = 0.069 \rangle & \text{if } CWR > 90\% \\ \langle \frac{(RT-PR)}{(W+PR)}, \alpha = 0.128 \rangle & \text{otherwise} \end{cases}$$

$$\approx \begin{cases} \langle \frac{(RJ+0.594845)(RT+PR)}{PR}, \alpha = 0.069 \rangle & \text{if } CWR > 90\% \\ \langle \frac{(RT-PR)}{PR}, \alpha = 0.128 \rangle & \text{otherwise} \end{cases}$$

The notation $\langle \cdot, \cdot \rangle$ indicates the dispatching rule and α value as in the middle and right subtrees of Fig. 3.5. Although both priority functions of $\mathcal{H}_{R_3}^{c2a}$ are variants of SPTtwk, the non-delay factor α for the case is smaller when CWR > 90%. One

```
(IF (>DJ 80%)
  (IF (>BWR 90%)
    (IF (≤ DJ 90%) (DISPATCH 0.426 WSPT)
      (IF (≤ MP 10%) (DISPATCH 0.436 WSPT) (DISPATCH 0.364 LPT)))
    (DISPATCH 0.065 WSPT))
  (IF (> BWR 20%)
    (IF (≤ DJ 30%) (DISPATCH 0.436 WSPT)
      (IF (≤ DJ 30%) (DISPATCH 0.364 LPT) (DISPATCH 0.389 WSPT)))
    (IF (≤ DJ 30%) (DISPATCH 0.436 WSPT)
      (IF (> DJ 80%) (DISPATCH 0.364 LPT) (DISPATCH 0.181 WSPT)))))
```

(a) Learned heuristic $\mathcal{H}_{R_1}^{t1a}$

```
(- (+ (*
      (* PR (* 0.614577 PR))
      (- ((-1*) RM) (/ RM W)))
      ((-1*) (* (* RT PR) (/ RT W))))
   (+ (-
      (/ (* RT PR) (- W 0.5214191))
      (* (/ RM W) (* 0.614577 PR)))
      (* (/ (* RT PR) (- W 0.5214191)) (+ (/ RM W) (/ RM W)))))
```

(b) Learned heuristic $\mathcal{H}_{R_2}^{t2a}$

```
(IF (≤ DJ 50%)
  (DISPATCH 0.331 ((-1*) (* DD (/ PR W))))
  (DISPATCH 0.163 ((-1*) (* (/ DD W) (/ RT W)))))
```

(c) Learned heuristic $\mathcal{H}_{R_3}^{t1a}$

Fig. 3.8 Selected learned heuristics for $Jm||\sum w_j T_j$

explanation is that when Ω' contains many critical operations, it is reasonable to start the available operations right away instead of waiting for the operations that will be ready after the ready time of m^*.

Learned Heuristics for $Jm||\sum w_j T_j$: Here, $\mathcal{H}_{R_1}^{t1a}$, $\mathcal{H}_{R_2}^{t2a}$, and $\mathcal{H}_{R_3}^{t1a}$ are selected to represent the learned rules for $Jm||\sum w_j T_j$. The three full heuristics obtained by the GP are shown in Fig. 3.8. For $\mathcal{H}_{R_1}^{t1a}$, it is quite interesting that this heuristic can obtain such a good result (as shown in Table 3.7) without any due-date-related components. The two main dispatching rules used in this case are WSPT and LPT. While WSPT can be considered as a suitable rule for $Jm||\sum w_j T_j$, it does not make sense to include LPT in this case. The result when we replace LPT with WSPT shows that the refined heuristic can still produce the results as good as $\mathcal{H}_{R_1}^{t1a}$. For this reason, the contribution to the success of the heuristic comes from WSPT and other factors instead of the combination of different rules as we observed in the previous section. It is noted that most values of α, in this case, are about 0.4. Using these values for the WSPT alone shows that WSPT with an appropriate choice of α can produce the results much better than the case when the non-delay scheduling strategy is used. In this heuristic, the contribution of the machine and system attributes are not very significant, and they are mainly employed to improve the worst-case performance.

For $\mathcal{H}_{R_2}^{t2a}$, we perform the following steps to simplify the learned function:

$$\mathcal{H}_{R_2}^{t2a} = 0.614PR^2(-RM - \frac{RM}{W}) - RT \times PR \times \frac{RT}{W} - (RT\frac{PR}{W} - 0.521)$$

$$- 0.614\frac{RM}{W}PR + 2RT\frac{PR}{(W - 0.5214191)}\frac{RM}{W}$$

$$\approx -0.614RM \times PR^2(1 + \frac{1}{W}) - RT\frac{PR}{W}(RT + 2\frac{RM}{(W - 0.5214191)})$$

$$\approx -k_1 \times PR^2(1 + \frac{1}{W}) - k_2 \times RT\frac{PR}{W}$$

The simplified function is a linear combination of two sophisticated variants of WSPT where the first part includes PR^2 instead of PR and the second part includes RT. Repeating the experiment on this simplified rule shows that it can perform better than sophisticated rules like ATC and COVERT (just slightly different from the full rule) regarding the average relative deviation with appropriate choice of k_1 and k_2 (with $k_2 > k_1$, similar to the original rule).

It is very surprising that the best-found heuristic for $Jm||\sum w_j T_j$ with the R_3 representation is also the smallest heuristic. Heuristic $\mathcal{H}_{R_3}^{t1a}$ can be formally described as follows:

$$\mathcal{H}_{R_3}^{t1a} = \begin{cases} \langle -\frac{DD \times PR}{W}, \alpha = 0.331 \rangle & \text{if } DJ \leq 50\% \\ \langle -\frac{DD \times RT}{W^2}, \alpha = 0.163 \rangle & \text{otherwise} \end{cases}$$

The dispatching rule following the first priority function is a combination of EDD and WSPT, and it is applied when the deviation of processing times of operations in Ω' is less than 50% (which means that the minimum processing time is less than half of the maximum processing time). When DJ is larger than 50% (which means that the gap between the minimum and maximum processing time is small), RT is used instead of PR and W^2 is used instead of W to increase the priority of jobs with small remaining processing times and high weights.

In general, even though the learned heuristics can be very complicated sometimes, they contain some good patterns which are very useful to create good dispatching schedules. While R_1 heuristics can be easily explained with the support of the machine and system attributes, R_2 heuristics are quite sophisticated and often possess very interesting properties. It is surprising that the R_3 heuristics presented here are quite straightforward even though they contain both machine attributes and composite dispatching rules. Heuristic $\mathcal{H}_{R_3}^{t1a}$ could be quite tricky to represent as a purely mathematical function, and it could be more difficult to discover its useful patterns; however, the use of attributes makes it much easier to identify and explain the rule.

3.3.4.2 Importance of Attributes in Learned Heuristics

Figure 3.9 shows the frequency that machine and system attributes have occurred in all of the learned R_1 and R_3 heuristics. For $Jm||C_{\max}$, critical machine-related attributes are the most frequent ones, while the bottleneck workload ratio has the lowest frequency. This result indicates that the information related to the critical machine is very important for the construction of a good heuristic to minimise the makespan. As shown in the previous section, suitable dispatching rules can be selected based on the idleness of the critical machine as well as the critical workload of m^*. In the case of $Jm||\sum w_j T_j$, the critical machine-related attributes are still employed very often, and CWR along with DJ is the ones with the highest occurrence frequency. However, the distribution of attributes in $Jm||\sum w_j T_j$ is more "uniform" than that in $Jm||C_{\max}$. This observation suggests that different attributes should be used to construct good heuristics for $Jm||\sum w_j T_j$.

For the priority functions (for composite dispatching rules) within R_2 and R_3 rules, Fig. 3.10 shows that RT and PR are the most used terms for $Jm||C_{\max}$. This explains the occurrence of SPTtwk variants in the best rules in the previous section. While W is the least used term for $Jm||C_{\max}$, it is the most used term in $Jm||\sum w_j T_j$. This emphasises the importance of weights for determining the priority of operations in $Jm||\sum w_j T_j$. It is quite surprising that the second most frequent term in learned heuristics for $Jm||\sum w_j T_j$ is PR instead of DD. Even though due dates of jobs also depend on the processing times in aggregate form, the results here suggest that such local information as PR is still very useful for dispatching tasks for due-date-related problems.

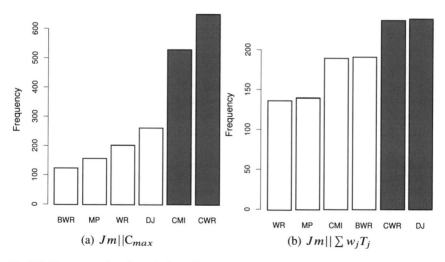

(a) $Jm||C_{max}$ (b) $Jm||\sum w_j T_j$

Fig. 3.9 Frequency of attributes in R_1 and R_3 heuristics

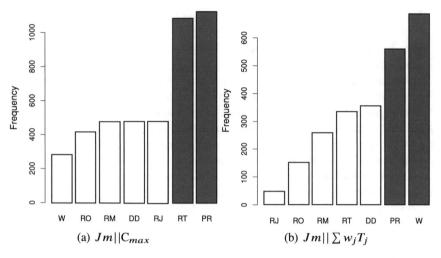

(a) $Jm||C_{max}$ (b) $Jm||\sum w_j T_j$

Fig. 3.10 Frequency of terminals in R_2 and R_3 heuristics

3.4 Chapter Summary

This chapter presents three GP representations for schedule construction heuristics. One point that we try to demonstrate here is the flexibility of GP, which allows us to evolve a wide range of heuristics with different levels of interpretability (i.e., understand how decisions are made) and comprehensiveness (i.e., ability to cover key components and different decision situations). Each representation has its own advantages and disadvantages as shown in the experiments above. In general, heuristics learned with these representations are significantly better than hand-crafted heuristics in the literature in terms of schedule quality. However, we should select a representation mainly based not only on its performance (e.g., quality of obtained schedules) but also on other practical requirements such as interpretability and ease of implementation. From the experiment results and analyses, we can see different computational challenges related to each representation such as overfitting and fitness evaluations. Many recent studies have been dedicated to addressing these challenges and have developed new GP systems that can discover more effective and interpretable scheduling heuristics.

Although the representations here are developed for JSS, they can be modified or expanded to cope with other production environments (interested readers can check [27] for more applications of these representations). In the JSS problems investigated in this chapter, we only need to focus on two main components: (1) dispatching rules and (2) non-delay factor. But in other environments such as flexible job shops, assembly shops, and wafer-fab factories, different meta-algorithms are needed and different sets of components will need to be designed. For example, we need to design both routing rules and sequencing rules for flexible job shops.

This chapter aims to help the readers get familiar with the fundamental issues in learning or evolving schedule construction heuristics with GP (and other machine learning techniques) and uses job shop scheduling as a case study. The later chapters in this part will show how this fundamental element can be utilised in wider and more sophisticated contexts.

Chapter 4
Learning Schedule Improvement Heuristics

Learning scheduling construction heuristics is a straightforward approach to handling production scheduling with GP and machine learning techniques. Since construction heuristics mainly utilise local information of jobs and machines, it is hard for construction heuristics to create optimal or near-optimal schedules for a problem instance. More appropriate use of construction heuristics is to quickly build a good schedule and further refine this schedule with some improvement heuristics. However, a smaller performance gap between schedules obtained by construction heuristics and optimal schedules is desirable. This chapter will explore how to improve the schedule construction heuristics to cope better with a wider range of decision situations.

4.1 Challenges and Motivations

Designing a universal heuristic to cope well with different problem characteristics and decision situations is very challenging, if not impossible. Two simple solutions to overcome this challenge are (1) using a set of heuristics or multi-pass heuristics to cover a wider range of decision situations, and (2) iteratively improving the schedules obtained by construction heuristics. The first solution can be easily done by running GP multiple times or using a diversification strategy to obtain a set of diverse heuristics. This solution usually works well with static scheduling problems, but the use of many heuristics makes it hard to understand why a scheduling decision is made (e.g., why job A is scheduled early this week by one heuristic but has to be delayed until the end of the week if another heuristic is used).

In this chapter, we investigate the second solution. The goal of this chapter is to explore how to design improvement heuristics that can logically (i.e., taking into account job and machine attributes and the information of current complete schedule) and efficiently (i.e., using a small number of iterations to achieve large improvements) modify a complete schedule to achieve better scheduling performance. Specially, we are interested in the following questions:

- How can GP be used to evolve improvement heuristics?
- What attributes are needed to develop improvement heuristics?
- How can ideas from optimisation search be incorporated to further improve the effectiveness of the evolved heuristics?

Section 4.2 shows how the representation and meta-algorithms introduced in the previous chapter are extended to help GP evolve improvement heuristics. Details of additional attributes/terminals to guide the improvements are also presented here. The similarity between the proposed improvement heuristics and other optimisation techniques in the literature is discussed. Section 4.3 compares proposed improvement heuristics and construction heuristics in the previous chapter.

4.2   Algorithm Design and Details

This section will introduce an improvement heuristic called iterative dispatching rules (IDR). IDR is an extension of the R_2 representation in the previous chapter with an improvement mechanism.

4.2.1   Meta-algorithm for Iterative Dispatching Rules

A composite dispatching rule $\mathcal{DR}$ uses a priority function to assign a priority for each job $\mathcal{J}$ in the queue of a considered machine $\mathcal{W}$ based on the information from $\mathcal{J}$ (e.g., processing time and due date) and $\mathcal{W}$ (e.g., ready time). After a priority has been assigned to all jobs in the queue, the one with the highest priority will be processed next. A limitation of these rules is that they can only provide sequencing decisions based on the available information of jobs and machines at the time the decision is made. Since these pieces of information come from a partial schedule, the future impact of the decisions made by a dispatching rule on the complete schedule is not considered in order to enhance the effectiveness of the rule. Different from these dispatching rules, an IDR is defined as $\mathcal{H}^I(\mathcal{J}, \mathcal{W}, \mathcal{R})$, where $\mathcal{R}$ is the recorded information of the previously generated schedule. The overall algorithm used to construct a schedule by $\mathcal{H}^I(\mathcal{J}, \mathcal{W}, \mathcal{R})$ is shown in Fig. 4.1.

In the beginning, the initial recorded information $\mathcal{R}^0$ (more details on how to determine $\mathcal{R}^0$ will be shown in the next section) is assigned to $\mathcal{R}$. Steps from 4 to 12 are used to construct a schedule. After a schedule is obtained, the objective is calculated and compared to the best objective Obj^* obtained in previous iterations. If the schedule obtained is improved, Obj^* will be updated and the information from scheduled jobs is used to update $\mathcal{R}$. This procedure will be repeated until no improvement is

1: $\mathcal{R} \leftarrow \mathcal{R}^0$
2: $Obj^* = +\infty$
3: $\pi \leftarrow \emptyset$ and $\Omega \leftarrow \{o_{11}, o_{21}, \ldots, o_{n1}\}$
4: **repeat**
5:     let $t(\Omega) = \min_{\sigma \in \Omega}\{\max\{r(\sigma), U_{M(\sigma)}\} + \delta(\sigma, M(\sigma))\}$ ▷ $U_{M(\sigma)}$ is ready time of machine $M(\sigma)$
6:     let σ^* be the operation that minimum is achieved, $M^* = M(\sigma^*)$, and $\Omega^* = \{\sigma \in \Omega | M(\sigma) = M^*\}$
7:     let $S(m^*) = \max\{\min_{\sigma \in \Omega^*}\{r(\sigma)\}, U_{M^*}\}$
8:     let $\Omega' = \{\sigma \in \Omega^* | r(\sigma) \leq S(M^*) + \alpha(t(\Omega) - S(M^*))\}$
9:     apply $\mathcal{H}^I(\mathcal{J}, \mathcal{M}, \mathcal{R})$ on Ω' to find the next operation σ' to be scheduled on M^*
10:     remove σ' from Ω' and include it into π
11:     include $next(\sigma')$ into Ω if $next(\sigma') \neq null$
12: **until** all operations have been scheduled
13: $Obj(\pi) \leftarrow$ calculate the objective value obtained from schedule π
14: **if** $Obj^* > Obj(\pi)$ **then**
15:     $Obj^* = Obj(\pi)$
16:     $\mathcal{R} \leftarrow get\_information(\pi)$
17:     go back to step 3
18: **else**
19:     end the procedure, and return $(Obj^*, \mathcal{R})$
20: **end if**

Fig. 4.1 Meta-algorithm for IDR

realised from the newly generated schedule. Different from traditional dispatching rules which only make sequencing decisions based on the partial schedule, IDRs can use the information from a complete schedule to correct the mistakes made in previous iterations.

Regarding the time complexity of IDRs, the maximum number of schedules (or iterations) $maxstep^{IDR}$ generated by IDRs is fewer than $\lceil \frac{Obj_I - LB}{\varepsilon} \rceil$, where Obj_I is the objective of the initial solution, LB is a lower bound of the problem, and ε is the smallest improvement in the objective values. Obviously, LB and Obj_I are always finite and ε is always larger than zero (for all instances with C_{max} as the objective function and integer processing times, ε is always larger than or equal to one). Therefore, $maxstep^{IDR}$ is always a finite value and IDRs can always stop in a finite time.

4.2.2 Representation of IDRs

The same GP algorithm introduced in the previous chapter will be used to evolve IDRs. The terminal set for IDRs is shown in Table 4.1. The terminals in the upper part are the same as those used by the R_2 representation. The three terminals in the lower part of this table are the recorded information $\mathcal{R}$ of scheduled jobs from the previous iteration. These terminals provide information about the previous schedule

Table 4.1 Terminal set used in IDR

Notation	Description
RJ	Operation ready time
RO	Number of remaining operations
RT	Work remaining of job
PR	Operation processing time
W	Weight
DD	Due date
RM	Machine ready time
#	Constant from Uniform [0, 1]
RFT	Recorded finish time
RWT	Recorded operation waiting time
RNWT	Recorded waiting time of the next operation

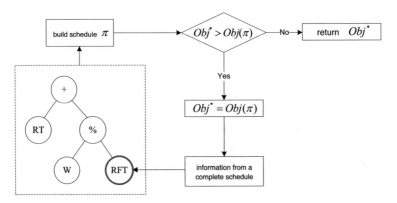

Fig. 4.2 Representation and evaluation of an evolved IDR

at the job level (RFT) and the operation level (RWT and RNWT). An illustration of how an IDR is represented and evaluated is shown in Fig. 4.2. In this case, a GP individual will play the role of a $\mathcal{H}^1(\mathcal{J}, \mathcal{W}, \mathcal{R})$ to assign the priorities to jobs in the queues, which is Step 9 in the schedule construction procedure in Fig. 4.1. The first time when a GP individual is used to construct the schedule, since there has been no recorded information $\mathcal{R}$ from the previous schedule, other pieces of information will be employed as the initial value of RFT, RWT, and RNWT, which represent the initial recorded information $\mathcal{R}^0$.

We employ two initialisation approaches for IDRS. In the first approach, the initial value for RFT is the total processing time of the considered job. For RWT and RNWT, they take half of the workload (total operation processing time of jobs waiting in the queue) of the considered machine as their initial recorded waiting times. This implies that a job has to wait for half of the jobs in the queue to be finished before it can be processed, given that all of those jobs have the same processing times.

Fig. 4.3 An example of a
pseudo-terminal

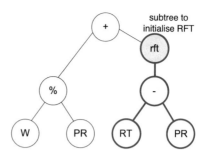

In the second approach, we let GP decide how RFT, RWT, and RNWT are initialised. Particularly, the three terminals RFT, RWT, and RNWT will be turned into *pseudo*-terminals **rft**, **rwt**, and **rnwt**. The difference between these pseudo- terminals and the original ones is that the pseudo-terminals are treated as a function that also includes a child node or a subtree. When the evolved IDRs are used in the first iteration, the values evaluated from the subtrees will be used as $\mathcal{R}^0$. In the later iterations, these subtrees will be ignored when the evaluation reaches the pseudo-nodes and $\mathcal{R}$ will be used in this case instead. An illustration of this approach is shown in Fig. 4.3. In this example, when evaluating node rft, we will check whether it is the first iteration of the scheduling construction procedure. If it is the first iteration, this node will return the output from its subtree which is (RT–PR) in this case. Otherwise, it will return the recorded finish time from the previous schedule.

4.2.3 Enhancing IDRs with Variable Neighbourhood Search

IDRs try to improve the quality of the schedule through some iterations. Different from a traditional dispatching rule that only provides a unique schedule π, an IDR will iteratively generate a sequence of schedules $\pi^0 \to \pi^1 \to \cdots \to \pi^n$ where $Obj(\pi^i) > Obj(\pi^{i+1})$ and π^{i+1} is a function $\mathcal{F}$ of π^i ($\mathcal{F} = \mathcal{H}$ in IDRs). This behaviour is very similar to that of a *local search*, which also tries to improve the quality of the solution iteratively. However, instead of using a neighbourhood structure $\mathcal{N}(\pi^i)$ to find a new improved solution, an IDR employs a function $\mathcal{F}(\pi^i)$ to generate a new schedule.

Being trapped at a local optimum is a big issue with local search methods. For this reason, many approaches to escaping from the local optima have been proposed in the literature. Some popular approaches used for JSS problems are simulated annealing [121], tabu search [166], large-step optimisation [138], and guided local search [12]. Even though these methods have been shown to be very effective when tested on benchmark JSS instances, they require extensive knowledge of the local search operators to make the approaches effective/efficient. Therefore, it would be hard to apply these approaches to evolve IDRs. For IDRs, it would be more suitable to consider high-level and more general approaches. Variable neighbourhood search

```
 1: select a set of iterative dispatching rules $\mathcal{F}_k$, $k = 1, \ldots, k_{max}$
 2: $Obj^{**} = +\infty$ and $k = 1$
 3: $\mathcal{R}^* \leftarrow \mathcal{R}^0$
 4: $improve \leftarrow false$
 5: $\mathcal{R} \leftarrow \mathcal{R}^*$
 6: $(Obj^*, \mathcal{R}) \leftarrow$ generate schedule with $\mathcal{F}_k$ as $\mathcal{H}^I (\mathcal{J}, \mathcal{W}, \mathcal{R})$
 7: if $Obj^{**} > Obj^*$ then
 8:      $Obj^{**} = Obj^*$
 9:      $\mathcal{R}^* \leftarrow \mathcal{R}$
10:      $improve \leftarrow true$
11: end if
12: if $k \neq k_{max}$ then
13:      $k = k + 1$ and return to step 5
14: else
15:      if $(improve = true)$ then $k = 1$ and return to step 4;
16:      else stop and return $(Obj^{**}, \mathcal{R}^*)$
17: end if
```

Fig. 4.4 IDR-VNS heuristic

(VNS) [89] seems a good candidate for this task since it mainly deals with the choice of neighbourhood structure or local search heuristics and their order in the search instead of low-level manipulations. One interesting idea in VNS is that a local optimum within one neighbourhood is not necessarily a local optimum within others, and a change of neighbourhoods has the potential to enhance the effectiveness of the search. In this study, we propose a new local search method based on IDR and VNS. An overview of the new heuristic IDR-VNS is shown in Fig. 4.4.

The key idea of this method is to employ different IDRs to explore better schedules since the best schedule obtained by an IDR is not necessarily the best solution for the other IDRs. However, instead of applying different IDRs independently, we will use the recorded information $\mathcal{R}^*$ from the best schedule obtained by an IDR to be the initial $\mathcal{R}^0$ of the next IDR to efficiently explore better schedules. Similar to VNS, in order to create effective IDR-VNS, we need to decide (1) how many $\mathcal{F}_k$ to be used (k_{max}), (2) what $\mathcal{F}_k$ to be used, and (3) the order in which these $\mathcal{F}_k$ are applied. While the first factor has to be decided experimentally, the next two factors can be handled by GP. A GP method is proposed in which a local search IDR-VNS is represented by a GP individual with multiple trees. An example of this representation is shown in Fig. 4.5. By evolving these GP individuals, GP can also help us find the effective IDRs and the order in which they will be applied.

The evolved IDRs are fundamentally more general than the construction heuristics examined in the previous chapter because they consider both construction steps and improvement steps rather than just construction steps. Given that the representation of IDRs in GP is very similar to that of R_2, we can expect that IDRs will always be as good as or even better than construction heuristics evolved with R_2 if GP can search the search space efficiently. The next section will show the experiment results to help us verify our hypotheses.

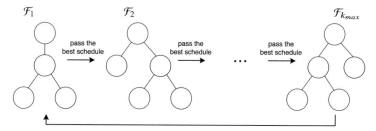

Fig. 4.5 Representation of an IDR-VNS heuristic

4.3 Empirical Study

To assess the performance of the learned improvement heuristics, we use the same datasets in the previous chapter. We run two sets of experiments to (1) compare the performance of IDRs and heuristics evolved with $R1$, R_2, and R_3 representations introduced in the previous chapter, and (2) examine the influence of the number of IDRs on the performance of IDR-VNS. In these experiments, the learned IDRs and construction heuristics evolved with R_1, R_2, and R_3 all use the non-delay factor $\alpha = 0$. All of the rules in this table are evolved using $dev_{average}(\Delta)$ as the fitness function with the crossover rate and mutation rate of 90% and 5%, respectively.

4.3.1 Comparing IDR Variants and Schedule Construction Heuristics

Figure 4.6 shows the performance of IDR variants and construction heuristics evolved in the previous chapter. In this figure, IDRs are evolved heuristics with a simple initialisation, IDR-P is used to indicate the evolved heuristics using pseudo- terminals, and IDR-VNS are evolved heuristics enhanced by variable neighbourhood search mechanism with $k_{max} = 2$.

It is easy to see that the evolved IDRs dominate the evolved schedule construction heuristics. For $Jm||C_{max}$, there is no significant difference between IDR and IDR-P. These results indicate that our approach to initialise R (as described in Sect. 4.2.2) is good enough for these problems. For $Jm||\sum w_j T_j$, IDR-P rules are significantly better than IDRs on the training set but there is no significant difference between these two types of heuristics on the test set. One explanation for this is that the initialisation of R can be useful when dealing with sophisticated objectives such as $\sum w_j T_j$; however, each problem instance may require different R^0, and it is really hard to create a general way to generate good initial R^0.

IDR-VNS is the best improvement heuristic in this comparison since the average relative performance of IDR-VNS is significantly smaller than those of all the other approaches. Obviously, the use of variable IDRs to search for good schedules is shown

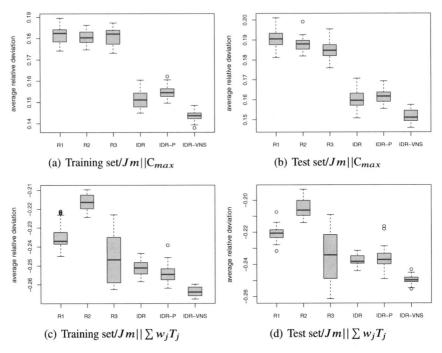

Fig. 4.6 Comparison between evolved schedule construction heuristics, IDR, IDR-P, and IDR-VNS

to be very effective. The fact that IDR-P has trouble improving the quality of IDRs and the success of IDR-VNS suggests two interesting points. First, the JSS instances in the datasets have very different characteristics, which cannot be handled easily by a single dispatching rule, even when the feedback from the previous schedule is used. Second, from the optimisation viewpoint, the search space of JSS problems is very complicated with many local optima. Therefore, the use of good $\mathcal{R}^0$ alone may not be useful because IDRs can be easily trapped at some local optimum. To enhance their performance, IDRs need to include some mechanism that helps escape from the local optima. VNS is an effective mechanism of this kind.

4.3.2 Influence of k_{max} on IDR-VNS

Given that IDR-VNS can effectively solve JSS problems, we now investigate how the number of IDRs, k_{max}, influences the performance of the IDR-VNS. Figure 4.7 shows the performance of IDR-VNS with $k_{max} = 2$, 3 and 4. It is quite clear that the performance of the IDR-VNS improves significantly when k_{max} increases. This again confirms the importance of using different IDRs to avoid being trapped in a local optimum. Another interesting observation is that while the average relative deviation

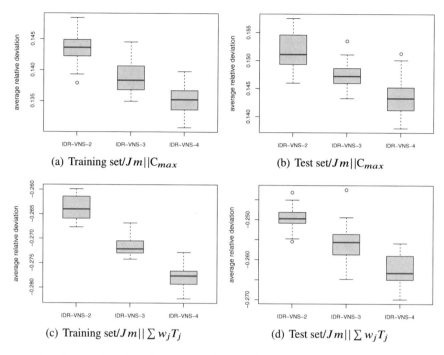

(a) Training set/$Jm||C_{max}$ (b) Test set/$Jm||C_{max}$

(c) Training set/$Jm||\sum w_j T_j$ (d) Test set/$Jm||\sum w_j T_j$

Fig. 4.7 Influence of number of trees k_{max} on IDR-VNS

tends to linearly decrease when k_{max} increases in the training set, the gap between IDR-VNS with different k_{max} tends to be larger in the test set. This indicates that a high k_{max} will also increase the generality of IDR-VNS. A reason is that a high k_{max} not only improves the quality of the schedules but also allows the evolved IDR-VNS rules to cover more situations in the training set.

4.4 Chapter Summary

This chapter introduces a simple way to transform the construction heuristics introduced in the previous chapter into improvement heuristics. The main difference between the two types of heuristics is the meta-algorithms they employ to construct and refine schedules. In our experiments, the learned improvement heuristics are significantly better than construction heuristics. Given that the computational efforts required by the improvement heuristics are not significantly larger than construction heuristics, they are a promising way to enhance the quality of obtained schedules.

Although we only investigate the R_2 representation in this chapter to evolve improvement heuristics, it is entirely possible to employ R_1 and R_3 by introducing appropriate attributes/terminals into these representations. Compared to construction heuristics, one of the additional challenges, when developing representations for improvement heuristics, is the selection of attributes from the complete schedule to improve the schedule decisions. Although GP can help to combine some primitive attributes (e.g., completion times of jobs or waiting times) to guide the improvement steps, identifying good primitive attributes is still challenging and usually requires problem-domain knowledge. The performance of the learned heuristics will depend on the choice of these primitive attributes similar to how neighbourhood structure influences the performance of a local search.

Both construction heuristics and improvement heuristics discussed in the first two chapters are useful to provide quick and good solutions for production scheduling problems. However, the performance of these heuristics relies on the choice of terminal sets, meta-algorithms, and existing problem-domain knowledge. Significant efforts are usually required to develop these components, and many low-level implementations are still needed to make GP efficient. The chapter will explore how combining GP and optimisation solvers can help to address these limitations.

Chapter 5
Learning to Augment Operations Research Algorithms

Production scheduling has been investigated by operations research and computer science communities for many decades. Nowadays, most well-established optimisation solvers such as CPLEX, Gurobi, and Google OR-Tools include production scheduling in their tutorials and even have some specialised modules for solving production scheduling problems. However, such generic solvers may not be efficient, especially, when dealing with large-scale problems. In those cases, optimisation tricks and problem-domain knowledge are useful to speed up the solving process. However, there are limitations to this solution. Optimisation tricks are complex and time-consuming to develop with problem knowledge. Meanwhile, problem-domain knowledge, if available, is not always straightforward to be incorporated into the solver. The ideas of developing a self-adaptive mechanism for optimisation solvers, i.e., learn to automatically adapt search strategies based on past experience, have been explored in recent years. Machine learning algorithms, from decision trees to deep learning, have been shown promising in achieving this goal. In this chapter, we will explore how GP can be employed to augment optimisation solvers.

5.1 Challenges and Motivations

The previous chapters show how GP can be used to evolve heuristics that solve production scheduling problems, either by constructing schedules or iteratively improving existing schedules. Nevertheless, there are several limitations that restrict the applications of these approaches. First, construction heuristics (i.e., build a solution from scratch) cannot guarantee optimality, and improvement heuristics require problem-domain knowledge to work effectively. These heuristics are very useful to generate quick solutions, but they are less attractive if decision-makers want to have some guarantee about the optimality. Second, meta-algorithms are needed for each problem to be solved. However, addressing all operational constraints in a meta-algorithm efficiently is not easy and time-consuming. Third, evolving heuristics for combinatorial optimisation problems are computationally expensive, especially when heuristics need to explore the solution search space. Although a number of techniques to overcome this limitation have been developed in the literature (e.g., using surrogate models [92, 161]), they are problem-specific and cannot be easily extended to cope with other problems.

This chapter introduces how GP can overcome the above challenges by learning efficient variable selectors for constraint programming (CP), a powerful optimisation approach to solving hard scheduling and combinatorial optimisation. The evolved programs or variable selectors use the attributes extracted from CP models to determine the orders in which variables are explored in CP. By providing good variable orderings to the CP solver, good solutions can be found quickly and potentially reduce the search space to efficiently find optimal solutions. In the proposed algorithm, CP plays the role of a meta-algorithm in which variable selectors are the key component to be evolved. This chapter will address the following questions:

- What are the roles and effects of variable selectors in CP solvers?
- How can variable selectors be represented and evaluated in GP?
- How can GP efficiently evolve variable selectors?
- What are the generalisation and interpretability of evolved variable selectors?

There are some key advantages when combining CP and GP. CP is a popular and powerful tool for combinatorial optimisation, and formulating problems with CP is straightforward and intuitive. However, to solve a complex problem efficiently, some modelling and optimisation tricks are usually needed. Having a mechanism to discover/learn these tricks automatically can help users focus on formulating the problem in order to address practical requirements. GP also benefits greatly from this combination. As CP can be used as a general meta-algorithm either for schedule construction (i.e., find feasible solutions) or schedule improvement (i.e., optimisation), GP users can significantly reduce the need for low-level implementations, which have to be efficient and addressing practical requirements. Also, the user-friendly modelling languages used in CP can help users quickly formulate and effectively communicate their problems.

Section 5.2 shows the CP formulation of job shop scheduling problems. Different from previous chapters in which the formulation is used to formally describe the problem, the formulation in this chapter is used by GP to evaluate variable selectors and solve problem instances. Section 5.3 describes the full GP algorithm to evolve variable selectors. Detailed comparisons and analyses of the proposed algorithm are shown in Sect. 5.4.

5.2 Constraint Programming Model for Job Shop Scheduling Problems

The problem considered here is the same as the one investigated in the previous chapter or one discussed in [185]. This section provides the CP model of the static JSS problem in CP as this model will be used by CP to search for optimal solutions, and the attributes for evolved variable selectors are also extracted directly from this model. Note that we have redefined and simplified some notations in this chapter to facilitate our discussions and descriptions of the algorithms.

Parameters:

- $\mathcal{J} = \{1, \ldots, j, \ldots, N\}$: the set of all jobs.
- n_j: the number of operations of job j.
- $route_j = (m_{j1}, \ldots, m_{jn_j})$: the sequence of machines that job j will visit where m_{ji} is the machine that processes the ith operation of job j.
- $time_j = (p_{j1}, \ldots, p_{jn_j})$: the processing times of all operations of job j where p_{ji} is the processing time of the ith operation of job j.
- r_j: the release time of job j.
- d_j: the due date of job j.
- w_j: the weight of job j.

Variables:

- s_{ji}: the starting time of the ith operation of job j.
- e_{ji}: the ending time of the ith operation of job j.
- C_j: the completion time of job j.
- T_j: the tardiness of job j calculated by $T_j = \max(C_j - d_j, 0)$.

Three scheduling objectives are considered in this chapter and their formal definitions are provided in Table 5.1.

The constraint programming formulation for JSS is as follows.

$$\forall j \in \mathcal{J}: \qquad\qquad\qquad\qquad s_{j1} > r_j \qquad (5.1)$$
$$\forall j \in \mathcal{J}, \ i \in \{1, \ldots, n_j\}: \qquad\qquad e_{ji} = s_{ji} + p_{ji} \qquad (5.2)$$
$$\forall j \in \mathcal{J}: \qquad\qquad\qquad\qquad C_j = e_{jn_j} \qquad (5.3)$$
$$\forall j \in \mathcal{J}: \qquad\qquad\qquad T_j = \max(C_j - d_j, 0) \qquad (5.4)$$

Table 5.1 Job shop scheduling objective functions

Objective	Description
Makespan	$C_{\max} = \max_{j \in \mathcal{J}} \{C_j\}$
Maximum tardiness	$T_{\max} = \max_{j \in \mathcal{J}} \{T_j\}$
Total weighted tardiness	$TWT = \sum_{j \in \mathcal{J}} (w_j \times T_j)$

To ensure that there is no overlap between operations on the same machine, disjunctive constraints are included. Specifically, if operations u and v from different jobs are to execute on the same machine $m_{ju} = o_{kv}$, the start time of one of these jobs must be greater than the end time of the other job.

$$\forall j, k \in \mathcal{J}, u \in \{1, \ldots, n_j\}, v \in \{1, \ldots, n_k\},$$
$$m \in route_j, o \in route_k : \quad\quad\quad\quad (5.5)$$
$$m_{ju} = o_{kv} \Rightarrow s_{ju} \geq e_{kv} \vee s_{kv} \geq e_{ju}$$

There are a number of precedence constraints between operations of a job:

$$\forall j \in \mathcal{J}, i \in \{1, \ldots, n_j - 1\} : \quad s_{j,i+1} \geq e_{ji} \quad\quad\quad (5.6)$$

The objective functions are defined as follows:

1. Makespan: we define a variable $C_{\max}$ which represents the latest completion time of any job. The objective is to minimise $C_{\max}$ subject also to Constraint (5.8):

$$\text{min} . \quad C_{\max} \quad\quad\quad\quad (5.7)$$
$$\forall j \in \mathcal{J} : \quad C_{\max} \geq e_{jn_j} \quad\quad\quad\quad (5.8)$$

2. Maximum tardiness: we define a variable $T_{\max}$ which represents the maximum tardiness of any job. The objective is to minimise $T_{\max}$ and we also add Constraint (5.10):

$$\text{min} . \quad T_{\max} \qu\quad\quad\quad\quad (5.9)$$
$$\forall j \in \mathcal{J} : \quad T_{\max} \geq T_j \quad\quad\quad\quad (5.10)$$

3. Total weighted tardiness (TWT): The objective is to minimise cumulative tardiness across all jobs:

$$\text{min} . \quad \sum_{j \in \mathcal{J}} w_j T_j \qu\quad\quad\quad\quad (5.11)$$

5.3 Algorithm Design and Details

Figure 5.1 shows the overview of the proposed genetic-based constraint programming (GCP) algorithm. Each individual/program in the GP population represents a variable selector (i.e., choosing an ordering of variables for labelling). Similar to the previous experiment, GCP starts by randomly initialising a population of programs, i.e., variable selectors, $\mathbb{S} = \{S_1, \ldots, S_{PS}\}$ where PS is the population size. To evaluate variable selectors in the population, they are applied with a CP solver

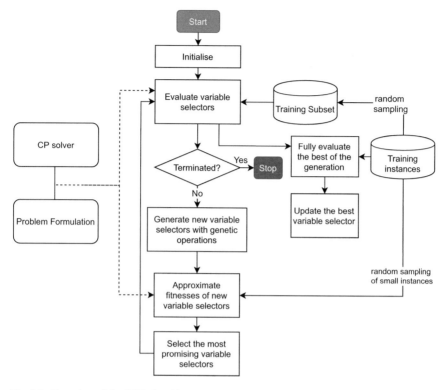

Fig. 5.1 Overview of the GCP algorithm

(i.e., Google OR-Tools: https://developers.google.com/optimisation in this chapter) to solve instances sampled from a training dataset $\mathcal{I} = \{I_1, \ldots, I_d\}$. Full evaluations with the complete training dataset are only applied to the variable selector S_g^* with the highest fitness in the generation. If improvements are made, it will be used to update the best-so-far selector S^*. After fitness evaluations, new programs are created by genetic operators. A pre-selection technique based on quick screening with reduced instances is used to filter potentially bad programs. This process is iterated until a stopping condition is met (i.e., the maximum number of generations in this paper). The details of key components of GCP are provided in the rest of this section.

5.3.1 Representation

To solve a JSS problem with constraint programming (CP), the user needs to build a CP model $\mathcal{M} = (X, \mathcal{D}, C)$. In this model, $X = \{x_1, \ldots, x_N\}$ is the set of variables to be optimised, i.e., start times of operations. $\mathcal{D} = \{\mathcal{D}_1, \ldots, \mathcal{D}_N\}$ is the set of domains of the variables where $x_k \in \mathcal{D}_k$ (e.g., start times are from 0 to an upper bound).

The set $C = \{C_1, \ldots, C_L\}$ represents the constraints, e.g., release times, disjunctive constraints, and precedence constraints. A constraint C_l includes a subset of variables $X_l \in X$ and a relation $\mathcal{R}_l$ that defines how values of variables $x \in X_l$ are assigned. The CP solver needs a search strategy to determine solutions for $\mathcal{M}$. A search strategy tells how the CP solver determines (1) how to select the next variable in a branching step and (2) how to assign value based on the variable domain to the selected variable. For variable selection, two sub-strategies can be applied: *domain-based variable selection* and (2) *variable ordering*.

Given a CP model, domain-based variable selection chooses the next variable or the starting time of an operation based on the values in its domain. There are a number of popular domain-based variable selection strategies. One of the simplest strategies is to select the variable with the smallest domain size or one with a minimum domain value (i.e., equivalent to operations with earliest starting times in JSS). Meanwhile, variable ordering is a tie-breaker in the case that domain-based variable selection cannot select a unique variable. The first sub-strategy influences the efficiency of CP in identifying feasible solutions. The second sub-strategy, if carefully designed, can help the solver find solutions with better objective values. The proposed GCP algorithm mainly emphasises the optimisation aspect, and feasibility will be handled by CP itself. In the evolved variable selector, we use *choosing-lowest-domain-value* as the fixed domain-based variable selection while variable orderings are evolvable. Meanwhile, *selecting-min-value* is used for the domain reduction strategy.

Similar to how sequencing decisions are made in the previous chapters, variable ordering is done by priority functions. The priority functions are represented by GP expression trees in which terminal nodes are feature values extracted from the CP model, and non-terminal nodes are arithmetic or logical operators. An example of a priority function is shown in Fig. 5.2. This example shows a simple priority function $W^2 \% PT$. In this example priority function, variables related to operations with a higher weight-to-processing times ratio will be selected first. Table 5.2 shows the terminal set and the function set to construct variable selectors in GCP. Although features here are similar to those used in the previous chapters, there are some key differences. Features here are obtained before any scheduling decisions are made, while features used in the previous chapters are obtained from partial schedules. The first five terminals PT, W, DD, RL, and ERC are job/operation data and the ephemeral random constants [190]. The last four terminals $ES, TPT, WL, MaxWL$ reflect some global (high-level) characteristics of jobs and machines. For example, ES determines if an operation is upstream or downstream. WL determines if a machine is busy, which is helpful to decide when an operation should start.

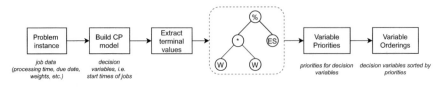

Fig. 5.2 Illustration of how an evolved variable selector generates variable orderings

Table 5.2 Terminal set (related to operation ith of job j) and function set in GCP

Notation	Description	Value
PT	Processing time	p_{ji}
W	Weight	w_j
DD	Due date	d_j
RL	Release time	r_j
ERC	Ephemeral constant	$[0, \ldots, 1]$
ES	Earliest start time	$r_j + \sum_{k=1}^{i-1} p_{jk}$
TPT	Total processing time	$\sum_{i=1}^{n_j} p_{ji}$
WL	Workload of m_{ji}	$\sum_{j=1}^{N} \sum_{k=m_{ji}} p_{jk}$
MaxWL	Maximum workload	$\max_{m=1}^{M} \{ \sum_{j=1}^{N} \sum_{k=m} p_{jk} \}$
Functions	$+, -, *, \%, \max, \min, \text{if\_then\_else}, \leq, \geq, =$	

5.3.2 Evaluation Mechanism and Fitness Function

The efficiency of an evolved variable selector S is determined by how much effort
(e.g., running times or numbers of branches) the CP solver is needed for S to identify
optimal solutions. In GCP, the computational effort is determined empirically by
applying the CP solver with S to find solutions for JSS instances. After an instance is
solved, some statistics are recorded: (1) objective values, (2) feasibility or optimality,
and (3) number of branches (reflecting the computational efforts). Ideally, we want
to have a variable selector that can find optimal solutions for all instances with the
least effort. The evaluation procedure of an evolved variable selector is presented in
Fig. 5.3. The fitness of a variable selector S can be calculated as follows:

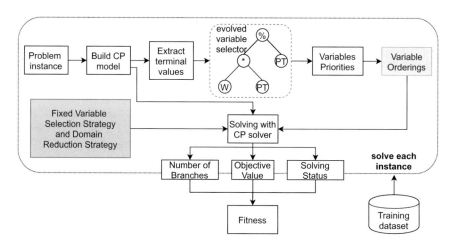

Fig. 5.3 Evaluating performance of evolved variable selectors

$$fitness(S, \mathcal{I}_g) = \frac{1}{|\mathcal{I}_g|} \sum_{I \in \mathcal{I}_g} nbr(S, I) \tag{5.12}$$

where $\mathcal{I}_g$ contains instances sampled from the training dataset $\mathcal{I}$ to evaluate S in generation g, and $nbr(S, I)$ is the number of branches to solve the instance I. A lower fitness value means that the CP solver needs less computational effort. Ideally, $nbr(S, I)$ is recorded when an instance is optimally solved. To avoid extreme running times caused by bad variable selectors, a time limit T is applied to stop the CP solver.

In the proposed algorithm, $fitness(S, \mathcal{I}_g)$ only reflects the efficiency of S on a subset of training instances $\mathcal{I}$ instead of the full dataset $\mathcal{I}_g$. When $\mathcal{I}_g = \mathcal{I}$, the $fitness(S, \mathcal{I}_g)$ is the true fitness of S, which is computationally expensive to evaluate. To make GCP more efficient, $\mathcal{I}_g$ is sampled from $\mathcal{I}$ in each generation for fitness evaluations. At the end of a generation g, S_g^*, the variable selector with the best $fitness(\cdot, \mathcal{I}_g)$, will be evaluated with the full training set $\mathcal{I}$. If $fitness(S_g^*, \mathcal{I}) < fitness(S^*, \mathcal{I})$, S_g^* will become the current best variable selector S^*.

5.3.3 Genetic Operations

To generate new priority functions, GCP uses the subtree crossover and the subtree mutation. Different from GP algorithms in the previous chapter, newly generated priority functions will not directly move to the next population but will form an intermediate population. Then, a pre-selection technique is applied to eliminate less promising individuals, and remained individuals will be kept for the next generation.

5.3.4 Pre-selection of Promising Programs

Pre-selection strategies have recently been employed to GP algorithms to converge faster towards good programs. This technique is very useful when it is computationally extensive to evaluate evolved programs. Pre-selection is used to identify potential programs and avoids wasting time evaluating bad programs. In GP, there are two common pre-selection techniques in previous studies: (1) surrogate models [92, 161] and (2) simplified models [165]. The first technique tries to predict the performance of newly generated programs by building a predictive model based on historical search data. The second technique assumes that the outcomes of solving simpler or auxiliary problems are good predictors of the original problems [204]. In GCP, the second technique is used as it is simple and general. The pre-selection algorithm in GCP is presented in Fig. 5.4. The main inputs for the pre-selection algorithm is a set $\mathbb{S}'$ of newly generated variable selectors. To maintain the diversity

---

Inputs: training instances $D \leftarrow \{I_1, I_2, \ldots, I_T\}$
Output: the best evolved rule Δ^*
    Input: A set of newly generated variable selectors $\mathbb{S}' = \{\mathcal{S}_1 \ldots, \mathcal{S}_{IPS}\}$, and training dataset $\mathcal{I}$
    Output: a new population $\mathbb{S}$
1: $\mathcal{I}' \leftarrow$ generate reduced instances based on $\mathcal{I}$
2: $\mathbb{S} \leftarrow \emptyset$
3: **for** $\mathcal{S} \in \mathbb{S}'$ **do**
4:     $fitness(\mathcal{S}, \mathcal{I}') \leftarrow$ evaluate $\mathcal{S}$ with new instances $\mathcal{I}'$
5: **end for**
6: sort $\mathbb{S}'$ based on $fitness(\cdot, \mathcal{I}')$ (lowest to highest)
7: $\mathbb{S} \leftarrow$ select the top PS selectors from $\mathbb{S}'$
8: return the new population $\mathbb{S}$

Fig. 5.4 Pseudo-code of the pre-selection algorithm

in the GP population, GCP will set the size of the intermediate population IPS much larger than the original population size PS.

GCP uses a set of reduced instances, generated from the distribution of the training dataset $\mathcal{I}$, to efficiently evaluate evolved variable selectors. Since JSS is a very challenging optimisation problem, the solving times can increase exponentially as the number of jobs and the number of machines increase. Solving reduced instances will be a lot faster than solving the original instances. For pre-selection, we only need to determine what the most promising individuals are rather than knowing exactly the fitness values. Therefore, ranking variable selectors based on their performance on reduced instances is sufficient. In GCP, a reduction ratio of $sr\%$ is introduced to determine the sizes of reduced instances. For example, if the original instances $I \in \mathcal{I}$ are 10×10, the reduced instance will be 5×5 with a reduction ratio of 50%. The empirical distributions [123] from $\mathcal{I}$ are used to generate processing times, weights, and routes of reduced instances. Similarly, the urgency and utilisation of machines in the original instances can be used to generate due dates and release times for reduced instances. If an instance generator is available (common for combinatorial optimisation), reduced instances can simply be created by setting the problem size with $sr\%$.

Since solving reduced instances is computationally cheap, a large number of such instances can be used to estimate the efficiency of evolved selectors and avoid overfitting issues. For simplicity, we choose $|\mathcal{I}'| = |\mathcal{I}|$.

5.4 Empirical Study

This section first provides details of the datasets for our experiments and performance metrics. Then results and analyses of GCP are provided.

5.4.1   Datasets

An instance generator is developed for experiments in this chapter. The generator here uses similar routines applied in other generators reported in the previous studies [100, 119]. For an instance $N \times M$, data related to a job j can be generated as follows:

- A route $route_j$ is a random sequence of machines.
- Processing times p_{ji} follow a discrete uniform distribution $\mathcal{U}[50, 100]$.
- Release times r_j follow a discrete uniform distribution $\mathcal{U}[0, TPT_j/2]$ where TPT_j is the total processing time of job j.
- Due dates $d_j = r_j + h \times TPT_j$ (total work-content rule [156]), where h is the due-date allowance factor.
- Weights w_j are randomly picked from the $\{4, 2, 1\}$ with the probabilities $\{0.2, 0.6, 0.2\}$, respectively.

In our experiments, the generator above will be used to generate training instances for GCP. Five variants of JSS problems based on the three objectives discussed earlier are investigated: (1) $C_{\max}$, (2) $T_{\max}$, $h = 1.5$, (3) $T_{\max}$, $h = 1.6$, (4) TWT, $h = 1.5$, and (5) TWT, $h = 1.6$. For training, only fixed problem size is used to reduce the variance of fitness values. To examine scalability of GCP, we will use three problem sizes $\{(5 \times 5), (7 \times 7), (10 \times 10)\}$ to generate training instances. For test, a larger and diverse set of instances are generated. Four problem sizes in the test set are $\{(5 \times 5), (10 \times 10), (15 \times 15), (20 \times 20)\}$. For each test set, 100 random instances are generated.

5.4.2   Parameter Settings

The parameters of GCP and the CP solver are provided in Table 5.3. In our experiments, a maximum depth of 7 is used to restrict the size of evolved selectors. It is noted that the maximum depth can be increased to explore a wide range of variable selectors without influencing the computational efficiency of GCP because variable orderings are only done once per instance. We only use the maximum depth to generate an interpretable selector. The reduction ratio is set to 50%, so solving reduced instances will be a lot faster than solving original instances.

For the CP solver, a time limit is used to avoid large running times caused by bad variable selectors. A longer time limit is set for test to help us accurately evaluate the performance of evolved selectors.

Table 5.3 Parameter settings of GCP and CP

	Parameter	Value
GCP	Population size	200
	Number of generations	50
	Terminal and function sets	See Table 5.2
	Initialisation	Ramped half-and-half
	Tournament selection size	5
	Subtree crossover	90%
	Subtree mutation	10%
	Maximum depth	7
	Intermediate population size	$= 2\times$ population size
	Reduce ratio	50%
	# of training instances $|\mathcal{I}|$	100
	# of instances per generation $|\mathcal{I}_g|$	5
	# of reduced instances	100
CP solver	Solver	Google OR-Tools
	Solving time limit (training)	5 (s)
	Solving time limit (test)	60 (s)
	Fixed variable selection	*choosing-lowest-domain-value*
	Fixed domain reduction	*selecting-min-value*

5.4.3 Performance Metrics

To comprehensively evaluate evolved variable selector, we use three performance metrics:

- *Computational effort*: average number of branches to optimally solve test instances or after the time limit is reached. Lower computational requirements are considered to be better.
- *Average objective value*: the average of the best objective values (optimal values if the instance is solved to optimality) found by the CP solver.
- *Number of optimal solutions*: the number of instances that the CP solver can solve to optimality.

For each experiment, 30 independent runs of GCP are conducted. The best selectors evolved in each run are recorded and tested.

5.4.4 Results

In this section, the evolved variable selectors are compared to the default search strategy of Google OR-Tools. Figures 5.5, 5.6, and 5.7 show the performance metrics of evolved selectors when 10×10 instances are used for training (results with other

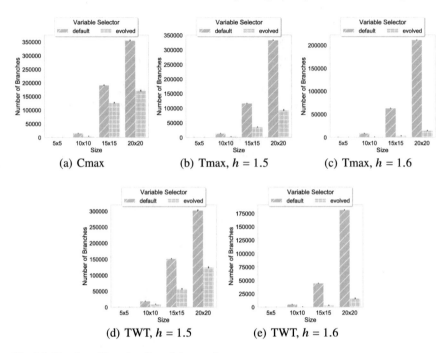

Fig. 5.5 Number of branches for solving test instances

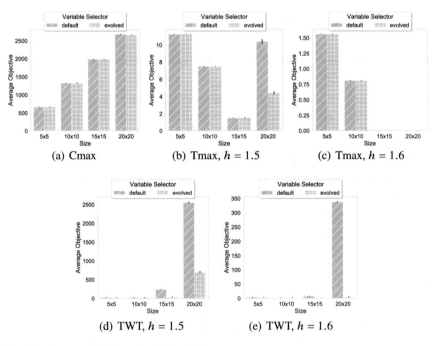

Fig. 5.6 Average objective values for test instances

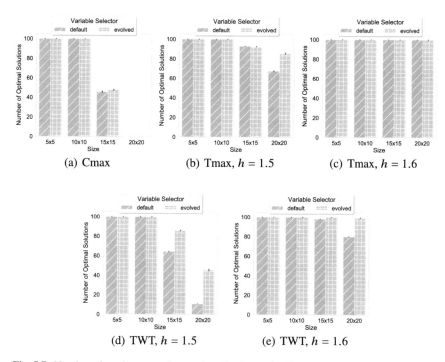

Fig. 5.7 Number of test instances that can be solved to optimality

training problem sizes are shown in the analysis section). It is clear from Fig. 5.5 that the evolved variable selectors can significantly reduce the number of branches required by the CP solver. For C_{max}, the computational efforts required for the largest test instances (i.e., 20×20) can be reduced by half with the evolved selectors. For T_{max} and TWT with $h = 1.5$, the computational efforts of the default settings are three times higher than those of evolved selectors. For T_{max} and TWT with $h = 1.6$, the evolved variable selectors are an order of magnitude faster than the default settings. In summary, GCP performs well across different JSS settings. Moreover, the evolved variable selectors can cope well with instances of sizes which they did not solve during training.

The quality of solutions generated by the CP solver is shown in Fig. 5.6. For C_{max}, there are only small differences in the average objective values between the evolved selectors and the default setting. This suggests that the evolved variable selectors can help the CP solver identify good or optimal solutions quickly, but proving optimality is still a challenge. For T_{max}, the performances with 5×5 and 10×10 are very similar for all settings because these instances can be solved very quickly. The differences in objective values become clearer when CP attempts to solve complex instances (large problem sizes or tight due dates). In such cases, evolved variable selectors are more successful in finding good solutions within the time limit. When due dates are loose ($h = 1.6$), optimal solutions can be found easily within the time limit, but the evolved variable selectors help the CP solver reduce the number of branches significantly as shown in Fig. 5.5c. For TWT, the evolved variable selectors are significantly better. When $h = 1.6$, the evolved variable selectors can help the CP solver find solutions that are 100 times better than those obtained by the default strategy.

Figure 5.7 again confirms the efficiency of the evolved variable selectors. For C_{max}, the number of optimal solutions of all strategies is similar, and it is much harder to prove optimality for larger instances. This is similar to other meta-heuristics in the literature. The evolved variable selectors are much more successful in terms of finding the optimal solutions for T_{max} and TWT. For the hardest problem with TWT and $h = 1.5$, the evolved variable selectors can find many more optimal solutions than the default strategy.

The experiments show that GCP, as a machine learning approach, is a very promising approach to improving the efficiency of the CP solver. GCP is very simple and the efforts to extract features for evolving variable selectors are minor. One key advantage of GCP as compared to other existing GP approaches for production scheduling is that the evolved programs can both determine good solutions (similar to schedule construction heuristics) and improve solutions (similar to scheduling improvement heuristics). In addition, the evolved variable selectors have good generalisation. The fact that variable selectors evolved with small/medium instances can be reused to efficiently solve much larger instances is very encouraging. This property allows GCP to train with solvable instances and to reduce training times and computational resources of GP.

5.4.5   Further Analyses

This section analyses the scalability and overfitting issues of GCP and the importance of features used to construct priority functions. This section also presents examples of evolved variable selectors and demonstrates the generalisation of these selectors using a set of popular benchmark datasets in the literature.

5.4.5.1   Scalability of GCP

Figures 5.8, 5.9, and 5.10 show the performance of evolved variable selectors when different instance sizes are used for training. For C_{max}, different instance sizes do not influence the performance of GCP across the three metrics. These results suggest that it is possible to use small instances for training when dealing with C_{max}. For this problem, using larger instances is less desirable. Meanwhile, the due-date-based objective functions are more sensitive to the choice of training instance sizes. It is clear that the larger training instances can help GCP discover better variable selectors in these cases. However, the improvements from increasing training problem sizes are not proportional.

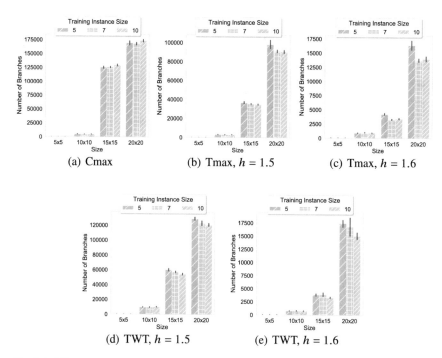

Fig. 5.8 Efforts to solve test instances with different training instance sizes

5.4.5.2 Overfitting Issues

Figures 5.11, 5.12, and 5.13 further examine the impact of training instance sizes on evolutionary processes. For C_{max}, good selectors can be found in early generations when larger training instances are used. GCP has trouble making progress with small training instances when the due-date-based objective functions are considered. For T_{max}, $h = 1.5$, Figs. 5.11b and 5.12b show serious overfitting issues.

Figure 5.13b shows that an instance size of 10 can improve the progress of GCP. For the easier problems T_{max}, $h = 1.6$, GCP again suffers from serious overfitting issues when smallest instances are used for training as shown in Fig. 5.11c. Meanwhile, relatively small progress is made in Fig. 5.12c. It is interesting that significant progress in early generations is observed in Fig. 5.13c but GCP still has overfitting issues in later generations. For TWT, large instances can help GCP find better variable selectors. These results are consistent with those in the previous section and suggest that optimal training instance sizes are sensitive to JSS problems to be solved.

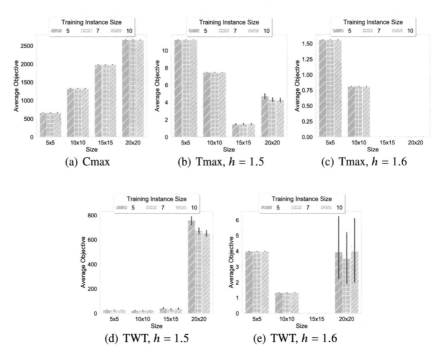

Fig. 5.9 Objective values of test instances with different training instance sizes

5.4.5.3 Feature Importance

To examine the general importance of features, we record the number of times those features appear in the best evolved variable selectors. The feature frequency is presented in Fig. 5.14. It is clear that ES is the most important feature in all problems. Since *choosing-lowest-domain-value* and *selecting-min-value* are, respectively, used as the fixed domain-based variable selection and the fixed domain reduction, using ES to guide the search of CP is natural. Surprisingly, DD is the second most used feature for C_{max} although this is not a due-date-based objective function. A possible explanation is that DD has a strong correlation with job completion times and total processing times because the total work-content rule is used to determine due dates in the instance generators. PT and RL are also often included in evolved selectors.

For due-date-based objective functions, TPT and PT are important features. It is interesting that features that are directly used to calculate objective functions such as DD and W appear less frequently.

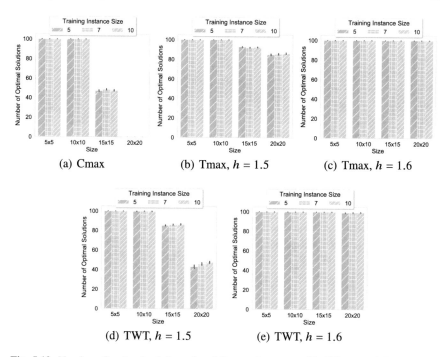

Fig. 5.10 Number of optimal solutions found for test instances with different training instance sizes

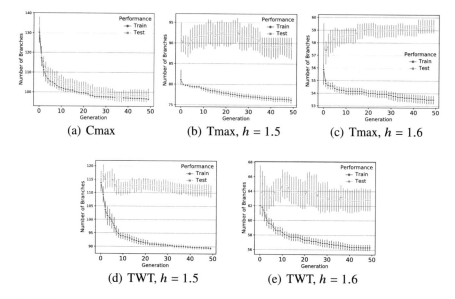

Fig. 5.11 Training with instance size = 5. y-axis shows the number of branches and x-axis shows the generation

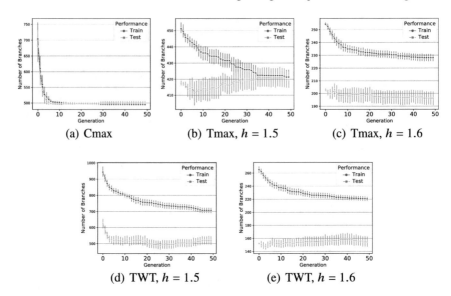

Fig. 5.12 Training with instance size = 7. y-axis shows the number of branches and x-axis shows the generation

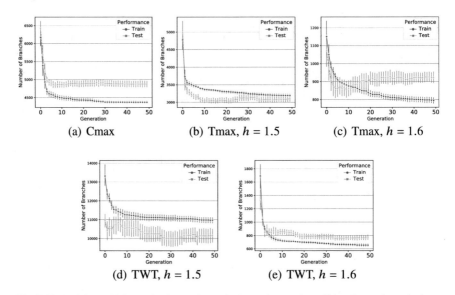

Fig. 5.13 Training with instance size = 10. y-axis shows the number of branches and x-axis shows the generation

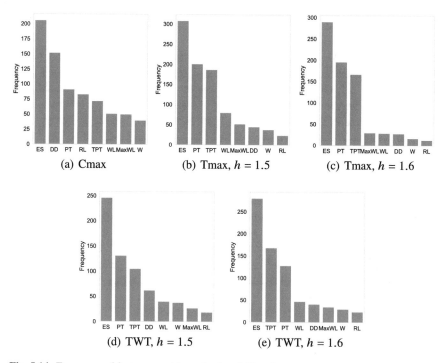

Fig. 5.14   Frequency of features used in evolved variable selectors

5.4.6   *Examples of Evolved Variable Selectors*

Three examples of the evolved priority function for variable ordering are presented here:

- $C_{\max}$: $(\mathsf{DD} + \mathsf{RL}) - 2 \times \mathsf{ES} = 2 \times \mathsf{RL} + h * \mathsf{TPT} - 2 \times \mathsf{ES}$
- $T_{\max}$: $-\mathsf{TPT} \times \mathsf{WL} \times \mathsf{ES}$
- TWT : $\mathsf{W} - \mathsf{PT} - \mathsf{WL} - \mathsf{ES}$

These three priority functions are compact and easy to interpret. Although DD is in the priority function of $C_{\max}$, we can easily remove it from the function. Priority functions for $T_{\max}$ and TWT are very straightforward. It is interesting that WL impacts priorities of operations. These examples show that workload, or indirectly, bottlenecks in the system are critical to determining the optimal start times for operations. Specifically, higher priorities are given to operations processed by machines with lower workload.

5.4.6.1   Generalisation of Example Variable Selectors

To further examine the generalisation of the above selectors, we apply them to five benchmark datasets in OR-library [20] for C_{max} and a set of instances used [119] for $TWT, h = 1.3$. Given that these instances are from different sources, they are excellent candidates in examining the generalisation of the evolved variable selectors.

Table 5.4 presents the computational efforts and solution quality for the example variable selector in the previous section for C_{max}. d-CP and e-CP show the performance of the default setting and the evolved variable selectors, respectively. $gap\%$ and #opt are the optimality gap (to the best upper bounds in the literature) and the number of optimal solutions obtained by the CP solver. For this experiment, the time limit for the CP solver is 10 minutes. The differences in the number of branches between d-CP and e-CP are measured by diffeff% (bolded if the evolved variable selector is more efficient). The results show that evolved variable selectors can reduce the gaps between the best found solutions and the upper bounds. For the ta dataset, e-CP can help the CP solver find more optimal solutions. In general, e-CP is much more efficient than d-CP in all five datasets. For the la dataset, e-CP requires about 44% fewer branches than d-CP.

Table 5.5 shows the results with benchmark datasets for $TWT, h = 1.3$. GCP is compared to two specialised optimisation algorithms, i.e., large-step random walk (LSRW) [119] and shifting bottleneck (SB) [185]. In the table, d-$gap\%$ and e-$gap\%$ (bolded if the evolved variable selector produces the best solutions) show the gaps compared to SB with the default setting and the evolved selector, respectively. In general, CP outperforms SB in most instances and evolved variable selector can also reduce the computational costs of CP. These are very promising results because these instances are very different from those used in the training datasets. The results with $h = 1.5$ and $h = 1.6$ are not shown here because these instances can be solved easily by all algorithms.

Table 5.4 Performance of evolved variable selectors for C_{max} with benchmark datasets

Dataset	#instances	d-CP		e-CP		diffeff%
		$gap\%$	#nopt	$gap\%$	#nopt	
ta	70	2.49	14	1.62	20	**−37.02**
la	40	0.00	39	0.02	39	**−44.02**
abz	5	0.55	2	0.62	2	**−4.85**
orb	10	0.00	10	0.00	10	**−29.04**
swv	20	3.23	6	3.11	6	**−17.43**

Table 5.5 Test performance with TWT, $h = 1.3$ with 10×10 benchmark instances

Instance	CP			LSRW	SB
	d-$gap\%$	e-$gap\%$	diffeff%	$gap\%$	
abz5	−4.17	−4.17	−3.11	−0.89	1464
abz6	0.00	0.00	−11.82	0.00	436
la16	−0.09	−0.09	−2.18	0.00	1170
la17	−0.11	−0.11	−20.87	0.00	900
la18	−0.75	−0.75	0.84	−0.75	936
la19	7.23	7.23	1.36	−0.42	955
la20	−8.31	−8.31	−0.14	−7.86	878
orb01	−5.92	−8.27	−1.09	−9.48	2890
orb02	1.56	1.56	11.88	1.56	1412
orb03	0.09	−0.09	−4.08	4.31	2113
orb04	10.29	0.00	6.66	3.14	1623
orb05	−4.44	−4.44	−2.31	−0.30	1667
orb06	−0.11	−0.11	−3.71	0.56	1792
orb07	4.07	3.05	1.64	4.75	590
orb08	−6.80	−6.80	−1.23	−2.41	2617
orb09	−11.26	−11.26	−4.22	−10.05	1483
orb10	−2.30	−8.10	2.73	−2.85	1827

5.5   Chapter Summary

This chapter introduces a way to integrate GP and constraint programming solvers. This integration has a number of key advantages. By applying CP for finding feasible solutions and optimising the objective functions, users no longer need to develop meta-algorithms (as introduced in the previous chapters for construction heuristics and improvement heuristics), which is a time-consuming process and requires problem-domain knowledge. Also, the availability of different and generic search strategies in CP can help us identify optimal solutions more efficiently. Users can now focus on formulating real problems with high-level modelling languages. From the CP viewpoint, users can speed up the search process automatically with GP. The methodology introduced in this chapter is general and can be extended to cope with a wide range of production scheduling problems. While this chapter only discusses the integration of genetic programming and constraint programming, similar principles can be applied to other optimisation solvers such as mixed-integer linear programming.

How to select an optimal set of attributes is one of the key elements for the effectiveness of GCP. The attributes used in this chapter are directly extracted from the instance data, but it is possible that high-level attributes (e.g., times to the bottleneck machine, and slack), if available, would be useful for building more powerful variable selectors. These high-level attributes can be provided as the inputs from the domain experts or automatically extracted by using advanced machine learning techniques.

There has been a growing interest in the applications of machine learning to scheduling and combinatorial optimisation in recent years. From traditional statistical learning (e.g., logistic regression, and decision trees) to deep learning (e.g., graph-embedding and deep reinforcement learning), machine learning has shown promises in boosting operations research algorithms. The GCP approach discussed here is an attempt towards that goal. Compared to other learning paradigms, GCP has a couple of advantages. First, GCP does not need to be supervised, i.e., datasets of optimal solutions (required by supervised learning techniques) are not needed to evolve variable selectors. Second, evolved variable selectors are partially interpretable, which is useful for users to understand the characteristics of optimal schedules. Finally, as an evolutionary computation approach, GCP can inherit different techniques to improve efficiency and address practical challenges such as multi-objective optimisation and transfer learning. Moreover, with its flexible representation and evaluation schemes, GCP can incorporate useful outputs from other machine learning techniques (e.g., high-level attributes extracted from optimal schedules) to boost its efficiency and effectiveness.

Part III
Genetic Programming for Dynamic Production Scheduling Problems

In the last part, we mainly focus on *static* production scheduling problems with GP. In this part, we will focus on how GP can tackle *dynamic* production scheduling problems. The literature has identified a number of key gaps and limitations of existing work in GP for DFJSS in terms of training efficiency (i.e., time-consuming due to the long simulation), search space (i.e., large search space with a large number of features), and search mechanism (i.e., no guidance for choosing the genetic materials from the parent(s) to generate offspring).

This part will cover efficiency improvement with multi-fidelity surrogates, search space reduction with feature selection, and search mechanism with specialised genetic operators for GP in DFJSS. We take DFJSS as an example for the investigation of how GP can be improved in this chapter. Before we move to these three studies, different ways to represent the scheduling heuristics of GP in DFJSS are introduced. Then, we will present new GP algorithms to address the mentioned limitations one by one.

Part III is organised into four chapters:

• Chapter 6 shows how GP is used to learn scheduling heuristics for making multiple decisions for DFJSS, i.e., machine assignment and operation sequencing. The advantages of GP with the cooperative coevolution strategy or multi-tree representation to learn multiple scheduling heuristics are discussed. A novel crossover operator is also developed to further improve the effectiveness of the proposed GP algorithm with multi-tree representation.

• Chapter 7 develops a GP algorithm with multi-fidelity surrogates to improve the effectiveness and efficiency for learning scheduling heuristics. This chapter also discusses how knowledge sharing between surrogates enhances the effectiveness of the proposed algorithm. The sensitivity of the proposed GP algorithm to the frequency of knowledge sharing is discussed.

• Chapter 8 proposes a two-stage GP feature selection algorithm to select important features for routing and sequencing rules simultaneously. This chapter also develops a novel GP algorithm that aims to learn effective scheduling heuristics with only selected features. The selected features and the learned scheduling heuristics are further analysed.

• Chapter 9 develops a GP algorithm with specialised genetic operators that can guide the search effectively. This chapter shows how to measure the importance of subtrees of a GP individual and how to use the subtree importance information to generate offspring by crossover effectively. The depths and chosen probabilities of the selected subtrees are further studied.

Chapter 6
Representations with Multi-tree and Cooperative Coevolution

The representation is an important factor for the success of evolutionary algorithms. For DFJSS, a key issue is how to evolve both the routing rule for machine assignment and the sequencing rule for operation sequencing simultaneously. This chapter aims to help the readers get the knowledge of how to represent individuals of GP for DFJSS.

6.1 Challenges and Motivations

Dealing with multiple interdependent decisions, especially in dynamic environments, is always difficult to find the global optimal solution. This is particularly challenging when multiple decisions need to be made simultaneously. First, an appropriate representation is definitely a fundamental factor for an evolutionary algorithm to solve a problem. An inappropriate representation is more likely to limit the searchability of the algorithms to find good solutions. Second, the representation determines the size of the search space and there is a clear trade-off between the complexity of the representation and the ability of GP to explore the search space. These two facts foster the motivation to develop a suitable representation for GP based on the characteristics of DFJSS. This chapter aims to investigate the representation or learning mechanism for GP to learn routing rule and sequencing rule simultaneously. Especially, this chapter tries to investigate the following questions:

- What representation or learning mechanism is suitable for evolving routing rule and sequencing rule simultaneously?
- What are the advantages and disadvantages of the investigated methods for learning routing rule and sequencing rule simultaneously?

F. Zhang et al., *Genetic Programming for Production Scheduling*, Machine Learning: Foundations, Methodologies, and Applications,
https://doi.org/10.1007/978-981-16-4859-5_6

In Sect. 6.2, details of the algorithms with a new representation or learning mechanism are provided. We suggest all the readers to check this section carefully, since the algorithms in other chapters will use one of the presented ways to evolve scheduling heuristics. Section 6.3 shows the details of the experiment design and parameter settings of GP. Readers interested in the chapters working on DFJSS are encouraged to read the details of this section. The results and discussions are also provided in this chapter.

6.2 Algorithm Design and Details

According to the characteristics of DFJSS, a routing rule and a sequencing rule are needed to make two types of decisions (i.e., machine assignment and operation sequencing) simultaneously. To the best of our knowledge, there are mainly two types of studies about the representation to handle DFJSS. The first one is that only one rule will be learned by fixing the other rule. Tay et al. [212] proposed to use GP to merely learn the sequencing rule while leaving the routing rule as a manually created rule for flexible JSS. However, only learning the sequencing rule may not be an effective way, if the fixed routing rule is not optimal. The second one is to learn the routing rule and the sequencing rule simultaneously. Yska et al. [228] developed a cooperative coevolution GP framework with two subpopulations to learn routing and sequencing rules simultaneously for the first time. Zhang et al. [239] introduced GP with multi-tree representation for learning routing rule and sequencing rule within an individual in one population. This chapter will focus on learning routing and sequencing rule simultaneously, and introduce the details of the cooperative coevolution and multi-tree representations.

6.2.1 Genetic Programming with Cooperative Coevolution

Within the cooperative coevolution framework, routing rule and sequencing rule are learned separately in different subpopulations. Each GP individual is either a routing rule or a sequencing rule. Figure 6.1 shows an example of the representation of learning routing and sequencing rule simultaneously for DFJSS with cooperative coevolution technique. Routing and sequencing rules are learned in two different subpopulations which are indicated by rectangles. It is noted that routing rule and sequencing rule work together to generate schedules in DFJSS. The learned best routing rule (i.e., indicated by the circle with dashed orange lines) and sequencing rule (i.e., indicated by the square with dashed blue lines) at the previous generation are saved for the rule evaluations in the current generation. All the routing rules are evaluated with the best sequencing rule at the previous generation, while all the sequencing rules are evaluated with the best routing rule at the previous generation. The fitness of an individual, either routing rule or sequencing rule, is calculated by the

Fig. 6.1 An example of the representation with cooperative coevolution for DFJSS

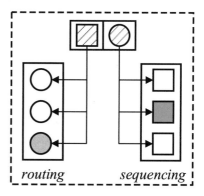

fitness function. The best routing rule and sequencing rule in the current generation will be kept to evaluate the new individuals in the next generation.

The offspring in different subpopulations are generated separately. Specifically, routing rules are used to produce routing rules, while sequencing rules are used to generate sequencing rules for the next generation. Since each individual represents a rule, the genetic operators implementation of GP with cooperative coevolution is the same as the traditional GP.

6.2.2 Genetic Programming with Multi-tree Representation

With the multi-tree representation, each GP individual is represented as multiple trees. Routing and sequencing rules can be learned together as different trees in one individual. Taking this into consideration, multi-tree representation naturally lends itself to DFJSS. In this book, we use the multi-tree representation that one individual contains two trees for DFJSS. To be specific, the first tree represents the routing rule and the second tree represents the sequencing rule. The quality of one individual depends on the effectiveness of two trees working together.

Figure 6.2 shows an example of the multi-tree representation with a routing rule and a sequencing rule of GP for DFJSS. The routing rule can be considered as a priority function to give priority values for all candidate machines of an operation. The corresponding priority function of the routing rule in Fig. 6.2 is WIQ * MWT + NIQ, where WIQ is the needed total processing time of operations in the queue of a machine, MWT is the waiting time for a machine to become idle, and NIQ is the number of operations in the queue of a machine. The machine with the highest priority assigned by the routing rule will be selected to allocate the operation. Similarly, the operation with the highest priority assigned by the sequencing rule (i.e., PT/W, where PT is the processing time and W indicates the importance of an operation) will be selected as the next operation to be processed on the idle machine.

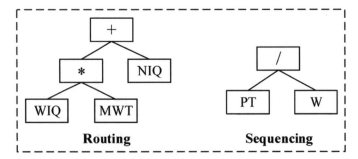

Fig. 6.2 An example of the multi-tree representation with a routing rule and a sequencing rule of GP for DFJSS

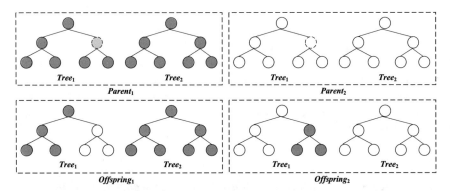

Fig. 6.3 Traditional crossover operator for multi-tree representation with two trees, i.e., *Tree$_i$* represents ith tree of a GP individual

6.2.2.1 Genetic Operators with Multi-tree Representation

For GP with multi-tree representation, it comes to a decision which trees the genetic operator should be applied. In multi-tree representation, the traditional genetic operators are defined to act upon only one tree in an individual at a time. Other trees are unchanged and copied directly from the parents to the offspring. Genetic operators are limited to a single type of tree at a time in the expectation that this will reduce the extent to which they disrupt "building blocks" of useful code. Figure 6.3 shows an example of the crossover in traditional GP with multi-tree representation with two trees. However, when coping with DFJSS, such a crossover operator has the following issues.

First, the crossover operator is only applied to one tree of the parents, and the offspring generated might not be substantially different from their parents. Thus, the population will lose its diversity and the ability of exploration will decrease. Second, the crossover operator limits the diversity of the combinations of routing and sequencing rules. Specifically, in DFJSS, a good rule cannot be "good" by itself,

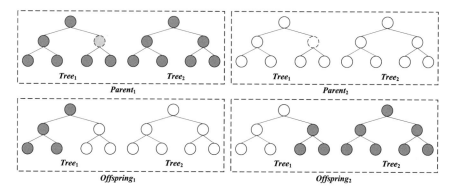

Fig. 6.4 Tree swapping crossover operator for multi-tree representation with two trees, i.e., *Tree*ᵢ represents *i*th tree of a GP individual

but should behave well when collaborating with the other rule. Thus, the diversity of combinations is an important factor for the success to achieve good schedules.

To handle these shortcomings and make the algorithm more in line with the properties of DFJSS, a new tree swapping crossover operator is presented. Figure 6.4 shows the tree swapping crossover operator, which is the same as the traditional crossover operator as shown in Fig. 6.3 except that the unselected trees (the same type) are also swapped with each other. Two parents (*Parent₁* and *Parent₂*) are selected to generate offspring and the second type (T_2) of trees is selected for crossover. The dotted circles mean that the subtrees are chosen and will be swapped. The standard crossover operator will stop here. But for the tree swapping crossover operator, the other tree is also swapped. Thus, two offspring (*Offspring₁* and *Offspring₂*) are generated. This will bring two benefits. The first is that useful blocks are not easily broken. The second is that more possible pairs or combinations of the routing and sequencing rules are examined.

6.3 Experiment Design

This section describes the simulation model (i.e., simulation configuration, training instance and test instance, and scenario configuration), comparison design and parameter settings (i.e., common parameter settings and specialised parameter settings), and experiment execution of GP for DFJSS.

6.3.1 Simulation Model

6.3.1.1 Simulation Configuration

For DFJSS in this book, the simulation model contains 6000 jobs that need to be processed by 10 machines. The jobs will arrive over time according to a Poisson process with rate λ. Each job has a different number of operations that are randomly generated from a discrete uniform distribution between 1 and 10. The importance of jobs are indicated by weights. The weights of 20, 60, and 20% jobs are set as 1, 2, and 4 following the setting in [92]. A job with a larger weight is more important. Each operation's processing time is drawn from a uniform discrete distribution with the range [212]. The number of candidate machines for a given operation is distributed uniformly between 1 and 10.

The first 1000 jobs are used as warm-up jobs to get typical situations occurring in a long-term simulation of a dynamic job shop system, and jobs arrive in a continual stream. We gather information from the next 5000 jobs. When the 6000th job is completed, the simulation comes to the end.

6.3.1.2 Training Instances and Test Instances

A problem instance is an instantiation of the problem/simulation that uses a specific seed for the pseudo-random number generator [27]. Instances distinguish each other by having different pseudo-random numbers for the simulations. The scheduling heuristics will be trained and tested using multiple different instances. Training instances are seen to GP for learning the scheduling heuristics, while the test instances are unseen to the training process. Test instances are used to measure the generalisation/effectiveness of the learned scheduling heuristics obtained from the training process.

To improve the training efficiency of GP at each generation, we only use one training instance to evaluate the quality of learned scheduling heuristics. To improve the generalisation of the GP method, the instance will be changed at each generation throughout the training phase by giving a different random seed. This strategy has been successfully used to improve the effectiveness and generalisation of learned scheduling heuristics of GP efficiently [93, 158].

6.3.1.3 Scenario Configuration

In this book, we use objectives and utilisation levels to build different scenarios, since these two factors affect the problem significantly. To verify the effectiveness and efficiency of the presented algorithm, the presented algorithms are examined in the scenarios with different settings. The *utilisation level* (p) is an essential factor for simulating various scenarios. It shows how much of the time a machine is likely to be in use. The rate of job arrival is controlled by the Poisson process with λ, which is calculated based on Eq. (6.1). Especially, μ is the average processing time of the machines. P_M is the likelihood of a job visiting a machine. If each job has two operations, for example, P_M is 2/10. A busier and more complex job shop environment results from a higher degree of utilisation level. New jobs will arrive over time according to a Poisson process with rate λ, where λ is adapted to keep a fixed utilisation level of a job shop.

$$\lambda = \mu * P_M / p \qquad (6.1)$$

In this part, we test the algorithms with three commonly used objectives, i.e., Max-flowtime, mean-flowtime, and mean-weighted-flowtime. To verify the effectiveness of the presented algorithm, we expect to test the presented algorithm on complex scenarios. We set the utilisation level as 0.85 and 0.95 since they can lead to complex job shop scenarios, and are commonly used to evaluate the performance of the presented algorithm. Overall, six different scenarios with three objectives (i.e., max-flowtime, mean-flowtime, and mean-weighted-flowtime) and the two utilisation levels (i.e., 0.85 and 0.95) are used to test the proposed algorithms.

It is noted that this scenario setting is used in this part. We also set different scenarios based on the characteristics of the algorithms in different chapters to investigate the performance of the algorithms if it is helpful.

6.3.2 Comparison Design

The goal of this chapter is to develop proper representations of GP for DFJSS. Three algorithms are involved in this chapter. The GP with cooperative coevolution is named CCGP [228]. The GP with multi-tree representation and traditional multi-tree crossover is named MTGP. The presented algorithm with multi-tree representation and the developed swapping crossover operator is named sMTGP. CCGP and MTGP are firstly compared to verify the effectiveness of these two GP representations for DFJSS. Then, the effectiveness of the developed tree swapping crossover operator for multi-tree representation is verified by comparing MTGP with sMTGP.

6.3.3 Parameter Settings

This section introduces the common parameter settings of GP which are also used in other chapters if they are not redefined. In addition, specialised parameter settings of GP are described.

Table 6.1 The common parameter settings of genetic programming

Parameter	Value
Method for initialising population	Ramped-half-and-half
Initial minimum/maximum depth	2/6
Terminal/function selection rate	10%/90%
Maximal depth of programs	8
The number of elites for a task	10
Parent selection	Tournament selection with size 7
Crossover/Mutation/Reproduction rate	80%/15%/5%
The number of generations	51

6.3.3.1   Common Parameter Settings of Genetic Programming

General Parameter Settings: Some common parameter settings of GP [114] are shown in Table 6.1. The individuals are initialised with the ramped-half-and-half method with a minimum depth of 2 and a maximum depth of 6. When generating a GP individual, the probabilities of selecting nodes from terminals and functions are 10% and 90%, respectively. The maximum depth of a GP individual is limited to 8. To maintain the quality of the obtained individuals in the evolutionary process of GP, the best 10 individuals (i.e., elites) from the previous generation are moved to the next generation directly. The remaining individuals for the next generation are produced by crossover, mutation, and reproduction (i.e., genetic operators) with a rate of 80%, 15%, and 5%, respectively. Tournament selection with size 7 is used to select the parent(s) for producing offspring with the genetic operators. The algorithm is stopped after 51 generations.

GP can select appropriate simple features from the terminals in the terminal set to build high-level features with functions in the function set automatically. The details of the terminal set and function set are introduced, as shown below.

Terminal Set: The GP terminals are used to capture the information about the problem. Following the suggestions in [27, 156, 244], the terminal set of GP in this book consists of a number of basic features of machines, jobs, and operations in the job shop. The routing terminal set is set the same as the sequencing terminal set in this book.

Machine-related features: machine states such as workload are important considerations for assigning operations to machines. A good schedule should not overload or underload a machine.

- NIQ denotes the number of operations in the machine's queue. Its purpose is to measure a machine's workload by counting the number of operations in its queue.
- WIQ indicates the total required processing time of the operations in the machine's queue. It calculates the time required for a machine to finish all the operations in its queue without any delay, which is to measure the workload of a machine.

- MWT stands for the waiting time for the machine to become idle again, which is equal to the completion time of the present processing on the machine minus the current time.

Job-related features: job states have a considerable impact on determining which job should be processed first. An effective schedule is expected to process key jobs earlier, and take into account both the current and future job information.

- W denotes the weight of a job. A larger weight indicates a more important job.
- NOR is the number of remaining operations for a job. It indicates the current processing stage of the job.
- WKR is the median amount of time required to complete the remaining operations. The median time is an estimation of the processing time, because the exact processing time of an operation in DFJSS is unknown in advance because the machine has not yet been determined. This feature estimates the processing stage of the job in terms of processing time.
- TIS is the amount of time that the job has been in the job shop since it arrives.

Operation-related features: the characteristics and states of operations are significant considerations for deciding which operation to process next. A decent schedule should take into account the time it takes to process the operation as well as the waiting time of the operation.

- PT represents the processing time of the operation on the candidate machine.
- NPT is the median processing time of the next operation of the candidate operation (i.e., it is 0 if the candidate operation is the last one of the job)
- OWT stands for the amount of time the operation has waited in the queue of the machine.

Function Set: The function set has been set to $\{+, -, *, /, \text{Max}, \text{Min}\}$ [143, 156]. Two arguments are required for the arithmetic operators. When divided by zero, the "/" operator is a protected division that returns one. The Max and Min functions return the maximum and minimum of their two arguments, respectively.

It is noted that, in this book, the settings of the GP parameters in Table 6.1 are the same in all chapters, if they are not redefined in the following chapters. It is noted that each chapter also has its own specialised parameter settings for GP according to the characteristics of the algorithms in the corresponding chapters.

6.3.3.2 Specialised Parameter Settings of GP

The specialised parameter settings of GP in this chapter are shown in Table 6.2. For CCGP, there are two subpopulations with 512 individuals, one for learning routing rules and the other for learning sequencing rules. For MTGP and sMTGP, we use one population with 1024 individuals. The total number of individual evaluations of CCGP, MTGP, and sMTGP is the same.

Table 6.2 The parameter settings of genetic programming

Parameter	Value
Number of subpopulations of CCGP	2
Subpopulation size pf CCGP	512
Number of (sub)populations of MTGP and sMTGP	1
Population size of MTGP and sMTGP	1024

6.3.4   Experiment Configuration

In this part, all experiments are run on an Arch Linux OS with an Intel (R) Core (TM) i7-4770 CPU at 3.40 GHz, with 8-GB RAM. The algorithms are encoded with the Java programming language, and ECJ package is used [140].

The learned scheduling heuristic is tested on 50 unseen instances, and the average objective value across the 50 test instances is reported as the test performance of the scheduling heuristic, which can be a good approximation of the true performance. This book works on the minimisation problem, and a smaller objective value indicates a better performance. Friedman's test with a significance level of 0.05 is applied to compare all the algorithms based on their performance. If Friedman's test gives significant results, we will further conduct Wilcoxon rank sum test with Bonferroni correction between the proposed algorithm and other algorithms with a significance level of 0.05 for pairwise comparisons with at least 30 independent runs (i.e., 30 runs and 50 runs). For all the results in this part, "−", "+", and "≈" indicate the corresponding result is significantly better than, worse than or similar to its counterpart. An algorithm is compared with the algorithm(s) before it one by one, as shown in tables in the Results and Discussions section. The results in other chapters are also compared in the same way.

6.4   Results and Discussions

6.4.1   Quality of the Learned Scheduling Heuristics

In this chapter, we use normalised objective value to indicate the quality of learned scheduling heuristics (i.e., the learned routing rule and sequencing rule are denoted by l_r and l_s). The normalised objective value ($norm\,Obj$), obtained by learned scheduling heuristics over the objective value obtained by benchmark rules, is used to indicate the quality of the learned scheduling heuristics. Due to the lack of the best objective values of the DFJSS instances, benchmark routing and sequencing rules [91] are chosen to normalise the objective values. The used benchmark routing and sequencing rules are shown in Table 6.3 according to objective functions. The equation of

Table 6.3 The benchmark rules used for different objectives

Rule type	Objectives	Rule	Description
Routing rule	Max-flowtime	LWIQ	Least work in queue
	Mean-flowtime		
	Mean-weighted-flowtime		
Sequencing rule	Max-flowtime	FCFS	First come first serve
	Mean-flowtime	SPT	Shortest processing time
	Mean-weighted-flowtime	FCFS	First come first serve

normalised objective value is shown in Eq. (6.2).

$$norm\,Obj\,(l_r, l_s) = \frac{Obj\,(S(l_r, l_s, \mathcal{I}))}{Obj\,(S(b_r, b_s, \mathcal{I}))} \qquad (6.2)$$

In Eq. (6.2), the instance set is represented by $\mathcal{I}$. $S(l_r, l_s, \mathcal{I})$ is the obtained schedule by the learned scheduling heuristics, while $S(b_r, b_s, \mathcal{I})$ is the obtained schedule by the benchmark rules (i.e., b_r and b_s indicate the benchmark routing rule and sequencing rule, respectively). The objective value obtained by the learned rules and benchmark rules are denoted by $Obj\,(S(l_r, l_s, \mathcal{I}))$ and $Obj\,(S(b_r, b_s, \mathcal{I}))$, respectively. A smaller normalised objective value (i.e., $norm\,Obj$) indicates a better scheduling heuristic. From $norm\,Obj\,(l_r, l_s)$, we can see how much better the learned scheduling heuristics are than the benchmark scheduling heuristics.

Figure 6.5 shows the boxplot of the average normalised objective values of CCGP, MTGP, and sMTGP on test instances according to 30 independent runs in six DFJSS scenarios. We can see that both the GPs with cooperative coevolution and multi-tree representations can learn comparable routing and sequencing rules. However, the GPs with multi-tree representation (i.e., MTGP and sMTGP) show its advantage of learning the routing rule and the sequencing rule simultaneously in terms of obtained objective values and standard deviation.

It is interesting that MTGP get better mean values than CCGP, but MTGP is not significantly better than CCGP in any scenario. When further looking into the boxplot in Fig. 6.5, one can see that CCGP has many more outliers than MTGP and sMTGP. This is because CCGP cannot handle well the interactions between routing and sequencing rules directly, thus can be stuck into poor local optima more often. The reason why there is no statistical significance between MTGP and CCGP is that the two algorithms show very similar performance except for the outliers. Figure 6.5 clearly shows that GP with multi-tree representation manages to reduce the probability of outliers dramatically.

According to these observations, the performance of GP with the multi-tree representation is more stable than GP with cooperative coevolution. Also, Wilcoxon rank

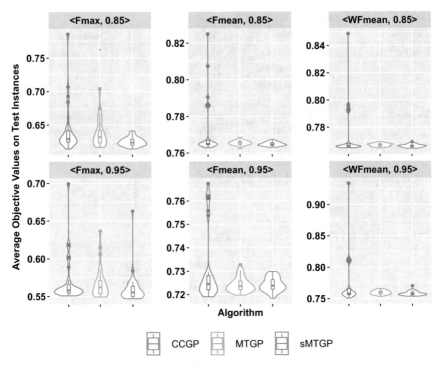

Fig. 6.5 The boxplot of the average normalised objective values of CCGP, MTGP, and sMTGP on unseen instances over 30 independent runs in six DFJSS scenarios

sum test results show that sMTGP is significantly better than MTGP in four scenarios (i.e., <Fmax, 0.85>, <Fmean, 0.85>, <WFmean, 0.85>, <WFmean, 0.95>). It means that the proposed tree swapping crossover operator can effectively improve the performance of MTGP.

6.4.2   Size of Learned Scheduling Heuristics

Figures 6.6 and 6.7 show the curves of the average best routing rule sizes and sequencing rule sizes of CCGP, MTGP, and sMTGP over 30 independent runs in six DFJSS scenarios, respectively. Figure 6.6 shows that the best routing rule sizes obtained by sMTGP and MTGP are smaller than that of CCGP. Figure 6.7 shows that the sizes of the learned best sequencing rules by sMTGP and MTGP are much smaller than the best rules learned by CCGP. These observations confirm the potential of using multi-tree-based GP to achieve a smaller scheduling heuristic.

To explore whether the best rule is smaller by chance or the rules in the whole population generally become smaller, the average rule sizes in the whole population

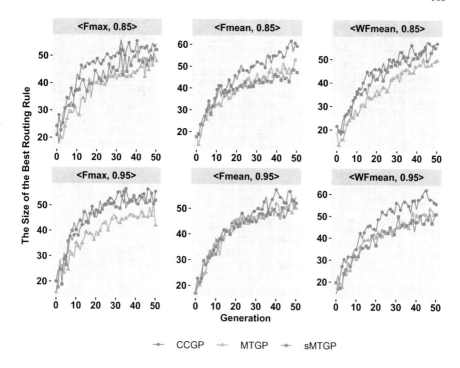

Fig. 6.6 The curves of the average best routing rule sizes of CCGP, MTGP, and sMTGP over 30 independent runs in six DFJSS scenarios

at each generation are investigated to get a clear vision of the changes in rule sizes. We take the scenario <WFmean, 0.95> as an example to further investigate the changes of rule sizes. As shown in Figs. 6.8 and 6.9, at the initial point, for all the three algorithms, the average sizes of both rules are about equal. However, the average sizes obtained by CCGP become larger than others over the generations. One possible reason is that, in multi-tree-based GP, effective and smaller rules are more likely to be well preserved because there is at least one rule structure will not be changed by the genetic operator at each time during the evolutionary process. In addition, the average sizes obtained by MTGP and sMTGP show the same trend and the routing rules are larger than the sequencing rules. This is consistent with the observations on the sizes of the best routing rule and the sequencing rule.

6.4.3 Training Time

Table 6.4 shows the mean and standard deviations of the training time of CCGP, MTGP, and sMTGP over 30 independent runs in six DFJSS scenarios. Table 6.4 shows that MTGP and sMTGP can learn scheduling heuristics for DFJSS more effi-

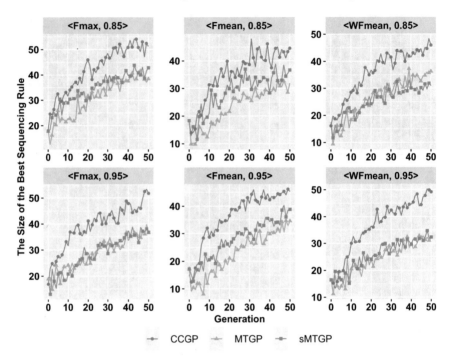

Fig. 6.7 The curves of the average best sequencing rule sizes of CCGP, MTGP, and sMTGP over 30 independent runs in six DFJSS scenarios

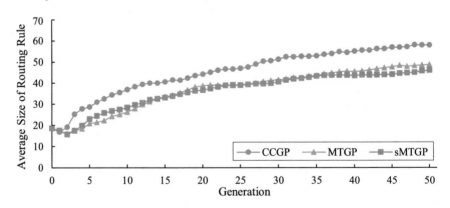

Fig. 6.8 The curves of average routing rule sizes of CCGP, MTGP, and sMTGP in population over 30 independent runs

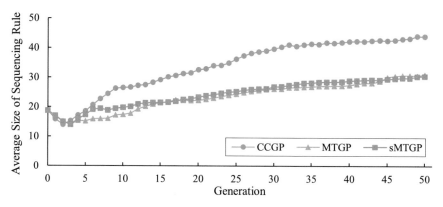

Fig. 6.9 The curves of average sequencing rule sizes of CCGP, MTGP, and sMTGP in population over 30 independent runs

Table 6.4 The mean and standard deviation of the training time of CCGP, MTGP, and sMTGP over 30 independent runs in six dynamic flexible job shop scheduling scenarios

Scenario	Training time (in minutes)		
	CCGP	MTGP	sMTGP
<Fmax, 0.85>	77(9)	71(7)(−)	74(11)(−)(≈)
<Fmax, 0.95>	86(9)	80(10)(−)	84(12)(−)(≈)
<Fmean, 0.85>	76(9)	71(12)(−)	70(9)(−)(≈)
<Fmean, 0.95>	81(10)	79(14)(≈)	78(9)(≈)(≈)
<WFmean, 0.85>	74(10)	70(8)(≈)	72(9)(≈)(≈)
<WFmean, 0.95>	83(9)	78(9)(−)	76(10)(−)(≈)

cient than CCGP in most of the scenarios (i.e., <Fmax, 0.85>, <Fmax, 0.95>, <Fmean, 0.85>, and <WFmean, 0.95>). There is no significant difference about the training time between MTGP and sMTGP which is consistent with our expectation, since there is no extra strategy for sMTGP to control the rule sizes compared with MTGP. This is a promising finding that GP with multi-tree representation is computationally cheaper than GP via cooperative coevolution for DFJSS.

6.5 Chapter Summary

This chapter presents two types of representations and a newly developed crossover operator for GP to learn the routing rule and the sequencing rule simultaneously. We can see that both GP with cooperative coevolution and multi-tree representation can help learn the routing rule and the sequencing rule simultaneously. The results show that GP with multi-tree representation is better than GP with cooperative coevolution for DFJSS. This indicates that the assumption in CCGP that the routing rule and the sequencing rule are independent and can be learned separately, might not be true. This suggests that, when we learn scheduling heuristics for DFJSS, it is better to take the interaction between the routing rule and the sequencing rule into consideration.

However, both GP with cooperative coevolution and GP with multi-tree representations for DFJSS have their advantages. The advantage of GP with cooperative coevolution is that, it can separate the routing and sequencing decisions properly, which makes it easy to implement new genetic operators and incorporate complex algorithm frameworks such as surrogate and feature selection algorithms. The representation with multi-tree makes it easy to manage the population, since the routing and sequencing rules are considered as an individual in one population. In this book, we will choose to use the representation with cooperative coevolution and multi-tree based on the characteristics of the designed algorithms.

This chapter shows how to learn multiple scheduling heuristics (i.e., routing rule and sequencing rule) with GP for DFJSS. In the next chapter, we will introduce how to use the surrogate technique to improve the training efficiency of GP for DFJSS.

Chapter 7
Efficiency Improvement with Multi-fidelity Surrogates

The fitness evaluation of GP for learning scheduling heuristics is time consuming, since it requires applying the scheduling heuristics to a long simulation process. Surrogate technique [103, 170, 205, 237] has been successfully used to reduce the computational cost of evolutionary algorithms. The success of surrogate technique lies in building computationally cheap models to approximate the original time-consuming fitness evaluation of individuals. The existing studies of surrogate-assisted GP for JSS can be grouped into two categories according to the number of surrogates used. One is to build one surrogate model [92, 162], the other is to build more than one surrogate model with multi-fidelities [165]. With one surrogate, the performance of the algorithms highly relies on the accuracy of the used surrogate model, either based on nearest neighbour [92, 162] or simplified simulation model [165]. With multi-fidelity surrogates, the performance of the algorithms depends on the interaction between surrogates, which relaxes the requirements for the accuracy of a specific surrogate model. This chapter will explore how to improve the training efficiency of GP with multi-fidelity surrogates for DFJSS.

7.1 Challenges and Motivations

Surrogate-assisted evolutionary algorithms [103] have been successfully used to reduce computational time in evolutionary optimisation for expensive problems such as drug design [35, 65] and surgical training optimisation [26, 135], where complex computational simulations are used. However, the studies related to surrogates for JSS are very limited. To the best of our knowledge, the existing two GP surrogate algorithms, either based on K nearest neighbour [92, 162] or simplified model [165, 240], are sensitive to the accuracies of the built surrogates. The reason is that, there is only a single surrogate model for the investigated problem.

© The Author(s), under exclusive license to Springer Nature Singapore Pte Ltd. 2021   107
F. Zhang et al., *Genetic Programming for Production Scheduling*, Machine Learning:
Foundations, Methodologies, and Applications,
https://doi.org/10.1007/978-981-16-4859-5_7

Generally, high-fidelity surrogate models can provide more reliable and accurate results than low-fidelity surrogate models, but at the cost of higher computational time [103]. A promising way to achieve a trade-off between prediction accuracy and computational cost is to integrate high-fidelity and low-fidelity surrogates by constructing a multi-fidelity surrogate model [251]. Multi-fidelity surrogate model can reduce the dependence on one single surrogate model. In addition, if the surrogate models in a multi-fidelity surrogate can help each other effectively, it can benefit the overall algorithm due to the reinforcement between them. However, the multi-fidelity surrogate model approach has not been studied in DFJSS.

This chapter will explore an effective GP algorithm with multi-fidelity-based surrogate models to evolve scheduling heuristics for the DFJSS problems efficiently. The algorithm is expected to speed up the convergence and reduce the training time of GP for DFJSS. Especially, this chapter is interested in the following questions:

- How are multiple surrogates developed?
- How can the surrogates help each other?
- What is the effect of the number of surrogates on the proposed algorithm?

Section 7.2 introduces the proposed algorithm with the details of how to build surrogates and how to share knowledge between surrogates. Section 7.3 gives the details of the experiment design followed by the results and discussion in Sect. 7.4. Section 7.5 gives further analyses of the proposed algorithm. Section 7.6 summaries this chapter.

7.2   Algorithm Design and Details

This section describes the framework of the presented algorithm first. Then, the key components of the presented algorithm are introduced.

7.2.1   Framework of the Algorithm

Algorithm 7.1 shows the main framework of the presented algorithm. The input includes k designed surrogate models with different fidelities. The surrogate models are developed using shorter DFJSS simulations. The last surrogate S_k is the original simulation which can be considered as a surrogate model of the original problem with an accuracy of 100%. The output of the presented algorithm is a set of best scheduling rules learned by GP and obtained from subpopulations with different surrogate models. The problems solved in different subpopulations can be considered as similar problems but with different problem scales which are indicated by different simulation lengths. The presented algorithm has three main differences compared with the traditional GP for JSS:

Algorithm 7.1 Framework of the Algorithm

Input: k multi-fidelity surrogate models $S_1, S_2, ..., S_k$
Output: The best learned heuristics with each surrogate model $ind_1^*, ind_2^*, ..., ind_k^*$
1: **Initialisation**: Randomly initialise the population with k subpopulations
2: set $ind_1^*, ind_2^*, ..., ind_k^* \leftarrow null$
3: set $fitness(ind_1^*), fitness(ind_2^*), ..., fitness(ind_k^*) \leftarrow +\infty$
4: $gen \leftarrow 0$
5: **while** $gen < maxGen$ **do**
6:     // **Evaluation**: Evaluate the individuals in the population
7:     **for** i = 1 to k **do**
8:         **for** j = 1 to $|subpopsize|$ **do**
9:             Run a DFJSS simulation (surrogate S_i) with ind_j to get the schedule $Schedule_j$
10:             $fitness(ind_j) \leftarrow Obj(Schedule_j)$
11:         **end for**
12:         **for** j = 1 to $|subpopsize|$ **do**
13:             **if** $fitness(ind_j) < fitness(ind_i^*)$ **then**
14:                 $ind_i^* \leftarrow ind_j$
15:                 $fitness(ind_i^*) \leftarrow fitness(ind_j)$
16:             **end if**
17:         **end for**
18:     **end for**
19:     **Selection**: Use tournament selection to choose parents for producing offspring
20:     **Evolution**: Generate offspring for each subpopulation by genetic operators with the proposed *knowledge transfer* mechanism
21:     $gen \leftarrow gen + 1$
22: **end while**
23: **return** $ind_1^*, ind_2^*, ..., ind_k^*$

---

1. At the initialisation stage, the population is formed with multiple subpopulations to incorporate multi-fidelity surrogate models into GP (line 1). Each subpopulation is associated with a surrogate model, respectively.
2. During the evaluation process, the individuals in different subpopulations are evaluated with the surrogate model associated with it (from line 6 to line 18).
3. During the evolution stage, each subpopulation's offspring are produced for the next generation using the proposed knowledge transfer mechanism (line 20).

The key idea of this chapter is to collaborate multi-fidelity-based surrogates for GP to learn scheduling heuristics for DFJSS. When using multi-fidelity surrogate models, some GP individuals are assessed using a simple surrogate, while others are evaluated using a more complicated surrogate. This chapter uses multiple subpopulations to solve multiple tasks simultaneously, one subpopulation for each task. Subpopulations can be thought of as a number of independent evolutionary processes evolving simultaneously.

The evolutionary framework of the GP algorithm with multi-fidelity surrogate models is shown in Fig. 7.1. Assuming k surrogate models with different fidelities (i.e., $S_1, S_2, \ldots, S_k$) from simple to complex are used to improve the efficiency of GP to learn effective scheduling heuristics, where the problem with S_k is the desired problem to be solved. The GP population is partitioned into k subpopulations (i.e.,

Fig. 7.1 The evolutionary
framework of the presented
algorithm

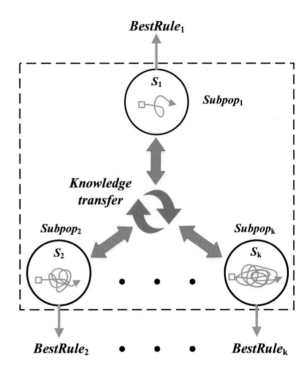

Subpop₁, Subpop₂, . . . , Subpopₖ) and each of which will learn scheduling heuristics based on the corresponding surrogate model. In addition, various subpopulations support each other through the evolutionary process by sharing their knowledge. It is noted that all the subpopulations are evolved in parallel. Therefore, the output of a GP run consists of k best learned rules (i.e., *Best Rule₁* (ind_1^*), *Best Rule₁* (ind_2^*), . . ., *Best Ruleₖ* (ind_k^*)). However, we only focus on the best learned rule *Best Ruleₖ* obtained from *Subpopₖ* with S_k, since it is the problem we aim to solve.

The advantages of using the presented evolutionary framework to realise collaborative multi-fidelity-based surrogate-assisted GP for DFJSS are shown as follows:

- The evolutionary framework makes it easier for surrogate models of various fidelities to collaborate.
- Because numerous surrogate models with diverse fidelities are used, the proposed algorithm is not sensitive to the correctness of any particular surrogate model.
- As a by-product, the proposed algorithm tackles the problems of various scales simultaneously. If one prefers to address numerous problems at different scales simultaneously, this is an efficient approach to use computational resources.

Algorithm 7.2 Offspring generation with knowledge transfer

Input: A population with
k subpopulations $P = \{Subpop_1, Subpop_2, \ldots, Subpop_k\}$
Output: A new population with
k subpopulations $P' = \{Subpop'_1, Subpop'_2, \ldots, Subpop'_k\}$
1: set $P, P' \leftarrow null$
2: $gen \leftarrow 0$
3: **while** $gen < maxGen$ **do**
4:   // **Evaluation**: Evaluate the individuals in each subpopulation, respectively
5:   // **Evolution**
6:   **for** i = 1 to k **do**
7:     **if** $rand <= tr$ **then**
8:       $parent_1 \leftarrow$ Select the first parent from $Subpops_i$
9:       $parent_2 \leftarrow$ Select the second parent from $Subpops_{\neg i}$
10:       $point_1$: the crossover point of $parent_1$
11:       $point_2$: the crossover point of $parent_2$
12:       $offspring$: replace $point_1$ of $parent_1$ by $point_2$
13:       $Subpop'_i \leftarrow Subpop'_i \cup offspring$
14:     **else**
15:       $parent_1 \leftarrow$ Select the first parent from $Subpops_i$
16:       $parent_2 \leftarrow$ Select the second parent from $Subpops_i$
17:       $point_1$: the crossover point of $parent_1$
18:       $point_2$: the crossover point of $parent_2$
19:       $offspring_1$: replace $point_1$ of $parent_1$ by $point_2$
20:       $offspring_2$: replace $point_2$ of $parent_2$ by $point_1$
21:       $Subpop'_i \leftarrow Subpop'_i \cup offspring_1 \cup offspring_2$
22:     **end if**
23:     $P' \leftarrow P' \cup Subpop'_i$
24:   **end for**
25:   $gen \leftarrow gen + 1$
26: **end while**
27: **return** $P' = \{Subpop'_1, Subpop'_2, \ldots, Subpop'_k\}$

7.2.2 Knowledge Transfer

The presented algorithm's essential characteristic is the inclusion of a collaborative approach for utilising the knowledge of surrogate models with multiple fidelities. The collaborative mechanism aims to share knowledge between different surrogates. The essential concerns of transfer learning include "how" and "when" to transfer knowledge, as well as "what" to transfer. In contrast to traditional transfer learning, this chapter does not involve the source and target problems. It suggests that the proposed algorithm has no knowledge extraction mechanism for obtaining knowledge from the source domain. The presented GP algorithm in this chapter conducts the knowledge transfer implicitly via the crossover operator [84].

How and When to Transfer. Crossover is an essential genetic operator in GP for producing offspring that can serve as effective carriers to transfer knowledge. Individuals from the same subpopulation or from different subpopulations can do crossover to produce offspring. Algorithm 7.2 shows the presented knowledge transfer mechanism. To govern when knowledge from other subpopulations is transferred at each generation, we define a transfer ratio tr. A larger (smaller) tr value suggests that knowledge transfer between subpopulations is promoted (or not). If the knowledge transfer mechanism is triggered, the first parent $parent_1$ from the current subpopulation will be chosen (line 8). The other parent $parent_2$ will be chosen from one of the other subpopulations (line 9). Only the offspring descended from $parent_1$ are maintained (line 10 to line 13). If the knowledge mechanism is not activated, both parents will be chosen from the current subpopulation (i.e., the same subpopulation) to generate two offspring for the next subpopulation (from line 14 to line 22).

It is worth highlighting that the knowledge can be transferred from a subpopulation with a lower-fidelity surrogate model to that with a higher-fidelity surrogate model and vice versa. From the perspective of knowledge transfer, the subpopulation with a lower fidelity surrogate (simpler problem) can identify promising individuals more quickly than the subpopulation with a higher fidelity surrogate (more complex problem). Introducing knowledge from a subpopulation with a lower-fidelity surrogate model can speed up the convergence of a subpopulation with a high-fidelity surrogate model. Learning knowledge from a subpopulation with a higher fidelity surrogate model can assist raise the quality of individuals in a subpopulation with a lower-fidelity surrogate model, because the learned rules with a higher-fidelity surrogate model are more dependable. In general, knowledge transfer collaboration is supposed to benefit all of the problems concerned.

What to Transfer. It is important to figure out what kind of information should be shared. Intuitively, the knowledge that promising individuals possess is advantageous. The knee- point technique [247] is used in this chapter to choose a group of promising individuals. Individuals with smaller fitness values (i.e., our problem is a minimisation problem) than the knee-point individual's fitness are chosen as promising individuals. The number of promising individuals across generations, and the knee-point technique can effectively find promising individuals. In addition, the number of selected promising individuals is not required to be defined in advance (i.e., the knowledge transfer mechanism is *parameter-free*). With the knee-point technique, the promising individuals can be decided adaptively. Only the knowledge of promising individuals is permitted to share knowledge to other subpopulations.

The pseudo-code for choosing promising individuals for knowledge transfer is shown in the Algorithm 7.3. The individuals in the subpopulation are first sorted in ascending order according to fitness values (line 1). Second, a line (L) is drawn by connecting the two points with the biggest and smallest fitness. Then, from line 6 to line 12, the distance between each individual and line L is calculated, and the knee point with the greatest distance to line L is identified. Finally, the individuals whose fitness values are smaller than the fitness of knee point are picked as *promising individuals* for knowledge transfer (line 13 to line 17).

An example of choosing promising individuals from a number of individuals with knee-point technique can be found in Fig. 7.2. First, the individuals in the population are sorted based on fitness values in ascending order, and a curve related to fitness values is obtained. Second, a line is generated by connecting the points with the

Algorithm 7.3 Framework of selecting promising individuals for knowledge transfer

Input: The current subpopulation with a set of individuals *Ind*
Output: Promising individuals *Ind** for knowledge transfer

1: *sort*(*subpopulation*)
2: minPoint(0, fitness(ind_0))
3: maxPoint(*subpopsize* - 1, fitness($ind_{subpopsize-1}$))
4: set *maxDistance* ← 0 and *kneePointIdx* ← 0
5: get a line *L* based on minPoint and maxPoint points
6: **for** i = 0 to *subpopsize* − 1 **do**
7:    calculate the distance (*d*) from *Point*(*i*, *fitness*(ind_i)) to line L
8:    **if** *d* > *maxDistance* **then**
9:        *maxDistance* ← *d*
10:       *kneePointIdx* ← *i*
11:   **end if**
12: **end for**
13: **for** i = 0 to *subpopsize* − 1 **do**
14:   **if** *i* <= *kneePointIdx* **then**
15:       *Ind** ← *Ind** ∪ *Ind*[*i*]
16:   **end if**
17: **end for**
18: return *Ind**

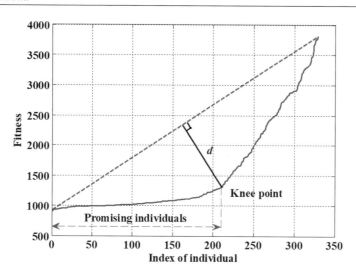

Fig. 7.2   The selected promising individuals based on knee-point method

smallest and the largest fitness value. Then, the distance between each point on the curve and the line is calculated. The point that has the largest distance to the line is selected as the knee point, and the individuals whose fitness value is smaller than that of the knee point are selected to be adapted. It is noted that if there is more than one knee point, we just randomly select one knee point.

7.2.3 Algorithm Summary

This chapter presents multi-fidelity surrogate models by shortening DFJSS simulations for GP to learn scheduling heuristics. The computational cost of GP will be reduced by using lower quality surrogate models with simplified DFJSS. Individuals assessed by poor fidelity surrogate models, on the other hand, are frequently evaluated incorrectly. By coordinating surrogate models with distinct fidelities, the given approach deftly handles this contradicting issue. The collaborative mechanism is realised via an effective knowledge transfer mechanism.

7.3 Experiment Design

7.3.1 Comparison Design

This chapter aims at reducing the training time of GP with collaborative multi-fidelity surrogate models for DFJSS. Two algorithms are involved in this chapter. The GP with multi-tree representation [239] (MTGP) algorithm is selected as the baseline algorithm because it can learn the routing rule and the sequencing rule simultaneously, and its framework is well suited to the use of multi-fidelity surrogate models in collaboration. The presented algorithm, with collaborative multi-fidelity surrogate models, is named M^3GP, since it involves multi-population framework, multi-tree representation, and multi-fidelity surrogate models. The surrogate models are used to test the algorithms first. It is worth emphasising that MTGP works with a single population of 1024 individuals, whereas M^3GP works with two subpopulations of 1024 individuals (i.e., 512 individuals for each subpopulation). M^3GP_1 and M^3GP_2 are used to indicate the effectiveness of M^3GP on task 1 and task 2, respectively. They can be regarded as the algorithms for measuring the performance of the learned scheduling heuristics with S_1 (i.e., the lower surrogate model) and S_2 (i.e., the original model) in M^3GP.

Because we are primarily concerned with solving the desired problem, the performance of the proposed algorithm is first assessed by comparing MTGP and M^3GP_2. The presented approach is compared to the state-of-the-art algorithms referred to as SGP_K [92] and SGP_H [165] in this chapter. In addition, the presented algorithm is investigated when there are more than two surrogates with varying fidelities. The performance of M^3GP_1 with and without the proposed knowledge transfer mechanism for GP is compared to each other in order to test the effectiveness of the suggested knowledge transfer mechanism for GP. Similarly, the impact of knowledge transfer mechanism on M^3GP_2 is investigated further.

Table 7.1 The parameter settings of M^3GP

Parameter	Value
Number of subpopulations	2
Subpopulation size	512
Transfer ratio tr	0.6
The number of jobs in surrogate model S_1	2500
The number of jobs in original model S_2	5000

7.3.2 Specialised Parameter Settings of Genetic Programming

Table 7.1 shows the specialised parameter settings of M^3GP in this chapter. For the sake of simplicity, only two models with varying fidelities are evaluated first. As a result, the GP population is divided into two subpopulations, with each subpopulation having 512 individuals. Using the idea from [165], the surrogate model (S_1) is created by building a "half shop" job shop with 2500 (i.e., $5000 * 0.5 = 2500$) jobs, which decreases the number of jobs to half of the original model but without reducing the number of machines to maintain the characteristics of DFJSS. The other model (S_2) is the original model with 5000 jobs, which is (i.e., can be regarded as a surrogate model with 100% accuracy). The only difference between S_1 and S_2 is the number of jobs, and the original model S_2 is more accurate than the surrogate model S_1, since S_2 accurately depicts the problem. The knowledge transfer ratio tr is set as 0.6, and the specifics of the sensitivity analysis are provided in Sect. 7.5.2.

The sizes of the intermediate population of SGP_H, SGP_K are set as two times of the population, as proposed in [92], to make fair comparison. The half shop surrogate model based on problem approximation is set to the same parameter as in [165], but with the maximum number of operations for a job with five to apply the idea properly in the investigated problem. In addition, for SGP_K, the number of neighbours for KNN is set to 1.

7.4 Results and Discussions

7.4.1 Training Time

According to 30 independent runs in six DFJSS scenarios, Table 7.2 shows the mean and standard deviation of the training time of MTGP and M^3GP. In all the studied scenarios, the training time of the proposed algorithm M^3GP is much less than that of MTGP. For instance, in scenario <Fmean, 0.95>, the training time of M^3GP

Table 7.2 The mean (standard deviation) of the **training time** (in minutes) of MTGP and M^3GP according to 30 independent runs in six dynamic flexible job shop scenarios

Scenario	MTGP	M^3GP
<Fmax, 0.85>	64(9)	51(9)(−)
<Fmax, 0.95>	67(12)	53(9)(−)
<Fmean, 0.85>	61(11)	48(8)(−)
<Fmean, 0.95>	64(13)	49(6)(−)
<WFmean, 0.85>	62(13)	49(7)(−)
<WFmean, 0.95>	63(11)	49(6)(−)

is lowered by 23.40%. In general, M^3GP is more efficient than MTGP in all the scenarios studied, and the training time of M^3GP is around 78.5% of that of MTGP.

The curve of the training time of MTGP and M^3GP during the training process in six different scenarios is shown in Fig. 7.3. It indicates that M^3GP takes less time in all scenarios to train scheduling heuristics than MTGP at all generations. M^3GP can save computational cost dramatically during the evolutionary process. In addition, the training time of M^3GP increases more slowly than that of MTGP. In all scenarios, the training time of M^3GP increases from about 40 to 70 s as the number of generations increases. However, the training time of MTGP increases from about 50 to 90 s in the scenarios with utilisation level of 0.85 (i.e., <Fmax, 0.85>, <Fmean, 0.85>, and <WFmean, 0.85>), while rises from 55 to 95 roughly in scenarios with utilisation level of 0.95 (i.e., <Fmax, 0.95>, <Fmean, 0.95>, and <WFmean, 0.95>).

Because the job shop scenarios with higher utilisation levels are more complicated than the scenarios with lower utilisation levels, more training time is typically required in scenarios with higher utilisation levels. In particular, for MTGP, compared with the training time required in scenarios with utilisation levels of 0.85, the scenarios with utilisation levels of 0.95 require additional training time. However, this is not the case for M^3GP. This demonstrates that the training time of M^3GP is not sensitive to the utilisation level of the job shop. One possible explanation is that the surrogate models with lower fidelities are shorter which require less time than the surrogates with higher fidelities. Even with a higher utilisation level, the simpler task has no significant influence on training time.

7.4.2 Quality of Learned Scheduling Heuristics

Based on 30 independent runs, the mean and standard deviation of the objective values on unseen instances of MTGP and M^3GP$_2$ with the same number of generations in six different scenarios are shown in Table 7.3. In five out of the six scenarios, it reveals that there is no statistical difference in the performance between MTGP and M^3GP$_2$. In addition, in scenario <Fmax, 0.95>, M^3GP$_2$ outperforms its counter-

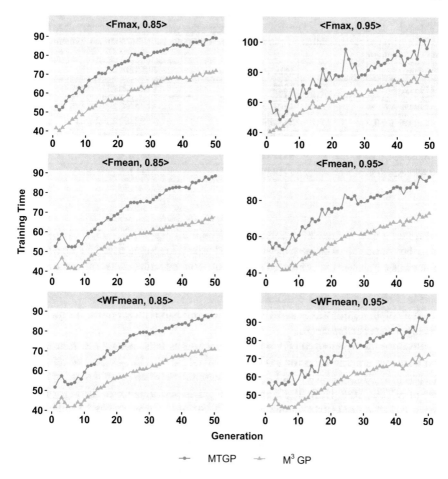

Fig. 7.3 The curve of the **training time** (in seconds) of MTGP and M³GP during the training process according to 30 independent runs in six dynamic flexible job shop scenarios

part significantly. This demonstrates that M³GP₂ can achieve equivalent or greater performance than MTGP at a lower computational cost than MTGP.

It would also be interesting to see if M³GP₂ can outperform MTGP when given the same amount of training time. To answer this question, we give all algorithms a fixed training time. The number of generations in each run for one algorithm may differ, and we are no longer to compare algorithms based on generations. To make a fair comparison, we first process the obtained results by carefully selecting the comparable data. We choose the data from each run of the compared algorithms at similar time points. In addition, the number of chosen data of each run for each algorithm is supposed to be the same. The training time budget is indicated by $time$, and the results are evenly divided into g groups. Each group's average period duration is $\frac{time}{g}$, and

Table 7.3 The mean (standard deviation) of the objective values on test instances of MTGP and M^3GP_2 **with the same number of generations** over 30 independent runs in six dynamic flexible job shop scenarios

Scenario	MTGP	M^3GP_2
<Fmax, 0.85>	1235.73(41.27)	1232.39(31.70)($\approx$)
<Fmax, 0.95>	1967.24(65.18)	1932.22(42.22)($-$)
<Fmean, 0.85>	384.55(1.04)	385.98(2.98) ($\approx$)
<Fmean, 0.95>	555.32(9.91)	552.50(5.07)($\approx$)
<WFmean, 0.85>	831.30(7.32)	831.24(5.22)($\approx$)
<WFmean, 0.95>	1114.93(13.80)	1114.36(8.88)($\approx$)

the training time demarcation points for g groups are $\frac{time}{g}*1$, $\frac{time}{g}*2$, ..., $\frac{time}{g}*g$. To map the contrasted data, the closest recorded time is identified based on the demarcation marks. Because the compared data is not obtained with exactly the same number of evaluations, there are bound to be inaccuracies when measuring the algorithm's performance in this way. Fortunately, because the sampling time duration for different runs of all algorithms is similar, these data are still representative for measuring the algorithm performance.

Because the baseline MTGP with 51 generations takes roughly 80 min to train, we limit the training time for MTGP and M^3GP to 80 min. We set the number of groups g to 20 so that we could acquire enough data to look into the objective values of MTGP, SGP_H, SGP_K, and M^3GP_2. The performance of the algorithms is measured using objective values of roughly 80 min. According to 30 independent runs, the mean and standard deviation of the objective values on unseen instances of MTGP, SGP_H, SGP_K, and M^3GP_2 with the same training time are shown in Table 7.4. First, we compare SGP_H, SGP_K, and M^3GP_2 with MTGP, respectively. Second, we compare M^3GP_2 with SGP_H and SGP_K, respectively. With the same training time, SGP_H and SGP_K performs worse than MTGP in three or one scenario(s), respectively. In four out of six scenarios (i.e., <Fmax, 0.85>, <Fmax, 0.95>, <Fmean, 0.95>, and <WFmean, 0.85>), M^3GP_2 outperforms MTGP. In addition, in all the other scenarios, M^3GP_2 is no worse than MTGP. This proves that the proposed algorithm is effective.

According to 30 independent runs on unseen instances, the curves of average objective values of MTGP, SGP_H, SGP_K, and M^3GP_2 with the same training time in the six DFJSS scenarios are shown in Fig. 7.4. Under the same training time, in all the scenarios, the proposed algorithm M^3GP_2 can converge quicker than the compared algorithms. M^3GP_2 outperforms the compared algorithms after about 10 min in scenario <Fmean, 0.85> and <Fmean, 0.95>. In addition, M^3GP_2 outperforms its counterparts after about 20 min in scenario <Fmax, 0.95> and <WFmean, 0.85>. We also see that SGP_K outperforms SGP_H, which is consistent with our expectations because the evaluations with half shop surrogate in [165] is much more computationally expensive than the KNN surrogate in [92].

Table 7.4 The mean (standard deviation) of the objective values on test instances of MTGP, SGP_H, SGP_K, and M^3GP_2 **with the same training time** of 80 min over 30 independent runs in six dynamic flexible job shop scenarios

Scenario	MTGP	SGP_H	SGP_K	M^3GP_2
<Fmax, 0.85>	1225.58(44.21)	1267.49(40.76)(+)	1239.48(40.73)($\approx$)	1212.25(28.60)($-$)($-$)($-$)
<Fmax, 0.95>	1963.85(61.53)	1981.81(52.60)($\approx$)	1956.72(29.60)($\approx$)	1925.87(28.98)($-$)($-$)($-$)
<Fmean, 0.85>	384.20(0.93)	387.24(4.22)(+)	386.74(3.32)(+)	384.61(1.25)($\approx$)($-$)($-$)
<Fmean, 0.95>	554.62(9.79)	554.75(6.71)($\approx$)	554.07(8.18)($\approx$)	550.56(3.32)($-$)($-$)($-$)
<WFmean, 0.85>	830.36(7.09)	834.79(8.13)(+)	830.40(5.52)($\approx$)	829.25(3.37)($-$)($-$)($\approx$)
<WFmean, 0.95>	1112.40(11.34)	1115.75(13.46)($\approx$)	1110.21(11.15)($\approx$)	1109.14(5.47)($\approx$)($-$)($\approx$)

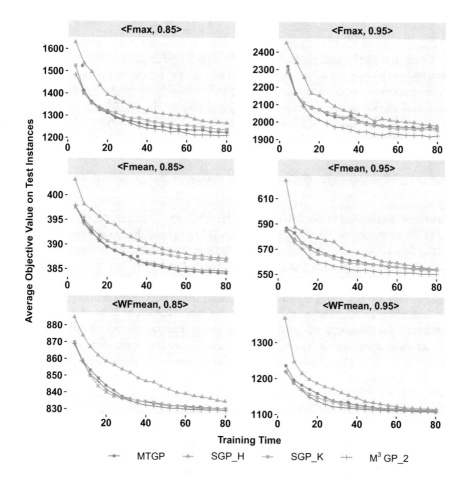

Fig. 7.4 The curve of average objective values according to 30 independent runs on test instances of MTGP, SGP_H, SGP_K, and M^3GP_2 **with the same training time** (in minutes) in six dynamic flexible job shop scenarios

Table 7.5 The mean (standard deviation) of the objective values of M^3GP_1 and M^3GP_2 **with and without knowledge transfer** with the same number of generations on test instances according to 30 independent runs in six dynamic flexible job shop scenarios

Scenario	M^3GP_1 (without)	M^3GP_1 (with)	M^3GP_2 (without)	M^3GP_2 (with)
<Fmax, 0.85>	1201.44(60.42)	1170.31(29.48)(−)	1261.85(65.40)	1232.39(31.70)(−)
<Fmax, 0.95>	1815.72(76.82)	1758.22(34.88)(−)	2001.56(80.43)	1932.22(42.22)(−)
<Fmean, 0.85>	390.24(4.21)	388.14(2.98)(−)	387.98(4.10)	385.98(2.98)(−)
<Fmean, 0.95>	566.98(7.81)	561.28(5.30)(−)	558.10(7.38)	552.50(5.07)(−)
<WFmean, 0.85>	840.19(9.50)	835.37(5.21)(−)	836.31(9.29)	831.24(5.22)(−)
<WFmean, 0.95>	1149.64(25.43)	1135.75(9.23)(−)	1130.06(25.95)	1114.36(8.88)(−)

Overall, the presented algorithm can improve the efficiency of GP for DFJSS by decreasing the computational cost without sacrificing algorithm performance, and thereby accelerating convergence of the algorithm. Given the same training time, the proposed algorithm outperforms the compared state-of-the-art GP algorithms with surrogate in most scenarios while doing no worse in all the scenarios.

7.4.3 Effectiveness of Knowledge Transfer Mechanism

Several experiments are carried out to test the effectiveness of the introduced knowledge transfer mechanism. To prevent knowledge sharing between different subpopulations, we set the transfer ratio tr to zero for M^3GP. M^3GP_1(without) and M^3GP_2(without) represent that there is no knowledge transfer between subpopulations for M^3GP. With a transfer ratio tr of 0.6, M^3GP_1(with) and M^3GP_2(with) are used to indicate that there is knowledge transfer between subpopulation.

According to 30 independent runs with the same number of generations on test instances, the mean and standard deviation of the objective values of M^3GP in the six scenarios are shown in Table 7.5. With knowledge transfer, the learned rules with multi-fidelity surrogate models outperform its counterpart without knowledge transfer. Specifically, M^3GP_1(with) outperforms M^3GP_1(without), while M^3GP_2(with) outperforms M^3GP_2(without) in all the scenarios. This suggests that the knowledge sharing can assist all the involved problems with various complexities. For more sophisticated surrogate models, the information gained with the simpler surrogate model is useful. The knowledge gained from the sophisticated surrogate model can be applied effectively to the simpler surrogate model as well. This demonstrates that the proposed knowledge sharing mechanism is effective.

The mean and standard deviation of the objective values on test instances of MTGP and M^3GP_2 without knowledge transfer in six scenarios are shown in Table 7.6. It reveals that M^3GP_2 without knowledge transfer performs significantly worse than MTGP. It validates the proposed knowledge sharing mechanism's effectiveness. This

Table 7.6 The mean (standard deviation) of the objective values on test instances of MTGP and M^3GP_2(without) according to 30 independent runs in six dynamic flexible job shop scenarios

Scenario	MTGP	M^3GP_2(without)
<Fmax, 0.85>	1235.73(41.27)	1261.85(65.40)(+)
<Fmax, 0.95>	1967.24(65.18)	2001.56(80.43)(+)
<Fmean, 0.85>	384.55(1.04)	387.98(4.10)(+)
<Fmean, 0.95>	555.32(9.91)	558.10(7.38)(+)
<WFmean, 0.85>	831.30(7.32)	836.31(9.29)(+)
<WFmean, 0.95>	1114.93(13.80)	1130.06(25.95)(+)

is consistent with our expectations, as more computational resources are used in MTGP to solve the target problem. Particularly, without knowledge transfer, the number of individuals to solve the desired problem in M^3GP_2(without) is 512, which equals half of the number of individuals in MTGP.

Based on 30 independent runs, the curve of average objective values on unseen instances of M^3GP_1(without) and M^3GP_1(with) in six different DFJSS scenarios are shown in Fig. 7.5. M^3GP_1(with) outperforms M^3GP_1(without) roughly after generation 5 in the max-flowtime related scenarios (i.e., <Fmax, 0.85> and <Fmax, 0.95>) and after 10 generations approximately in mean-flowtime and weighted mean-flowtime related scenarios (i.e., <Fmean, 0.85>, <Fmean, 0.95>, <WFmean, 0.85>, and <WFmean, 0.95>). It is possible that this is because the individuals in the population before generation 5 or generation 10 have not yet attained a high level of quality, and the knowledge transferred has not yet made a significantly contribution to the other subpopulation. Fortunately, the transferred knowledge before generation 5 or 10 has no negative impact on the other subpopulations. The same trend can also be seen between M^3GP_2(without) and M^3GP_2(with), as illustrated in Fig. 7.6.

7.5 Further Analyses

To deeply understand the effectiveness of the presented algorithm, we have conducted in-depth analyses. This section delves deeper into the presented algorithm, which includes multiple surrogate models with varied fidelities as well as a sensitivity analysis of the knowledge transfer ratio.

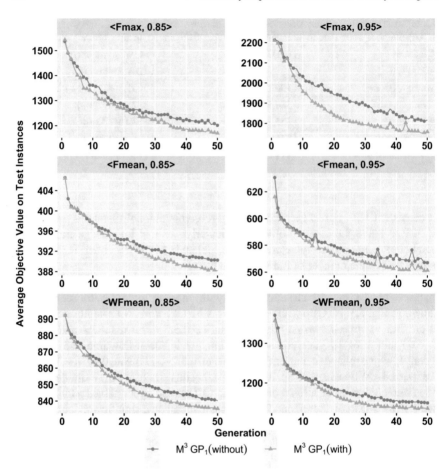

Fig. 7.5 The curve of average objective values on test instances of M^3GP_1(without) and M^3GP_1(with) according to 30 independent runs in six dynamic flexible job shop scenarios

7.5.1 Number Analysis of Multi-fidelity Surrogate Models

It would be interesting to see if incorporation more surrogate models with diverse fidelities can improve problem-solving. Half of the population is preserved for opti-mising the desired problem in order to assure the performance of the algorithm for the problem to be addressed. The remaining half of the population is divided equally for handling easier problems using simpler surrogate models. The number of jobs of surrogate models follows an arithmetic sequence with an upper bound of 5000. The corresponding algorithms on the desired problem are represented by M^3GP_2, M^3GP_3, M^3GP_4, and M^3GP_5, respectively. The number of individuals and jobs of the proposed algorithm with one, two, three, and four surrogates is shown in Table 7.7.

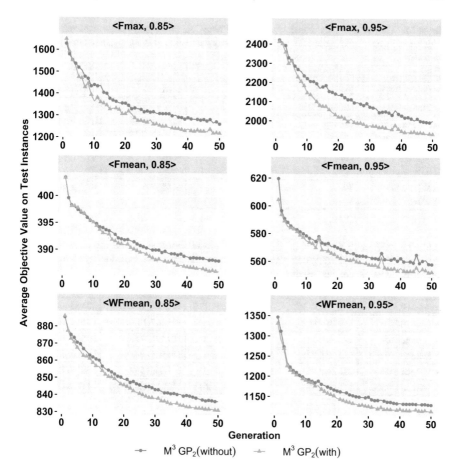

Fig. 7.6 The curve of average objective values on test instances of M^3GP_2(without) and M^3GP_2(with) according to 30 independent runs in six dynamic flexible job shop scenarios

Table 7.7 The number of individuals/jobs of the presented algorithm with two, three, four, and five surrogates

Algorithm	$Subpop_1$	$Subpop_2$	$Subpop_3$	$Subpop_4$	$Subpop_5$
M^3GP_2	512/2500	512/5000	–	–	–
M^3GP_3	256/1667	256/3333	512/5000	–	–
M^3GP_4	170/1250	171/2500	171/3750	512/5000	–
M^3GP_5	128/1000	128/2000	128/3000	128/4000	512/5000

Table 7.8 The mean (standard deviation) of the training time (in minutes) of the involved algorithms with the same number of generations based on 30 independent runs in six dynamic flexible job shop scenarios

Scenario	M^3GP_2	M^3GP_3	M^3GP_4	M^3GP_5
<Fmax, 0.85>	51(9)	48(9)($\approx$)	46(8)($\approx$)	48(8)($\approx$)
<Fmax, 0.95>	53(9)	48(7)($-$)	47(7)($-$)	49(7)($-$)
<Fmean, 0.85>	48(8)	43(4)($-$)	47(8)($\approx$)	47(7)($\approx$)
<Fmean, 0.95>	49(6)	47(7)($\approx$)	48(6)($\approx$)	46(7)($\approx$)
<WFmean, 0.85>	49(7)	48(7)($\approx$)	48(6)($\approx$)	47(8)($\approx$)
<WFmean, 0.95>	49(6)	47(9)($\approx$)	47(8)($\approx$)	47(7)($\approx$)

Table 7.9 The mean (standard deviation) of the objective values on test instances of M^3GP_2, M^3GP_3, M^3GP_4, and M^3GP_5 with the same number of generations according to 30 independent runs in six dynamic flexible job shop scenarios

Scenario	M^3GP_2	M^3GP_3	M^3GP_4	M^3GP_5
<Fmax, 0.85>	1232.39(31.70)	1228.03(37.98)($\approx$)	**1222.44(29.73)**($\approx$)	1230.54(33.47)($\approx$)
<Fmax, 0.95>	**1932.22(42.22)**	1948.53(45.67)($\approx$)	1960.67(120.22)($\approx$)	1954.50(68.28)($\approx$)
<Fmean, 0.85>	385.98(2.98)	385.33(2.06)($\approx$)	**384.93(1.73)**($\approx$)	385.59(3.11)($\approx$)
<Fmean, 0.95>	**552.50(5.07)**	553.43(5.30)($\approx$)	554.02(4.84)($\approx$)	555.82(6.76)($+$)
<WFmean, 0.85>	831.24(5.22)	830.83(6.50)($\approx$)	831.36(6.56)($\approx$)	**830.79(4.27)**($\approx$)
<WFmean, 0.95>	**1114.36(8.88)**	1119.21(15.79)($+$)	1122.82(17.39)($+$)	1117.59(12.98)($\approx$)

According to 30 independent runs, with the same number of generations, the mean and standard deviation of the training time of M^3GP_2, M^3GP_3, M^3GP_4, and M^3GP_5 in six different scenarios is shown in Table 7.8. As the number of surrogate models increases, compared with M^3GP_2, M^3GP_3, M^3GP_4, and M^3GP_5 does not differ much in most scenarios in terms of training time.

According to 30 independent runs, with with the same number of generations, the mean and standard deviation of the objective values of M^3GP_2, M^3GP_3, M^3GP_4, and M^3GP_5 on unseen instances in six different scenarios are shown in Table 7.9. There is no substantial difference between the compared algorithms in terms of objective values on unseen data in most scenarios. In one of the scenarios, the performance of M^3GP_3, M^3GP_4, and M^3GP_5 is much worse than that of M^3GP_3. In half of the scenarios (i.e., Fmax, 0.95>, Fmean, 0.95>, and WFmean, 0.95>, M^3GP_2 outperforms the other compared algorithms in terms of mean and standard deviation, as highlighted in bold. Furthermore, the investigated problems are not particularly sensitive to the number of surrogates, as the performance of the proposed algorithm with various numbers of surrogates is comparable.

Overall, the results show that with the settings in this chapter, the proposed algorithm with two surrogates obtains the best performance. One probable explanation

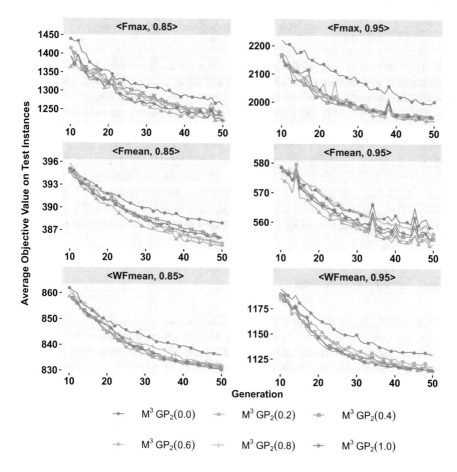

Fig. 7.7 The curve of average objective values on test instances of M^3GP_2 with different transfer ratios over 30 independent runs in six dynamic flexible job shop scenarios.

is that the additional surrogate models introduce more noise than the algorithm only with two surrogate models. Surrogates' accuracy varies depending on the complexity of the problem. In our scenario, we find that utilising two surrogate models is the best option. There are a number of intriguing but challenging research questions that should be investigated more in the future. First, how to choose the best number of surrogates? Second, based on domain knowledge or information from the evolutionary process, how to create efficient surrogate models? Finally, how to build successful knowledge transfer methods for a large number of surrogates? The interaction between more than two surrogates is complex.

7.5.2 Sensitivity Analysis of Knowledge Sharing Ratio

This section delves deeper into the transfer ratio, which determines the frequency of information transfer between distinct subpopulations at each generation. According to 30 independent runs, the curves of average objective values on unseen instances of M^3GP_2 with different transfer ratios in six different scenarios are shown in Fig. 7.7. The performance of M^3GP_2 with different transfer ratios are comparable in the scenarios <Fmax, 0.95>, <WFmean, 0.85>, and <WFmean, 0.95>. There are some slight differences between the algorithms with different transfer ratios in the scenarios <Fmax, 0.85>, <Fmean, 0.85>, and <Fmean, 0.95>. M^3GP_2 with a transfer ratio of 0.6 has a slightly better performance than its counterparts from an overall perspective. This is the reason why this chapter sets the transfer ratio to 0.6 for M^3GP. However, in general, the performance of M^3GP_2 is not sensitive to the transfer ratio.

7.6 Chapter Summary

The goal of the presented algorithm in this chapter is to develop an effective strategy that collaborates multi-fidelity surrogate models to improve the efficiency of GP to evolve scheduling heuristics automatically for DFJSS. The purpose is successfully achieved with an effective collaboration framework in GP that allows the subpopulations with different surrogate models to learn from each other, and developing an effective knowledge transfer mechanism.

The results show that the presented algorithm M^3GP_2 can dramatically reduce the computational time of GP without losing its performance. Within the same training time, M^3GP_2 can achieve significantly better performance in most of the scenarios, while no worse than its counterpart in all the scenarios. The efficiency of the presented algorithm is verified by comparing the training time. The effectiveness of the presented algorithm is verified by comparing the quality of learned scheduling heuristics and the analysis of knowledge transfer mechanism. In summary, M^3GP_2 can successfully improve the efficiency of GP, and achieve effective scheduling heuristics automatically for DFJSS. The presented algorithm shows its superiority compared with the state-of-the-art surrogate-assisted GP algorithms for the JSS problems.

This chapter mainly uses surrogate techniques to improve the training efficiency and effectiveness of GP for DFJSS. In the next chapter, surrogate techniques will be used with other techniques to improve the efficiency of feature selection in GP for DFJSS.

Chapter 8
Search Space Reduction with Feature Selection

A GP individual in this book can be considered as a priority function, which is typically represented as a tree. GP uses a terminal set and a function set to learn a population of such tree. The terminal set is vital to the success of GP [213]. In DFJSS, a wide range of features about the job shop state (e.g., the processing time of each operation and the idle time of each machine) can be considered as features to be included in the terminal set. The importance of a feature, on the other hand, is determined by job shop scenarios [232]. In fact, it is often difficult to tell which features are useful to learn scheduling heuristics. Previous studies often put all potential features into the terminal set. As a result, learned scheduling heuristics have a lot of different features, which makes them difficult to interpret. Additionally, a huge terminal set with redundant or unrelated features results in an exponentially vast and noisy search space, reducing GP's search power. It has been shown in the literature that the searchability of the evolutionary algorithms can be improved by reducing the number of features with feature selection technique [227].

8.1 Challenges and Motivations

Feature selection [86] is an important machine learning technique to select relevant and complementary features to reduce the search space of evolutionary algorithms. Feature selection has been successfully used for different tasks such as regression [43], classification [56, 216, 227], and clustering [126]. However, there are some challenges which make traditional feature selection approaches not directly applicable to DFJSS. First, the task (i.e., prioritising operations or machines) in DFJSS and the training instances are different from the traditional machine learning tasks (e.g., regression) and training data. The available data are only generated with the simulation execution in DFJSS, while the training data are accessible directly in the

F. Zhang et al., *Genetic Programming for Production Scheduling*, Machine Learning: Foundations, Methodologies, and Applications, https://doi.org/10.1007/978-981-16-4859-5_8

traditional machine learning tasks. Thus, filter-based feature selection approaches used for traditional machine tasks cannot be applied for DFJSS. The reason is that it is impossible to measure the importance of each feature based on the filter measures such as entropy [105] and Pearson's correlation [200]. Second, it is much more computationally expensive to apply the wrapper-based feature selection approaches in DFJSS than in traditional machine learning tasks, due to the long simulation execution of DFJSS with GP. Specifically, running a GP process to obtain a reliable estimation of the best objective value of a feature set is much slower than that in traditional machine learning tasks. Last, in most embedded feature selection approaches, although GP has been successfully used to handle both the feature selection and the supervised machine learning tasks (e.g., regression [42, 43] and classification [41, 154]) simultaneously, feature selection is rarely used for GP in JSS.

In this chapter, we investigate feature selection in DFJSS. One goal of this chapter is to explore how to design feature selection algorithms to select features efficiently for routing rule and sequencing rule simultaneously. The other goal is to investigate how to use the selected features to improve the interpretability of learned scheduling heuristics. Specially, we are interested in the following questions:

- How to measure the feature importance in tree-based representation?
- How to measure the feature importance efficiently?
- How to use the selected features to improve the interpretability of learned scheduling heuristics?

Section 8.2 introduces a two-stage genetic programming with feature selection to realise the framework of doing feature selection and using the selected features. Section 8.3 describes the experiment design followed by the results in Sect. 8.4. Section 8.5 further discusses the selected features and the learned scheduling heuristics. Section 8.6 makes the conclusion of this chapter.

8.2   Algorithm Design and Details

This section will show how the important features are selected and how to use the information carried by the learned scheduling heuristics obtained from the feature selection process. The proposed algorithm's framework is described first, followed by the details of its essential components.

8.2.1   Two-stage Genetic Programming with Feature Selection

Figure 8.1 shows the flowchart of the GP feature selection algorithm for DFJSS. The two-stage GP algorithm contains two consecutive phases. The first stage is mainly for selecting important features. Using the niching and surrogate techniques in stage 1 helps GP quickly reach a diverse set of reasonably good individuals for investigating

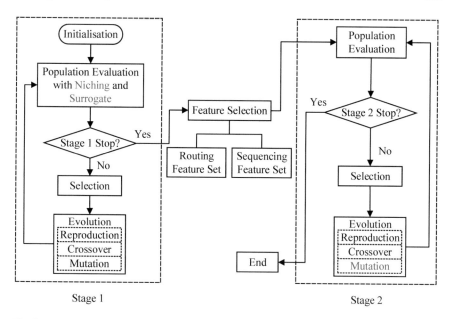

Fig. 8.1 The flowchart of two-stage GP with feature selection for dynamic flexible job shop

the importance of features. In the second stage, the obtained information in the first stage is utilised by inheriting the final population from stage 1 to stage 2. In addition, the selected features are used to guide the search space via the mutation operator during the subsequent evolutionary search.

The main steps in stage 1 and 2, which are evaluation, selection, and evolution, are the same as the traditional GP algorithm. The difference is that, there is a checkpoint (i.e., generation 50) for separating the whole GP process into two stages. The details of the presented two-stage GP with feature selection algorithm are shown as follows:

1. In the first stage (i.e., before generation 50), GP proceeds with a niching evaluator and a surrogate model. The output of stage 1 is the number of individuals for investigating the importance of features.
2. At generation 50, the feature selection mechanism will be used to select two subsets of informative features for the routing rule and the sequencing rule, respectively. Then, the terminal sets for learning routing and sequencing rules are reset to the selected feature subsets.
3. In the second stage (i.e., after generation 50), the population will be evaluated with the original evaluation of GP in DFJSS. The mutation operator generates the new subtree only with the selected features.

Algorithm 8.1 Selecting features based on a diverse set of promising individuals

---

Input: A diverse set of promising individuals ($baseInds$) obtained from stage 1
Output: Selected feature set F

1: set $F \leftarrow \{\}$
2: **for** i = 1 to $|features|$ **do**
3:      $vote_{f_i} \leftarrow 0$ // the number of votes obtained for feature f_i
4:      **for** j = 1 to $|baseInds|$ **do**
5:          Measure the contribution C_{f_i} of feature f_i
6:          $ind \leftarrow baseInds_j$
7:          **if** $C_{f_i}^{ind} > 0$ **then**
8:              $vote_{f_i} \leftarrow vote_{f_i} + 1$
9:          **end if**
10:     **end for**
11:     **if** $vote_{f_i} > 0.5 * |baseInds|$ **then**
12:         $T \leftarrow T \cup f_i$
13:     **end if**
14: **end for**
15: **return** F

8.2.1.1 Niching and Surrogate

The niching technique is used to obtain a diverse set of good individuals for feature selection. In a nutshell, the niching strategy preserves individual diversity by dividing individuals into niches and regulating the number of individuals in each niche. Involving feature selection mechanism increases the computational cost of GP for DFJSS. To reduce the extra computational cost, the surrogate technique is applied in GP with feature selection [142]. The surrogate technique was founded on the premise that knowledge gained through tackling simpler or auxiliary problems could be utilised to the original problems [165]. This chapter uses the idea but adapts it to the DFJSS. There are 5000 jobs and 10 machines in the original simulation. The surrogate model is created by reducing the simulation to 500 jobs and 5 machines.

8.2.1.2 Feature Selection

Algorithm 8.1 shows the pseudo-code of feature selection. The feature selection consists of three main steps. First, the top 10 individuals from the last population in stage 1 are chosen as a diversified group of promising individuals (i.e., represented by $baseInds$) based on fitness values. Second, each feature's importance is determined by its contribution to the fitness of the individuals in $baseInds$, and an individual in $baseInds$ will vote for a feature if it contributes to their fitness. Finally, a feature will be selected and added to the selected feature set with important features, if a feature can get more than half of the votes from the individuals in $baseInds$.

The Importance of Features: The contributions of a feature f to a group of individuals $baseInds$ are used to determine its importance. To determine the contribution

Fig. 8.2 An example of how to examine the contribution (denoted as C_y) of a feature y for an individual r

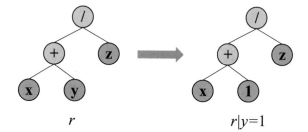

of a feature f to an individual r (i.e., the contribution is denoted by C_f^r), the feature f in the investigated individual is first substituted with the constant of one. A GP individual r with three features (i.e., x, y, and z) is shown in Fig. 8.2. Feature y is replaced with 1 to investigate the significance of feature y.

A feature's contribution is calculated as the difference between the fitness before and after the replacement, as shown in Eq. (8.1).

$$C_f^r = fitness(r|f = 1) - fitness(r) \tag{8.1}$$

The contribution of feature y, as shown in Fig. 8.2 is defined as $C_y^r = fitness(r|y = 1) - fitness(r)$.

Since the DFJSS problem investigated in this book is a minimisation problem, thus, if $C_f^r > 0$, it signifies that without the measured feature f, fitness deteriorates, and the measured feature f is important. The individual r votes for the measured feature f due to its contribution to the fitness.

Selecting Features: The feature selection technique in [142] is extended in this chapter to fit the DFJSS problems in two ways. To begin, two groups of individuals from each subpopulation are chosen to learn routing and sequencing rules, respectively. Second, based on the two groups of individuals, the feature selection algorithm is used to pick the important routing feature set and the sequencing feature set, respectively. If a feature f contributes positively to at least 50% of the selected individuals *baseInds*, it will be chosen (from line 11 to line 13).

8.2.2 Individual Adaptation Strategies and Genetic Programming Feature Selection

Based on the two-stage GP feature selection algorithm mentioned in the last section, to eliminate unselected features in scheduling heuristics, the information from both the selected features and the observed individuals from the feature selection process is used in this chapter to introduce individual adaptation strategies.

Figure 8.3 shows the flowchart of the two-stage GP feature selection algorithm with individual adaptation strategy. Stage 1 is the same as shown in Fig. 8.1. The

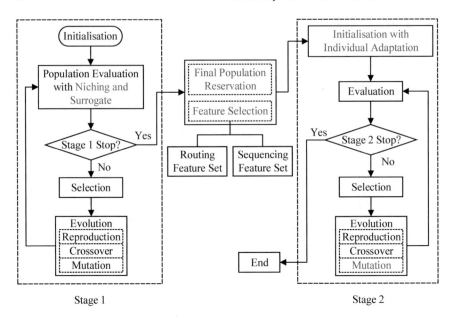

Fig. 8.3 The flowchart of two-stage GP feature selection algorithm with individual adaptation strategies (i.e., the reddish font parts are the main steps of the presented algorithm)

goal of *Stage 1* is to create an informative population of individuals for feature selection and adaptation. The unique feature of the GP feature selection algorithm with individual adaptation strategy is the initialisation in stage 2. The goal of *Stage 2* in the GP feature selection algorithm with individual adaptation strategies is to use the information (i.e., final population and selected features) obtained in stage 1 to use only selected features to learn effective scheduling heuristics. This chapter discusses a variety of innovative individual adaptation strategies for initialising the population in stage 2 with only the selected features while maintaining the performance of the algorithm. In stage 2, the chosen features will be applied in two cases. One is, during the population initialisation, to generate new individuals in order to form a new population. The other is, via mutation, to create a new subtree with only the selected features to replace a selected subtree of a parent. Readers interested in the genetic operators of GP can refer to [190]. Readers can find that [190] is helpful to learn about genetic operators of GP.

8.2.2.1   Individual Adaptation Strategies

Individual adaptation strategy aims at inheriting the promising individuals' information of stage 1's final population to keep the quality of learned scheduling heuristics, but with only selected features in stage 2. This chapter presents two strategies to adapt the individuals at the previous generation of stage 1 to stage 2.

1. The first individual adaptation strategy is to simply substitute each unselected feature in the individuals with a constant of one. This can completely remove the unselected features from the individuals, while still keeping the structures of individuals as much as possible. If a feature is not selected, it is expected to have little contribution to a majority of individuals, and thus replacing them by one would not change the fitness much.

   A potential drawback is that the average quality of individuals in the first generation of stage 2 might not be as well as the previous generation of stage 1. One possible reason is that replacing a number of unselected features in an individual by one is likely to change the behaviour of the individual in certain ways.

2. The second individual adaptation strategy is developed according to the idea of "mimicking" the behaviour of individuals. To be specific, it generates a large number of individuals at random using only the given features. Each promising individual in stage 1's final population is replaced by the randomly produced individual with the most similar behaviour. The behaviour of a GP individual is measured based on the phenotypic characterisation [92], which is indicated by a numeric vector. The details of the phenotypic characterisation calculation are shown as follows:

Phenotypic Characterisation: The phenotypic characterisation of a GP individual is a decision vector based on a set of decision situations [92]. In this chapter, decision situations are sampled from preliminary simulation runs with 5000 jobs on 10 machines using the reference rules (e.g., WIQ, work in the queue for routing, and SPT, shortest processing time for sequencing). The preliminary simulation generates about 50,000 routing and 50,000 sequencing decision situations. Following the steps in [92], we randomly sample decision situations from the generated decision situations that contain 2 and 20 jobs. To balance the accuracy and complexity of the phenotypic characterisation, the number of candidates, i.e., machines for routing and operations for sequencing, in each decision situation, is set to 7 in this chapter. In other words, from all the generated decision situations, a subset of 20 routing situations and 20 sequencing situations with the length of 7 is sampled for measuring the phenotypic characteristic of a GP individual. A smaller distance between the phenotypic characterisations of two individuals suggests that the two individuals are more similar.

The phenotypic characterisation of a rule is a vector of rank numbers, where the dimension of phenotypic characterisation of a GP individual equals the number of decision situations. The element in the ith dimension of a phenotypic characterisation indicates the rank of the most prior candidate (i.e., operation or machine) by the characterised rule in the rank list of the reference rule. Table 8.1 shows an example of calculating the phenotypic characterisation of a routing rule with four decision situations, and each decision situation consists of three candidate machines. In the first decision situation, M_2 is the most prior machine by the characterised routing rule. When looking at the rank value of M_2 by the reference rule, we find that the rank value is 2. Therefore, the value of the phenotypic characterisation in the first situation PC_1 is set to 2. Similarly, PC_i can be obtained in other decision situations. The

Table 8.1 An example of calculating the phenotypic characterisation of a routing rule with four decision situations (i.e., each with three candidate machines)

Decision situation	Reference rule	Characterised rule	PC_i
1 (M_1)	1	3	
1 (M_2)	2	1	2
1 (M_3)	3	2	
2 (M_1)	1	3	
2 (M_2)	3	1	3
2 (M_3)	2	2	
3 (M_1)	2	2	
3 (M_2)	3	1	3
3 (M_3)	1	3	
4 (M_1)	2	3	
4 (M_2)	3	2	1
4 (M_3)	1	1	

PC_i indicates the ith dimension of phenotypic characterisation

Routing PC				Sequencing PC			
2	3	3	1	1	3	2	1

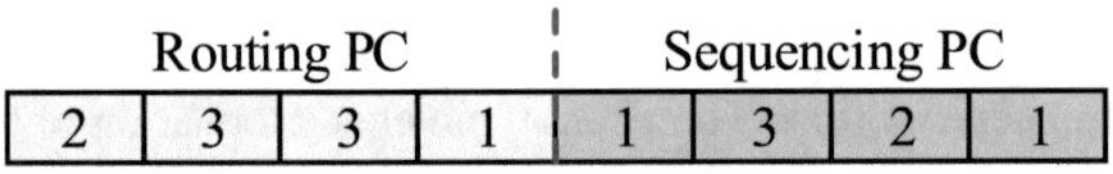

Fig. 8.4 An example of the phenotypic characterisation of an individual in dynamic flexible job shop scheduling (PC indicates phenotypic characterisation)

corresponding observation indicators for finalising the phenotypic characterisation are underlined in each decision situation. Finally, the phenotypic characterisation of this routing rule is [213, 227, 232]. It is noted that the way to calculate the phenotypic characterisation of the sequencing rule is the same as the routing rule, except that the sequencing rule is examined with sequencing decision situations.

This chapter extends the idea in [160, 161] to calculate the phenotypic characterisation for an individual in DFJSS by concatenating the phenotypic characterisations of routing and sequencing rules. An example of the phenotypic characterisation of an individual in DFJSS is shown in Fig. 8.4. The phenotypic characterisation of an individual consists of the decision vectors of routing and sequencing heuristics. The individuals with both similar routing (left part) and sequencing (right part) phenotypic characterisations are considered to have similar behaviour.

Mimic the Behaviour of Individuals:
Figure 8.5 depicts an example of the process of mimicking individuals with a two-dimensional phenotypic characterisation. To develop phenotypic characteriszation, two decision situations are used. A decision situation is a small part of decision marking processes in the simulation, either when a machine becomes idle to choose the next operation to process or an operation becomes ready to be assigned to a machine. The stars indicate the promising individuals in stage 1 that need to be

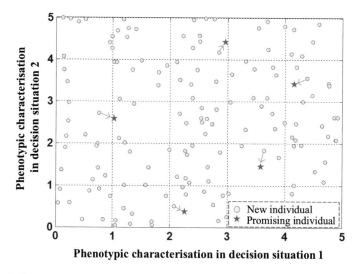

Fig. 8.5 Individual mimicking process by producing new individuals with only selected important features

mimicked. A large number of new individuals (i.e., denoted as circles) are generated with only the selected features. The new individuals whose phenotypic vectors are closest to the mimicked individuals will replace the promising individuals in stage 1. It is noted that it is not always possible to find individuals with the same behaviour (i.e., the distance between two individuals equals zero). The new individuals with the most similar behaviours with promising individuals will be saved in the initial population for stage 2.

Adapting all of the individuals from stage 1 is not necessary. There will be too many randomly generated individuals if the number of adapted individuals is too small, and the performance will be close to just re-initialisation. If it is too big, it will make a lot of noise and reduce diversity. Only the "promising" individuals are adapted in this chapter. The knee point, which is a parameter-free technique, is used in this chapter to identify promising individuals. The knee point can be used to draw a line between two areas. Individuals with fitness smaller (i.e., minimising problem) than the knee point are chosen as promising individuals. The details of detecting promising individuals with the knee point technique can be found in Chap. 7. In the initial population of stage 2, the rest of the individuals, will be randomly initialised with only the selected features. In this way, we can get a new population containing individuals with only the selected features.

8.2.3   Algorithm Summary

This chapter presents two GP feature selection algorithms for DFJSS. The first algorithm is a two-stage GP feature selection algorithm for DFJSS where both the routing terminal set and sequencing terminal set are obtained simultaneously. The output of the feature selection process is two feature subsets, one is the important feature set for routing rule and the other is the important feature set for sequencing rule. The selected features are used during the evolutionary process of GP via mutation in stage 2. The second algorithm further extends the two-stage GP feature selection algorithm, where the proposed individual adaptation strategies are developed to initialise the population to eliminate the unselected features in stage 2. The learned scheduling heuristics are expected to be effective and contain only the selected features.

8.3   Experiment Design

8.3.1   Comparison Design

GP with two subpopulations for incorporating cooperative coevolution learning mechanism is used to learn routing and sequencing rules simultaneously for DFJSS [228]. This can separate the routing rules and sequencing rules into different subpopulations to make it easy to measure the importance of features for different types of scheduling heuristics.

In this chapter, a total of five algorithms are compared. The baseline algorithm is the cooperative coevolution genetic programming (CCGP) [228] (i.e., without feature selection and individual adaptation strategies). $CCGP^2$ [244] is the name of the first algorithm described in this chapter. $CCGP^2$ only employs the selected features from stage 2 in mutation, and it is used to see if the new mutation operator has any effect on the performance. The proposed algorithms with the new individual adaptation strategies are called as $CCGP^{2a}$(mimic) [236] (i.e., mimicking the behaviour of promising individuals) and $CCGP^{2a}$(rep) (i.e., replicating the behaviour of promising individuals) (i.e., directly replacing the unselected features with one). The algorithm (i.e., named $CCGP^{2a}$(rand)) that randomly initialises all the individuals in the new population in stage 2 is also compared to check the effectiveness of $CCGP^{2a}$(mimic) and $CCGP^{2a}$(rep). Using the selected features to randomly re-initialise the new population in stage 2 is the most straightforward way to eliminate unselected features.

Table 8.2 The specialised parameter settings of genetic programming

Parameter	Value
Number of subpopulations	2
Subpopulation size	512
Number of generations in stage 1 and stage 2	50/50

8.3.2 Specialised Parameter Settings of Genetic Programming

Table 8.2 shows the specialised parameter settings of GP in this chapter. Two subpopulations are involved. One subpopulation is responsible for learning routing rules, while the other is responsible for learning sequencing rules. The GP with feature selection contains 100 generations. The first 50 generations are designed for feature selection, and the second 50 generations aim to learn scheduling heuristics for DFJSS.

8.4 Results and Discussions

8.4.1 Quality of Learned Scheduling Heuristics

Based on 30 independent runs, in six DFJSS scenarios, the mean and standard deviation of the involved five algorithms are shown in Table 8.3. $CCGP^2$, which uses the selected features in the mutation operator, obtains similar results to CCGP. One reason could be that the GP can automatically discover important features. The other is that there is not that much change when only a few individuals are mutated at a low rate (i.e., 0.15). The disadvantage is that learned scheduling heuristics by $CCGP^2$ still contain unselected characteristics. It does not make the rules easier to interpret because all/most of the features are still present, making the rules complex.

In most scenarios, the performance of $CCGP^{2a}$(rand) is much worse than that of CCGP. One cause might be that for stage 2, a completely new random population with poorer individual productivity (i.e., a fresh start) is formed. It is difficult to achieve good scheduling heuristics as CCGP (i.e., actually evolved for 100 generations). In most scenarios, $CCGP^{2a}$(mimic) and $CCGP^{2a}$(rep) (i.e., just with selected features) can achieve comparable performance to $CCGP^2$. It suggests that the proposed individual adaptation strategies are effective in utilising the population information from stage 1.

In the scenarios <Fmax, 0.85> and <Fmax, 0.95>, all the algorithms with feature selection (i.e., $CCGP^2$, $CCGP^{2a}$(rand), $CCGP^{2a}$(rep), and $CCGP^{2a}$(mimic)) have bigger variances. One reason for this could be that the maximum flowtime is more sen-

Table 8.3 The mean (standard deviation) of the objective values of the five algorithms according to 30 independent runs for six dynamic flexible job shop scheduling scenarios

Sce.	CCGP	CCGP^2	CCGP^{2a}(rand)	CCGP^{2a}(rep)	CCGP^{2a}(mimic)
1	1223.83(41.78)	1225.41(43.51)($\approx$)	1314.87(121.35)(+)	1237.53(81.28)($\approx$)	1238.34(99.27)($\approx$)
2	1959.24(46.63)	1998.09(115.26)($\approx$)	2054.56(204.36)(+)	2032.85(145.16)($\approx$)	2034.08(153.61)($\approx$)
3	385.42(2.65)	384.77(1.32)($\approx$)	387.34(2.23)($\approx$)	385.07(1.24)($\approx$)	385.14(1.87)($\approx$)
4	553.65(7.89)	552.88(6.78)($\approx$)	559.21(8.21)(+)	553.07(6.31)($\approx$)	551.20(6.11)($\approx$)
5	830.74(6.89)	829.58(5.56)($\approx$)	833.02(6.15)(+)	830.11(5.42)($\approx$)	831.51(6.52)($\approx$)
6	1109.89(13.07)	1110.86(12.01)($\approx$)	1112.35(12.91)($\approx$)	1109.58(7.96)($\approx$)	1112.94(14.62)($\approx$)

* Sce. 1: <Fmax, 0.85>, 2: <Fmax, 0.95>, 3: <Fmean, 0.85>, 4: <Fmean, 0.95>
* Sce. 5: <WFmean, 0.85>, 6: <WFmean, 0.95>

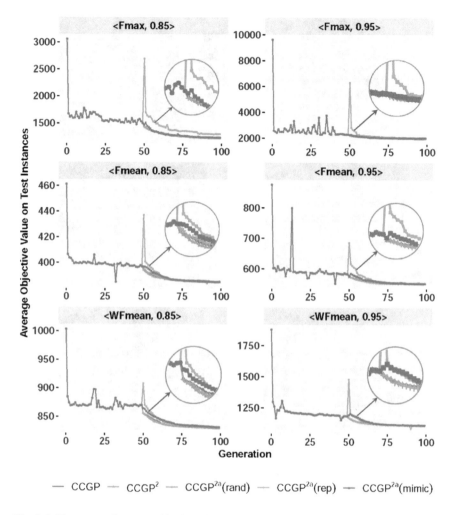

Fig. 8.6 The curves of average objective values on test instances of the five algorithms according to 30 independent runs in six dynamic flexible job shop scheduling scenarios

sitive to the worst-case situation of processing jobs than the mean-flowtime. Another possible explanation is that feature selection depends on the individuals obtained in stage 1, which is not always accurate, and inaccurate selected features can lead to some outliers.

According to 30 independent runs, in different scenarios, the convergence curves of the average objective value on unseen instances of CCGP, $CCGP^2$, $CCGP^{2a}$ (rand), $CCGP^{2a}$ (rep), and $CCGP^{2a}$ (mimic) are shown in Fig. 8.6. In all of the scenarios, $CCGP^{2a}$ (mimic), and $CCGP^{2a}$ (rep) can reliably mimic the behaviours (i.e., shown at generation 50), and their performance does not degrade significantly. Compared with $CCGP^2$, $CCGP^{2a}$ (mimic) can achieve comparable results in the scenario <Fmax,

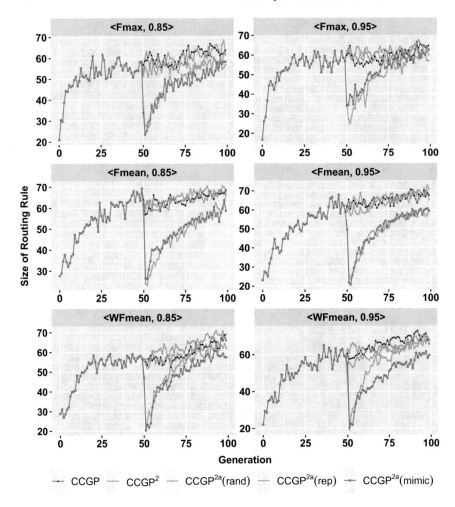

Fig. 8.7 The curves of **the best routing rule sizes** of the population of the five algorithms according to 30 independent runs in six dynamic flexible job shop scheduling scenarios

0.95> after generation 50. In all scenarios, however, $CCGP^{2a}$(rep) can get comparable results with $CCGP^2$ in terms of effectiveness. This suggests that $CCGP^{2a}$(rep) has a better ability to do individual inheritance. In general, $CCGP^{2a}$(mimic) and $CCGP^{2a}$(rep) are more effective than $CCGP^{2a}$(rand) from the perspective of objective values. From stage 1 to stage 2, both $CCGP^{2a}$(mimic) and $CCGP^{2a}$(rep) can successfully inherit the information of the individuals.

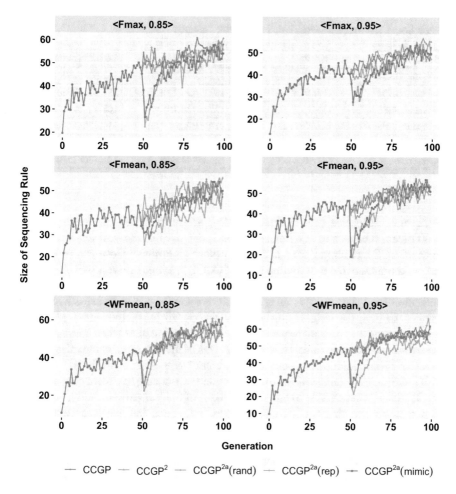

Fig. 8.8 The curves of **the best sequencing rule sizes** of the population of the five algorithms according to 30 independent runs in six dynamic flexible job shop scheduling scenarios

8.4.2 Rule Size

Based on 30 independent runs, the curves of the mean values of the routing and sequencing rule sizes are shown in Figs. 8.7 and 8.8. From the figures, at the beginning of stage 2, the rule sizes of routing rules and sequencing rules of $CCGP^{2a}$(mimic) and $CCGP^{2a}$(rand) decrease dramatically. This is caused by the individual adaptation strategies of re-initialising the population with smaller individuals, especially for the routing rules. We also found that the sizes of the routing rules of $CCGP^{2a}$(rep) are not smaller than $CCGP^{2a}$(mimic) and $CCGP^{2a}$(rand). This is because the structures of the large rules of $CCGP^{2a}$(rep) in stage 1 are kept to stage 2.

Table 8.4 The mean (standard deviation) of **the rule sizes** obtained by CCGP, CCGP^2 and CCGP^{2a}(mimic) over 30 independent runs in six dynamic flexible job shop scheduling scenarios

Scenario	CCGP	CCGP^2	CCGP^{2a}(mimic)
<Fmax, 0.85>	122.07(30.60)	123.80(26.72)($\approx$)	112.60(25.75)($\approx$)
<Fmax, 0.95>	117.27(25.22)	117.27(28.72)($\approx$)	112.27(31.38)($\approx$)
<Fmean, 0.85>	115.73(24.91)	125.07(18.18)($\approx$)	108.93(26.32)($\approx$)
<Fmean, 0.95>	121.27(17.60)	118.53(28.08)($\approx$)	110.47(22.31)($\approx$)
<WFmean, 0.85>	116.87(23.88)	115.67(26.53)($\approx$)	115.73(26.31)($\approx$)
<WFmean, 0.95>	127.40(24.40)	125.67(23.84)($\approx$)	121.47(22.80)($\approx$)

In most scenarios (i.e., <Fmax, 0.85>, <Fmean, 0.85>, <Fmean, 0.95>, <WFmean, 0.85>, and <WFmean, 0.95>), the sizes of the best learned routing rules of CCGP^{2a}(mimic) are smaller than other algorithms. However, in <Fmax, 0.95>, the sizes of the best routing rules of CCGP^{2a}(mimic) are comparable to those of other compared algorithms. This may be owing to the fact that Fmax with a higher utilisation level (i.e., 0.95) is more difficult to optimise due to its sensitivity to the worst case (i.e., the longest finished time among all jobs). Individual adaptation strategies have a significant impact on the sizes of routing rules when compared with the sizes of sequencing rules. However, the sizes of the best sequencing rules learned by CCGP^{2a}(mimic) have no significant difference with other algorithms.

It is worth mentioning that the routing rule and sequencing rule work cooperatively for DFJSS. It is reasonable to measure the sizes of the learned scheduling heuristics by combining a routing rule and its corresponding sequencing rule. According to our preliminary observations, a small routing rule and a large sequencing rule can have the same ability compared with a large routing rule and a small sequencing rule for DFJSS. We find that with small scheduling heuristics, CCGP^{2a}(mimic) can achieve comparable performance with CCGP^{2a}(rep). For the rest of this section, the rule sizes of three algorithms (i.e., CCGP, CCGP^2, and CCGP^{2a}(mimic)) are compared in more details.

According to 30 independent runs, the mean and standard deviation of the learned rule sizes (i.e., routing rule size plus sequencing rule size) by CCGP, CCGP^2, and CCGP^{2a}(mimic) in six DFJSS scenarios are shown in Table 8.4. It demonstrates that the rule sizes produced by all three algorithms are comparable. The main distinction between the learned rules is that the rules learned by CCGP^{2a}(mimic) only contain selected features (i.e., a small number of features), whereas the rules learned by siCCGP and CCGP^2 include all possible features. The following subsection delves deeper into the distinctive feature.

Table 8.5 The mean (standard deviation) of **the average number of unique features of routing rules** obtained by the five algorithms over 30 independent runs in six dynamic flexible job shop scheduling scenarios

Scenario	CCGP	CCGP^2	CCGP^{2a}(rand)	CCGP^{2a}(rep)	CCGP^{2a}(mimic)
\<Fmax, 0.85\>	8.40(1.33)	7.80(1.47)($\approx$)	6.27(1.68)(–)	6.57(1.74)(–)	6.67(1.81)(–)
\<Fmax, 0.95\>	8.67(0.92)	8.37(1.33)($\approx$)	6.73(1.72)(–)	6.83(1.64)(–)	6.70(1.66)(–)
\<Fmean, 0.85\>	8.03(1.03)	7.67(1.03)($\approx$)	5.70(1.62)(–)	5.83(1.46)(–)	5.63(1.52)(–)
\<Fmean, 0.95\>	8.40(1.10)	7.87(1.14)($\approx$)	5.57(1.33)(–)	5.73(1.53)(–)	5.83(1.56)(–)
\<WFmean, 0.85\>	8.27(1.23)	7.93(1.23)($\approx$)	5.60(1.63)(–)	6.17(2.09)(–)	5.70(1.99)(–)
\<WFmean, 0.95\>	8.20(1.13)	7.50(1.57)($\approx$)	5.70(1.47)(–)	5.90(1.73)(–)	5.70(1.99)(–)

8.4.3 Number of Unique Features

According to 30 independent runs, the mean (standard deviation) of the number of unique features about routing rules in six DFJSS scenarios is shown in Table 8.5. In all scenarios, in terms of the number of unique features about the routing rules, there is no significant difference between CCGP and CCGP^2. This suggests that applying selected features to mutation alone is not a good way to reduce the number of unique features of routing rules. In addition, the number of unique features in routing rules of CCGP^{2a}(mimic), CCGP^{2a}(rep), and CCGP^{2a}(rand) is significantly smaller than CCGP, regardless of the type of individual adaptation strategy used.

Based on 30 independent runs, the mean (standard deviation) of the number of unique features in the sequencing rules in different DFJSS scenarios is shown in Table 8.6. The number of unique features in the sequencing rules is significantly smaller in all scenarios for all the algorithms (i.e., CCGP^{2a}(mimic), CCGP^{2a}(rep), CCGP^{2a}(rand), and CCGP^2) that involve feature selection, especially the algorithms with individual adaptation strategies (i.e., CCGP^{2a}(mimic), CCGP^{2a}(rep), and CCGP^{2a}(rand)).

The goal of simplification is to make the complicated equation easy to understand by performing some algebraic operations. For example, if we have a rule (A − B)/(A − B), it will be simplified to one. In addition, the rule A + B + A + A will be simplified to 3 * A + B, and the rule A * B/B will be simplified as A.

According to the results from 30 independent runs, the violin plot of rule size after the simplification taking routing and sequencing rules as a pair obtained by CCGP, CCGP^2, and CCGP^{2a}(mimic) in six DFJSS scenarios is shown in Fig. 8.9. It demonstrates that the sizes of simplified rules acquired by CCGP^{2a}(mimic) are much smaller than those learned by CCGP and CCGP^2. It suggests that CCGP^{2a}(mimic) has a promising ability to get smaller rules which are critical for rule interpretation.

Table 8.6 The mean (standard deviation) of **the average number of unique features of sequencing rules** obtained by the involved five algorithms according to 30 independent runs in six dynamic flexible job shop scheduling scenarios

Scenario	CCGP	CCGP^2	CCGP^{2a}(rand)	CCGP^{2a}(rep)	CCGP^{2a}(mimic)
<Fmax, 0.85>	7.13(1.59)	6.53(1.14)(–)	5.23(1.41)(–)	5.00(1.20)(–)	5.20(1.27)(–)
<Fmax, 0.95>	7.40(1.57)	6.53(1.41)(–)	4.97(1.25)(–)	5.03(1.27)(–)	5.17(1.39)(–)
<Fmean, 0.85>	6.57(2.10)	5.47(1.36)(–)	3.53(1.07)(–)	3.40(1.00)(–)	3.70(1.06)(–)
<Fmean, 0.95>	6.90(1.60)	6.03(1.43)(–)	4.00(1.23)(–)	3.97(1.19)(–)	3.70(0.99)(–)
<WFmean, 0.85>	6.53(1.59)	5.17(1.05)(–)	4.00(0.79)(–)	3.93(0.69)(–)	4.00(0.79)(–)
<WFmean, 0.95>	6.80(1.52)	5.70(1.47)(–)	4.33(0.92)(–)	4.17(0.95)(–)	4.27(0.87)(–)

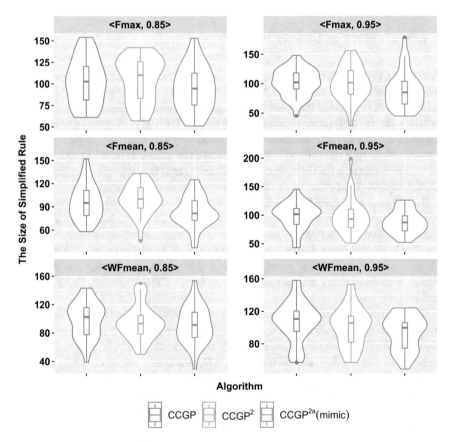

Fig. 8.9 The violin plot of rule sizes (i.e., routing rule plus sequencing rule) obtained by CCGP, CCGP^2, and CCGP^{2a}(mimic) after simplification over 30 independent runs in six dynamic flexible job shop scheduling scenarios

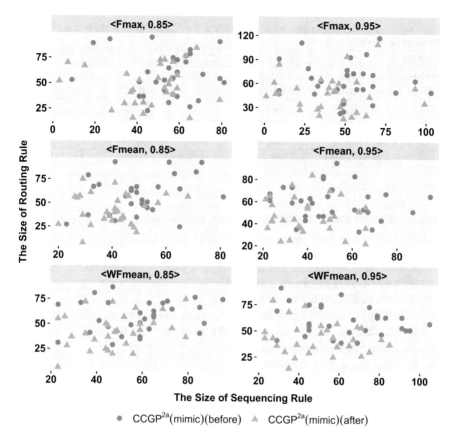

Fig. 8.10 The scatter plot of the sizes of routing rules and sequencing rules before and after simplification obtained by CCGP^{2a}(mimic) over 30 independent runs in six dynamic flexible job shop scheduling scenarios

The scatter plot of the sizes of routing and sequencing rules before and after simplification is shown in Fig. 8.10. In every scenario, the routing rule sizes get substantially smaller (i.e., positioned downwards). In addition, in all scenarios, the sizes of the sequencing rule are smaller (i.e., to the left) than that without simplification. In summary, CCGP^{2a}(mimic) can obtain smaller routing rules and sequencing rules than CCGP and CCGP^2 after simplification.

8.4.4 Training Time

Table 8.7 shows the training time (in minutes) of the algorithms CCGP, CCGP^2, CCGP^{2a}(rand), CCGP^{2a}(rep), and CCGP^{2a}(mimic). CCGP^2 has no significant dif-

Table 8.7 The mean (standard deviation) of **training time** (in minutes) by the five algorithms in six DFJSS scenarios

Scenario	CCGP	$CCGP^2$	$CCGP^{2a}$(rand)	$CCGP^{2a}$(rep)	$CCGP^{2a}$(mimic)
<Fmax, 0.85>	100(17)	96(14)($\approx$)	80(15)(–)	83(13)(–)	75(10)(–)
<Fmax, 0.95>	107(15)	111(20)($\approx$)	88(13)(–)	92(15)(–)	88(12)(–)
<Fmean, 0.85>	98(12)	99(13)($\approx$)	78(11)(–)	87(12)(–)	77(12)(–)
<Fmean, 0.95>	109(15)	108(16)($\approx$)	86(11)(–)	94(16)(–)	85(12)(–)
<WFmean, 0.85>	94(11)	98(11)($\approx$)	82(9)(–)	88(15)($\approx$)	83(15)(–)
<WFmean, 0.95>	113(16)	109(16)($\approx$)	90(13)(–)	98(16)(–)	88(13)(–)

ference compared with CCGP in terms of the training time. The training time of $CCGP^{2a}$(rand), $CCGP^{2a}$(rep), and $CCGP^{2a}$(mimic) decrease dramatically compared with CCGP and $CCGP^2$. The training time of CCGP, $CCGP^2$ and the algorithms with individual adaptation strategies (i.e., $CCGP^{2a}$(rand), $CCGP^{2a}$(rep), and $CCGP^{2a}$(mimic)) has difference due to the individual adaptation strategy. Intuitively, for $CCGP^{2a}$(rand), $CCGP^{2a}$(rep), and $CCGP^{2a}$(mimic), a longer training time may be resulted from the extra algorithm operators, i.e., individual adaptation. However, it turns out that the time of the presented algorithms with the individual adaptation strategies are smaller than CCGP and $CCGP^2$, especially $CCGP^{2a}$(rand) and $CCGP^{2a}$(mimic). This is because, the sizes of the learned scheduling heuristics obtained by $CCGP^{2a}$(rand) and $CCGP^{2a}$(mimic) are smaller than CCGP and $CCGP^2$. In summary, $CCGP^{2a}$(mimic) is more efficient in terms of the mean values of the training time despite more algorithm operators are used such as selecting features and mimicking individuals' behaviours.

The average rule sizes over the population of $CCGP^{2a}$(rand) and $CCGP^{2a}$(mimic) are smaller than their counterparts, as shown in the curve of average rule sizes in the population in Fig. 8.11. This is because, at generation 50, all of the individuals in the population are re-initialised which have smaller sizes. Smaller rules tend to reduce computing time by shortening the time it takes to evaluate individuals. As a result, $CCGP^{2a}$(rand) and $CCGP^{2a}$(mimic) can greatly shorten computational time.

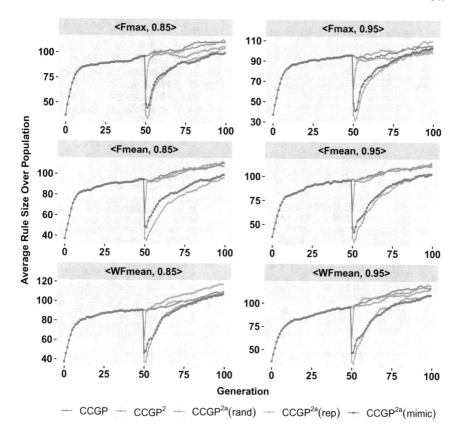

Fig. 8.11 The **average rule (routing rules plus sequencing rules) sizes over population** of the five algorithms in six dynamic flexible job shop scheduling scenarios

8.5   Further Analyses

The selected features and learned scheduling heuristics are further examined in this section to better understand the effect of the proposed algorithm.

8.5.1   Selected Features

According to 30 independent runs, the selected and unselected features of the sequencing rules and routing rules in the six DFJSS scenarios are shown in Figs. 8.12 and 8.13, respectively. A larger blue area for a feature (i.e., the associated feature is selected more frequently) indicates that the feature is more essential. Because the evolutionary processes are the same in stage 1, the selected features of $CCGP^2$,

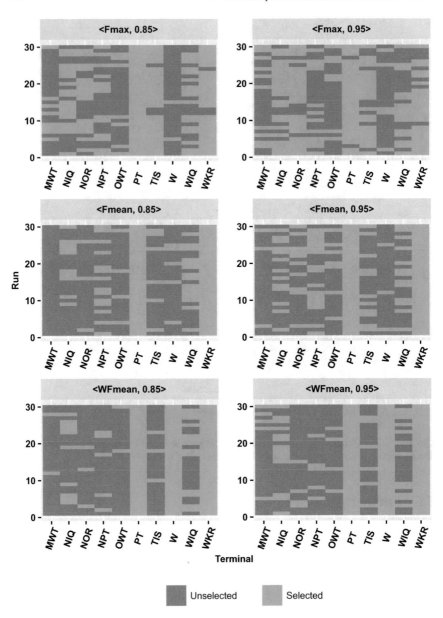

Fig. 8.12 Selected and unselected features of **sequencing rules** of 30 independent runs in six dynamic flexible job shop scheduling scenarios

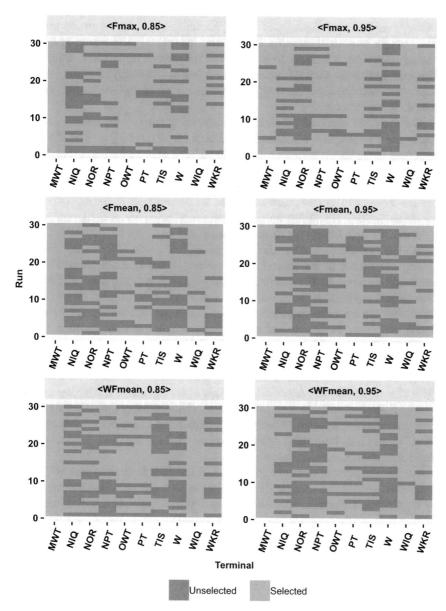

Fig. 8.13 Selected and unselected features of **routing rules** of 30 independent runs in six dynamic flexible job shop scheduling scenarios

$CCGP^{2a}$(rand), $CCGP^{2a}$(rep), and $CCGP^{2a}$(mimic) are the same in the same run (i.e., with the same random seed). The selected features for the sequencing rules and routing rules change from run to run based on the given feature selection algorithm. This implies that the chosen features can be adjusted adaptively using the two-stage framework described, and the chosen features are based on the specific problems (i.e., different runs).

The top three significant features for sequencing rules in the scenario <Fmax, 0.85> are PT, TIS, and WKR, as illustrated in Fig. 8.12. NIQ and NOR are also important in the scenario <Fmax, 0.95> in addition to these three features. In comparison to <Fmax, 0.85>, Fig. 8.12 illustrates that in the scenario <Fmax, 0.95>, more features are selected. It is possible that a higher utilisation level makes it more challenging to optimise the problem. PT and WKR have an essential role in minimising mean-flowtime (i.e., <Fmean, 0.85> and <Fmean, 0.95>. (i.e., they are selected in all the 30 runs). Except for PT and WKR, W is a significant feature when considering mean-weighted-flowtime (i.e., <WFmean, 0.85> and <WFmean, 0.95>.

The importance of PT (processing time) and WKR (median amount of work remaining for a job) for the flowtime-related objectives are congruent with our intuition. Additionally, W is frequently chosen over max-flowtime and mean-flowtime for minimising the mean-weighted-flowtime. It is in line with our expectations, given that W is not used in the calculations of mean-flowtime and max-flowtime.

MWT, OWT, and WIQ, as shown in Fig. 8.13, are important for updating routing rules in all scenarios. Because the new operation has a higher possibility of being processed early, the machine with less workload (WIQ) and earlier ready time (MWT) is recommended. In addition, if a new operation is ready, it will be assigned to a machine immediately (i.e., OWT, the operation waiting time equals zero). In comparison to the selected features in the sequencing rules, the routing rules use more features (i.e., larger blue areas). It suggests that learned routing rules are more sophisticated than learned sequencing rules.

The distributions of test objective values of the 30 independent runs of $CCGP^{2a}$ (mimic) in the scenario <Fmax, 0.85> are shown in Fig. 8.14, which are categorised by whether each feature is selected or not in the sequencing rules. TIS is not picked in three runs in which the test performance is significantly lower. Nonetheless, even if NIQ is a pretty important feature (i.e., it is generally selected in half runs), the test performance is still very good. In three runs, WKR is not chosen. In two runs, WKR has a big impact on the quality of the learned scheduling heuristics, while it has no influence on the third. This shows that, although the features utilised in scheduling heuristics are crucial, the scheduling heuristics' quality also depends on other aspects, such as the structures of rules.

8.5.2 Insights of Learned Scheduling Heuristics

The scenario <WFmean, 0.95> is used as an example for rule analysis in this section. This is due to the fact that the objective in the scenario <WFmean, 0.95> is more

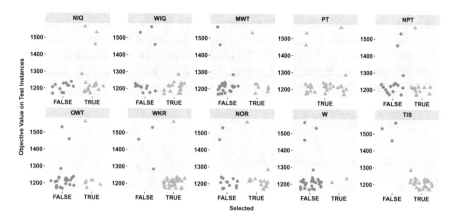

Fig. 8.14 The scatter plot of the test objective values of the 30 independent runs of CCGP^{2a}(mimic) in scenario <Fmax, 0.85>, categorised by whether each feature is selected or not in sequencing rules

difficult to optimise than the objectives in other scenarios. We analyse the best routing rule and the accompanying sequencing rule with an objective value of 1096.02. The rules select the candidates (i.e., either machines or operations) with the lowest priority value (i.e., operation for sequencing or machine for routing).

The routing terminal set for CCGP^2 comprises NIQ, WIQ, MWT, PT, NPT, OWT, WKR, NOR, and WKR (i.e., nine features). NIQ, WIQ, MWT, PT, WKR, and W are the six features chosen as the sequencing terminal set. All of the features in the routing terminal set (i.e., seven features, NIQ, WIQ, MWT, PT, OWT, WKR, and W) are picked for the routing rules obtained by CCGP^{2a}(mimic). Five of them (NIQ, WIQ, PT, WKR, and W) are included in the sequencing rule obtained by CCGP^{2a}(mimic). Compared with CCGP^2, CCGP^{2a}(mimic) employs a smaller number of features.

However, the learned scheduling heuristics acquired by CCGP^{2a}(mimic) perform better in tests than those obtained by CCGP^2. As indicated in Eq. (8.2), the routing rule provided by CCGP^{2a}(mimic) can be simplified.

$$R_1 = \min\{2 * NIQ, \max(\frac{NIQ * PT}{MWT},$$
$$PT * WIQ * \min(WIQ, WKR))\}+ \tag{8.2}$$
$$NIQ * \frac{PT}{W} - MWT$$

After simplification, it is clear that this routing rule is fairly modest small. In terms of machine characteristics, this routing strategy chooses to select the machine with the longest waiting time (MWT) and the smallest NIQ (number of operations in the queue) and WIQ (i.e., the workload of machines). This routing strategy favours machines that can process an operation more efficiently with a smaller PT (i.e., processing time) in terms of operation characteristics. In addition, when selecting a machine from the standpoint of operation, W (i.e., the weight used to signify the importance of operation) is a constant. From a machine's point of view, a more important operation with a larger W has more priority to select a machine.

As demonstrated in Eq. (8.3), the corresponding sequencing rule derived by CCGP^{2a}(mimic) is simplified.

$$
\begin{aligned}
S_1 = & \frac{PT + WKR}{W} \\
& - \frac{W}{PT}(W * WIQ - W + WIQ) - \frac{W}{PT} * \\
& (W * WIQ + WKR + \frac{PT + WKR}{W * WIQ - W + WKR})
\end{aligned}
\tag{8.3}
$$

It is worth mentioning that the machine-related features such as WIQ (i.e., a machine's workload) are the same for all operations in a machine's queue. This indicates that they are not a critical feature to make an operation sequencing decision. This rule favours operations having a smaller PT (processing time) and a higher W (i.e., weight, the importance of an operation). It is noteworthy to note that, based on the first line of S_1, this sequencing rule prefers operations with smaller WKR (i.e., the median amount of work remaining for a job), while partially preferring operations with greater WKR based on the third line of S_1. It implies that, although humans can read rules to some extent, fully comprehending rule behaviour remains challenging. In the future, we will continue to work on this issue.

As illustrated in Eqs. (8.4) and (8.5), the corresponding routing and sequencing rules generated by CCGP^2 can be simplified.

$$
\begin{aligned}
R_2 = & Max\{NIQ^2, (NIQ + NPT) * Min(NIQ, NOR)\} \\
& - Min(MWT, \frac{WKR}{MWT * WKR - 1}) \\
& * Max\{WKR, -MWT * WKR + \\
& \frac{NIQ * PT * Max\{WIQ, \frac{Min(MWT,PT)}{(W+WKR)}\}}{Max\{NIQ^2, \frac{NOR-W+WKR}{W}, \frac{NIQ+NPT}{(Min(NIQ,NOR))^{-1}}\}}\}
\end{aligned}
\tag{8.4}
$$

$$
\begin{aligned}
S_2 = & NIQ(PT - W)(PT + \frac{WKR}{W} * Max\{PT, \frac{WIQ}{W}, \\
& \frac{Max\{WIQ, \frac{WKR}{W}\}}{W}\}) + Max\{\frac{WIQ}{W^2}, \frac{NIQ}{W^2} + WIQ\} \\
& * (MWT + W + Max\{\frac{WKR}{W^2}, NIQ - 1\}) \\
& * \frac{Max\{WIQ, \frac{WKR}{W}\}}{W}
\end{aligned}
\tag{8.5}
$$

Even after simplification, the structures of the routing rule and sequencing rule are more complex than those of siCCGP2a(mimic). The functions Max and Min are frequently utilised and nested inside each other from the perspective of the components. Because it is dependent on several elements, it is difficult to determine which component played a real role. From a human standpoint, it is difficult to comprehend.

In conclusion, $CCGP^{2a}$(mimic) can learn comparable scheduling heuristics efficiently with fewer unique characteristics and sizes. This can help real-world applications because learned scheduling heuristics are easier for humans to understand.

8.6 Chapter Summary

This chapter introduces an effective GP feature selection algorithm to evolve scheduling heuristics for DFJSS without compromising any performance. This chapter firstly presents the two-stage feature selection algorithm $CCGP^2$ for DFJSS. The presented algorithm is based on the idea that promising features can benefit more for the evolutionary process and can be used to guide the process effectively. In addition, this chapter presents the GP feature selection algorithm with novel individual adaptation strategies to eliminate unselected features in the learned scheduling heuristics. In the feature selection process, both the information of the selected features and the investigated individuals are used to maintain the quality of the scheduling heuristics.

The results verify the effectiveness of the presented GP feature selection algorithm $CCGP^2$. Moreover, the results show that the learned rules by $CCGP^{2a}$(mimic) are smaller and contain fewer unique features due to feature selection. These kinds of rules tend to be interpreted easier. The semantic assessments of the routing and sequencing rules learned by $CCGP^{2a}$(mimic) corroborate this. In terms of training time, $CCGP^{2a}$(mimic) is more efficient than the baseline algorithm, since it can reduce the average rule size of the population. In summary, the presented algorithm $CCGP^{2a}$(mimic) can evolve effective scheduling heuristics with a smaller number of unique features and smaller size efficiently.

It is noted that the surrogate technique is used to improve the training efficiency of GP in both the last chapter and this chapter. The last chapter focuses on the design of surrogate models for DFJSS. However, this chapter focuses on using the individual information and selected features to maintain the quality of individuals only with the selected features. The surrogate technique used in this chapter is to improve the efficiency of obtaining individuals for feature selection, and the surrogate itself is not the focus in this chapter.

This chapter uses the phenotypic characterisation technique to measure the behaviour/performance of GP individuals for finding their similar individuals with only the selected features. In the next chapter, phenotypic characterisation will be used to measure the subtree importance for improving the search mechanism of GP for DFJSS.

Chapter 9
Search Mechanism with Specialised Genetic Operators

For evolutionary algorithms, genetic operators play an essential role in producing offspring. An effective crossover is expected to generate offspring of good quality. In essence, the crossover operator is a recombination of different genetic materials from the parents. In traditional GP, subtrees are randomly chosen from two parents to produce two offspring by swapping the selected subtrees. However, the importance of subtrees for a GP individual can be different. Some subtrees are redundant or less important, and removing them might not affect the fitness of an individual too much. On the other hand, some subtrees play essential roles for an individual, and losing them will cause considerable loss to the fitness. The random way of recombination may disrupt beneficial building blocks. Although there are some studies [98, 176, 245] related to genetic operators for GP, little research has been conducted on the crossover operator to improve the quality of offspring by investigating the importance of subtrees. This chapter will present an effective algorithm to improve the effectiveness of crossover by designing an effective and adaptive recombinative guidance mechanism based on the importance of subtrees.

9.1 Challenges and Motivations

A GP individual (i.e., a tree) contains a number of subtrees. Taking the crossover operator as an example, in a typical GP crossover, subtrees are randomly chosen from the two parents to produce two offspring by swapping the selected subtrees. However, the subtrees can have different importance for a GP individual. Some subtrees play important roles for an individual, and removing them may worsen the fitness of the individual. Some other subtrees may be redundant (unimportant) [62], and removing them does not affect the fitness of an individual. In general, if the important subtrees of one parent can be reserved and the unimportant subtrees can be replaced by the important subtrees from the other parent, it is more likely to generate offspring with good fitness. Theoretical studies in [191, 192] support that introducing biases to

F. Zhang et al., *Genetic Programming for Production Scheduling*, Machine Learning: Foundations, Methodologies, and Applications, https://doi.org/10.1007/978-981-16-4859-5_9

GP operators can be helpful for different purposes such as improving the quality of newly generated offspring and controlling the size of offspring. For offspring quality improvement, this chapter aims at incorporating biases for the crossover operator by measuring the importance of subtrees, since crossover is the most important genetic operator for GP [116]. However, there are no existing studies related to subtree importance measures for a GP individual. In addition, it is not clear how to apply the expected "biases" to GP for DFJSS.

Even with slight modifications of a GP individual, its behaviour is likely to cause significant alterations. What kinds of genetic operators can improve GP's performance is still not clear. To our knowledge, how to improve the recombinative effectiveness of GP via the DFJSS crossover has not been comprehensively investigated. In [191, 192], Riccardo et al. presented a comprehensive generic schema theory for GP with subtree-swapping crossover. According to the idea, GP operators' biases can be advantageous for various goals, including boosting offspring quality and managing offspring size and shape. One of the most difficult challenges is determining how to assess subtrees based on the desired goal. The other difficulty is figuring out how to apply the expected "biases" to GP for a specific problem. The purpose of this chapter is to create a guided crossover operator for boosting offspring generation effectiveness. Specially, we are interested in the following questions:

- What information is suitable for measuring the subtree importance?
- How to measure the importance of subtrees?
- How to efficiently guide the search via crossover in GP?

Section 9.2 shows two strategies to measure the importance of subtrees. Section 9.3 provides the details of the experiment design. Section 9.4 discusses the results followed by further discussion in Sect. 9.5. Section 9.6 summaries this chapter.

9.2   Algorithm Design and Details

This section will show how the effectiveness of the GP crossover operator can be improved with subtree importance measure. This section will first introduce the framework of the GP algorithm with specialised genetic operators. Then, this section will describe the important components of the algorithm including two ways of measuring the importance of the subtree and the crossover with recombinative guidance mechanism.

9.2.1   Framework of the Algorithm

The flowchart of the proposed algorithm is shown in Fig. 9.1. The main procedure is similar to that of a traditional GP. It starts by randomly initialising the population and then evaluating the individuals inside it. There are two subpopulations. One

Fig. 9.1 The flowchart of
proposed genetic
programming algorithm with
specialised genetic operators

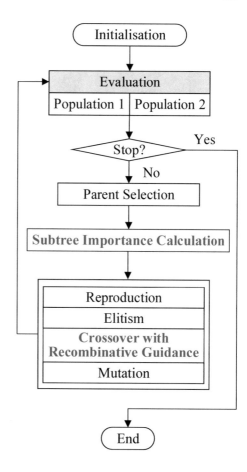

subpopulation is responsible for evolving routing rules, while the other is responsible for evolving sequencing rules. In Fig. 9.1, there are two new components that differ from the traditional GP and are marked in red. Before the mating procedure, the importance of each subtree of the parents is calculated. Second, the crossover is performed during the mating process using the proposed recombinative guidance mechanism. Two research questions need to be answered.

- How to measure the importance of subtrees?
- How to apply the subtree importance information to guide the recombination between the parents via the crossover operator?

This chapter proposes two strategies for subtree importance calculation for GP. The details are shown as follows.

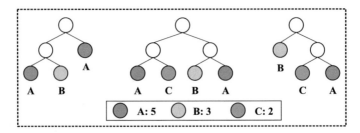

Fig. 9.2 The occurrences of features in the top three individuals

9.2.2 Subtree Importance Measure Based on Feature Importance

The ability of GP to automatically choose important features to build individuals during the evolutionary process is one of its advantages. The features which appeared in the individuals with good fitness are more likely to be important. Individuals who have key features are more likely to be promising. As a result, the features of promising individuals can be utilised to assess the value of subtrees.

9.2.2.1 The Occurrences of Features

Because preliminary studies demonstrate that the top 10 individuals have potential fitness, which is effective for detecting feature properties [236, 244]. The occurrences of features in the top 10 individuals are utilised to estimate the value of subtrees of an individual in this chapter. Another benefit of employing feature occurrence information is that, we do not have to put in much extra work to get information because it has already been generated during the evolutionary process.

Using three individuals as an example, Fig. 9.2 shows how to extract feature occurrence information. These three individuals have various structures and have varied numbers of features. Assume they are the top three individuals in the population in terms of fitness. The occurrences of features in all three individuals are counted according to the three individuals. Feature A, B, and C occur 5 times, 3 times, and 2 times, respectively. This information will be utilised to determine the relative importance of a GP's subtrees.

9.2.2.2 The Importance of Subtrees

A single individual (i.e., a tree) can be divided into several subtrees. The subtree is determined once a function node is chosen. Subtree importance is measured from bottom to top, and we utilise the concept score to reflect a subtree's importance value. Each feature's occurrence information is collected, and the parent node's score (i.e.,

Fig. 9.3 The importance
(i.e., scores) of subtrees of a
GP individual

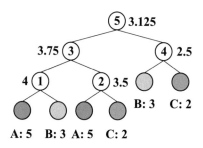

the importance of the subtree) is determined by the average occurrence number of its child nodes. Assume that the importance (i.e., occurrence) of features A, B, and C are ranked in the following order: $A > B > C$ (i.e., $5 > 3 > 2$). There are three possible choices for the subtree if just the simplest subtrees are considered (i.e., depth is two) and only two features are considered: A and B, A and C, and B and C. Subtrees should be prioritised as follows: $subtree(A, B) > subtree(A, C) > subtree(B, C)$.

An example of how to measure the importance of each subtree for an individual is shown in Fig. 9.3. For example, the $subtree1$ (in the bottom-left corner) has two features (A and B), and the parent node's score is set to 4 (i.e., $(5 + 3)/2$). The average score of its two subtrees (i.e., $subtree1$ and $subtree2$) is used to determine the relevance of $subtree3$ (i.e., 3.75, $(4 + 3.5)/3$. As indicated in Fig. 9.3, the scores of all the subtrees will be calculated by analogy. There are three subtrees (i.e., indicated by subtree 1, 2, and 4) whose importance values are 4, 3.5, and 2.5, respectively, when considering the subtrees with a depth of two. The importance values of subtrees 1, 2, and 4 are rated as follows: $subtree1 > subtree2 > subtree4$, which matches the importance measurement design. The importance rank of all subtrees in this individual is $subtree1 > subtree3 > subtree2 > subtree5 > subtree4$.

9.2.3 Subtree Importance Measure Based on the Behaviour Correlation

Intuitively, if a subtree can make more consistent decisions with the entire tree, it is deemed more significant to the tree. Figure 9.4 depicts an example of a GP individual with five subtrees. Each subtree can be thought of as a self-contained "individual", which is capable of making its own decisions. This chapter uses a decision vector $\mathbf{d_i}$ (i.e., phenotypic characterisation) to characterise the behaviour of a subtree T_i in a decision situation. The vector is a list of the ranks of the candidates (i.e., machines for routing decision situations, or operations for sequencing decision situations).

The decision vectors of subtrees of a GP individual are calculated in the manner shown in Table 9.1. Assume the individual is a routing rule with four subtrees. The decision situation is to assign one of the three candidate machines for a ready operation. The numbers in the machine columns (i.e., M_1, M_2, and M_3) represent priority

Fig. 9.4 An example of a
labelled tree-based GP
individual

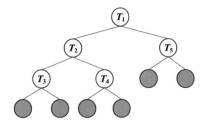

Table 9.1 An example of the calculation for **decision vector** of the subtrees of a genetic programming individual

Subtree (T_i)	M_1	M_2	M_3	Decision vector (d_i)
T_1	100 ①	150 ②	200 ③	(1, 2, 3)
T_2	300 ①	320 ②	350 ③	(1, 2, 3)
T_3	140 ③	120 ②	110 ①	(3, 2, 1)
T_4	100 ①	160 ③	130 ②	(1, 3, 2)

values (i.e., real numbers) based on the corresponding subtrees (i.e., routing rules) and machine rankings based on the priority values. Machines with a lower priority value have a higher priority than machines with higher priority values. Finally, the ranks make up the decision vectors. It demonstrates that distinct subtrees can have the same (T_1 and T_2) or opposite (T_1 and T_3) or partially same (T_1 and T_4) decisions. Because a decision is made exclusively on the basis of the candidates' rankings rather than their specific priority values, this chapter concentrates on the relationship in terms of ranks rather than priority values.

Pearson and Spearman correlation coefficients are two regularly used indicators of the relationship between the two variables [55]. Pearson's correlation coefficient examines linear relationships, whereas Spearman's correlation coefficient analyses monotonic correlations [155] (i.e., regardless of whether they are linear or not). The Spearman correlation coefficient, in particular, assesses the statistical relationship between two variables' rank values. In the DFJSS, subtree decision-making is based on the ranks of machines or operations. As a result, the Spearman correlation coefficient is a natural candidate for determining the relationship between subtree behaviour. To quantify the importance of a subtree T_i, this chapter employs the correlation c_i between the decisions (i.e., $\mathbf{d_i}$ and $\mathbf{d_1}$) made by T_i and T_1 (i.e., the entire tree). The values range from -1 to 1. If $|c_i|$ is close to 1, T_i will behave similarly to T_1 (either positively or negatively), and Ti will be an important subtree for an individual. If $|c_i|$ is near to 0, T_i's behaviour is virtually irrelevant to that of T_1, and so T_i is not an essential subtree for T_1.

The calculations for the correlation of subtrees of the individual depicted in Fig. 9.4 are shown in Table 9.2. The correlations between subtrees differ (i.e., either positive or negative values). T_2 makes the exact same judgements as T_1, making it a key subtree of T_1. T_5, on the other hand, has a correlation of -1, indicating that

Table 9.2 An example of the calculations for **correlation** of subtrees of a genetic programming individual in a decision situation

Subtree (T_i)	Decision vector (d_i)	Correlation (c_i)
T_1	(1, 2, 3, 4, 5, 6)	1
T_2	(1, 2, 3, 4, 5, 6)	1
T_3	(1, 3, 2, 6, 4, 5)	0.77
T_4	(6, 5, 1, 2, 3, 4)	−0.43
T_5	(6, 5, 4, 3, 2, 1)	−1

Algorithm 9.1 Calculation of a subtree's importance

Input: An individual T, a subtree T_i of T, and a set of decision situations
Output: The importance of the subtree T_i
1: $S(T_i) \leftarrow null$, $\mathbf{d_i} \leftarrow null$
2: $c_i \leftarrow 0$, $sum(c_i) \leftarrow 0$
3: **for** j = 1 to $|decisionSituations|$ **do**
4:     Based on the subtree T_i, calculate the priority values of machines or operations
5:     Prioritise the machines or operations and rank them accordingly
6:     $\mathbf{d_i} \leftarrow$ get the decision vector of subtree T_i based on the ranks
7:     $c_i \leftarrow$ calculate the correlation of $\mathbf{d_i}$ and $\mathbf{d_1}$
8:     $sum(c_i) \leftarrow sum(c_i) + |c_i|$
9:     $S(T_i) \leftarrow \frac{sum(c_i)}{|decisionSituations|}$
10: **end for**
11: **return** $S(T_i)$

---

its behaviour is entirely opposite that of T_1. In this case, T_5 is an essential subtree of T_1 since its behaviour can be transformed to that of T_1 with a minor adjustment, i.e., "$0 - T_1$". T_3 and T_4, on the other hand, have a weaker association with T_1, and hence are deemed less significant than T_2 and T_5.

The pseudo-code for assessing the importance of a subtree is shown in the Algorithm 9.1. A subtree becomes much more important as the correlation's absolute value approaches 1. Because the features of jobs (e.g., processing time) and machines (e.g., the workload) can differ, the correlation between the behaviours of two trees can change across distinct decision scenarios. To obtain an accurate measure of the relationship, we sample a set of representative decision situations [92] and define the relationship between the behaviours of two trees as the average correlation values across all of the sampled decision situations. This chapter uses the WIQ (work in the queue) rule for routing and the SPT (shortest processing time) rule for sequencing to sample a set of representative decision situations, and runs a preliminary simulation with 5000 jobs on 10 machines, resulting in about 50,000 routing and 50,000 sequencing decision situations. Decision situations, including between 2 and 20 jobs, were generated randomly in [92], which have been demonstrated to be good for monitoring the behaviour of individuals by phenotypic characterisation. In this chapter, the number of candidates, either machines or operations, for both the routing and sequencing decisions is 7, taking into account the complexity of DFJSS. Then, at random, we choose 50 routing and sequencing choice circumstances with a length of 7 from the created routing and sequencing decisions. This indicates that each subtree has a seven-dimensional decision vector. The fixed dimension length seeks to set the same length decision vectors for all individuals in order to achieve a feasible Spearman's correlation coefficient value. The priority values of machines or operations are determined (line 4) in each decision situation to determine their ranks (line 5). The decision made by T_i is represented by a vector $\mathbf{d_i}$ (line 6). The importance of T_i is determined by the correlation c_i between T_i and T_1 (line 7). The average c_i overall decision scenarios is the final importance of subtree T_i (line 10).

9.2.4   Crossover with Recombinative Guidance

A GP crossover operator is normally performed on two parents (i.e., $parent_1$ and $parent_2$), both of them are deemed promising members of the population (e.g., selected by tournament selection). For each parent, it desires to choose the unimportant subtree and replace it with an important subtree from the other. Two probability calculations are intended for distinct reasons based on the importance of subtrees $S(T)$. One is for important subtree selection, while the other is for unimportant subtree selection. The crossover operator with recombinative guidance is then designed based on the probabilities.

The Probability for Each Subtree. This chapter employs the concept of roulette wheel selection to select the required subtrees based on the subtree importance information. Each subtree's likelihood is proportionate to its importance. To demonstrate how to calculate the probability of each subtree, we utilise the example provided in Table 9.2 with the correlation. We suppose that there is just one decision situation and that the calculated importance of subtrees from T_1 to T_5 is 1, 1, 0.77, 0.43, and 1. The example of two distinct strategies to calculate the likelihood of each subtree in a GP individual for crossover is shown in Fig. 9.5. The strategy for selecting unimportant subtrees is shown in Fig. 9.1a. A subtree with a higher score has a reduced chance of getting chosen, as indicated by the caption "$\downarrow$". The strategy for selecting important subtrees is shown in Fig. 9.1b. A subtree with a higher score has a higher chance of getting chosen, as indicated in the caption by the "$\uparrow$" symbol. The roulette wheel selection based on correlation values or transformed correlation values is used to calculate the probabilities of subtrees. As illustrated in Fig. 9.5 "(a) Correlation $\downarrow$" and Fig. 9.5 "(b) Correlation $\uparrow$", the correlation values of subtrees are the same at the beginning. However, different from Fig. 9.5b, the correlation value of each subtree will be converted to $1 - S(T)$, as shown in Fig. 9.5 "(a) Converted Correlation $\downarrow$", since we are choosing unimportant subtrees. In the last row of Fig. 9.5, the probabilities of subtrees are presented beside the function nodes. In Fig. 9.5a, the rank of the probability of subtrees is $T_4 > T_3 > T_1 = T_2 = T_5 > T_3 > T_4$, and in Fig. 9.5b, it is

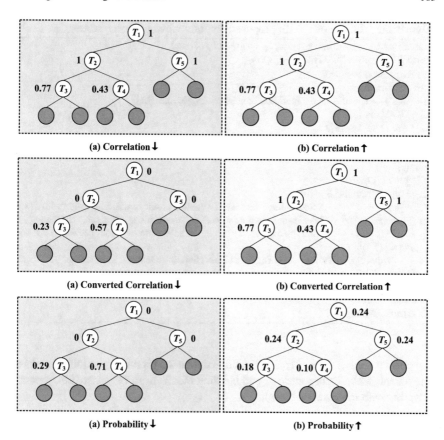

Fig. 9.5 An example of obtaining the probabilities for subtrees. Figure 9.5a tends to choose unimportant subtrees, while Fig. 9.5b tends to choose important subtrees

$T_1 = T_2 = T_5 > T_3 > T_4$. This chapter can guarantee that important and unimportant subtrees are chosen in accordance with the requirements in this way.

The Recombinative Guidance Mechanism The pseudo-code of the presented crossover operator is shown in the Algorithm 9.2. Before determining significant and unimportant subtrees based on roulette wheel selection (line 2 to line 8 for $parent_1$, line 9 to line 13 for $parent_2$), the importance of subtrees of an individual is computed. Finally, by replacing the unimportant subtree $parent_1(T^*)^n$ from $parent_1$ with the important subtree $parent_2(T^*)^p$ from $parent_2$, one offspring is formed (line 14). The unimportant subtree $parent_2(T^*)^n$ from $parent_2$ is replaced with the important subtree $parent_1(T^*)^p$ from $parent_1$ to produce the other offspring (line 15).

Continuing the example from Figs. 9.5 and 9.6 illustrates an example of the created offspring using the recombinative guidance presented. The unimportant subtree with an unhappy face from $parent_1$ ($parent_2$) is expected to be replaced by the important subtree with a happy face from $parent_2$ ($parent_1$), resulting in a better offspring for

---

Algorithm 9.2 Crossover with proposed recombinative guidance strategy

---

Input: Two parents for the crossover ($parent_1$ and $parent_2$)
Output: The generated offspring (*offspring*)
1: set *offspring* $\leftarrow$ null
2: **if** $parent_1$ **then**
3:      $S(T) \leftarrow$ Calculate the importance of subtrees (**Algorithm 9.1**)
4:      $S(T)^p \leftarrow |S(T)|$
5:      $S(T)^n \leftarrow 1 - |S(T)|$
6:      $parent_1(T^*)^p \leftarrow$ Selected important subtree based on roulette wheel selection with $S(T)^p$
7:      $parent_1(T^*)^n \leftarrow$ Selected unimportant subtree based on roulette wheel selection with $S(T)^n$
8: **end if**
9: **if** $parent_2$ **then**
10:     repeat from line 3 to line 5
11:     $parent_2(T^*)^p \leftarrow$ Selected important subtree based on roulette wheel selection with $S(T)^p$
12:     $parent_2(T^*)^n \leftarrow$ Selected unimportant subtree based on roulette wheel selection with $S(T)^n$
13:     $offspring_1 \leftarrow$ Produce offspring by replacing the subtree chosen from $parent_1(T^*)^n$ with $parent_2(T^*)^p$
14:     $offspring_2 \leftarrow$ Produce offspring by replacing the subtree chosen from $parent_2(T^*)^n$ with $parent_1(T^*)^p$
15:     *offspring* $\leftarrow$ *offspring*$_1$ $\cup$ *offspring*$_2$
16: **end if**
17: **return** *offspring*

$parent_1$ ($parent_2$). The offspring should maintain the promising building blocks of one parent, while incorporating good building blocks from the other (i.e., produce offspring with more happy faces).

9.2.5 Algorithm Summary

Rather than choosing subtrees at random, the given algorithm uses a recombinative guidance mechanism to improve the crossover effectiveness. We presume that deleting an unimportant subtree from an individual has very little impact on its fitness. Placing an important subtree to the position where the unimportant subtree is removed, on the other hand, has a good chance of improving the individual. It should be highlighted that the idea presented in this chapter does not limit to DFJSS, but can be applied to GP in general. Designing a good measure for subtree importance based on the specific problem to be solved is a crucial challenge. Using sampling semantics, the subtree importance may be quantified in the symbolic regression problem [215].

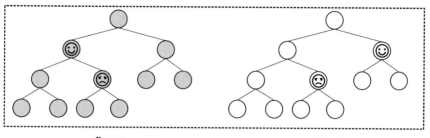

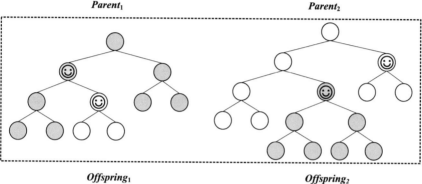

Fig. 9.6 An example of generated offspring from two parents with the presented recombinative guidance mechanism

9.3 Experiment Design

9.3.1 Comparison Design

The purpose of this chapter is to improve the crossover operator of GP with recombinative guidance mechanism in learning effective scheduling heuristics for DFJSS. The comparison in this chapter includes three algorithms. The basic algorithm is [228], a cooperative coevolution genetic programming (CCGP) algorithm. CCGP uses the uniform crossover operator. The algorithm that measures the subtree importance based on feature importance is named CCGP^f [233], while the algorithm that measures the subtree importance based on the correlation between subtrees and the whole tree is named CCGP^c. Additionally, we compare the proposed subtree importance measure and recombinative guidance mechanism to a reverse algorithm named CCGP$^{!c}$, which employs unimportant subtrees to replace important subtrees in GP.

Table 9.3 The specialised parameter settings of genetic programming

Parameter	Value
Number of subpopulations	2
Subpopulation size	512
The number of elites for each subpopulation	5

9.3.2   Specialised Parameter Settings of Genetic Programming

The specialised parameter settings of GP are shown in Table 9.3. This chapter adopts the representation with cooperative coevolution along with two subpopulations, and the crossover for the routing rules and the sequencing rules is independent. This makes it convenient for validating the effectiveness of the presented crossover operator.

9.4   Results and Discussions

9.4.1   Quality of the Learned Scheduling Heuristics

Table 9.4 shows the test performance (mean and standard deviation over 50 independent runs) of the four compared algorithms in six DFJSS scenarios. From the table, we can observe that $CCGP^f$ and CCGP performed very similarly in all the scenarios. This suggests that the number of times each feature is used may not be an accurate measure of the subtree importance, since there may be many redundant branches in GP that introduce a lot of noise. $CCGP^c$ managed to outperform CCGP significantly in three scenarios, and no worse in all the scenarios (including <Fmax, 0.85> and <Fmean, 0.95>, where $CCGP^c$ still obtained insignificantly better mean

Table 9.4 The mean (standard deviation) of the objective values of CCGP, $CCGP^f$, $CCGP^c$ and $CCGP^{lc}$ on unseen instances over 50 independent runs in six dynamic flexible job shop scheduling scenarios

Scenario	CCGP	$CCGP^f$	$CCGP^c$	$CCGP^{lc}$
<Fmax, 0.85>	1212.05(34.68)	1215.55(32.62)($\approx$)	1211.83(27.45)($\approx$)	1291.96(48.23)(+)
<Fmax, 0.95>	1941.98(29.93)	1939.84(32.97)($\approx$)	1942.09(29.16)($\approx$)	2026.88(80.15)(+)
<Fmean, 0.85>	385.95(3.22)	384.66(1.19)($\approx$)	384.68(1.92)(−)	389.79(3.96)(+)
<Fmean, 0.95>	551.18(5.78)	551.11(3.81)($\approx$)	550.30(3.72)($\approx$)	563.76(10.14)(+)
<WFmean, 0.85>	831.41(6.08)	829.89(4.76)($\approx$)	828.98(3.57)(−)	841.22(9.78)(+)
<WFmean, 0.95>	1111.01(12.02)	1109.52(11.27)($\approx$)	**1105.84(7.21)**(−)	1141.54(23.04)(+)

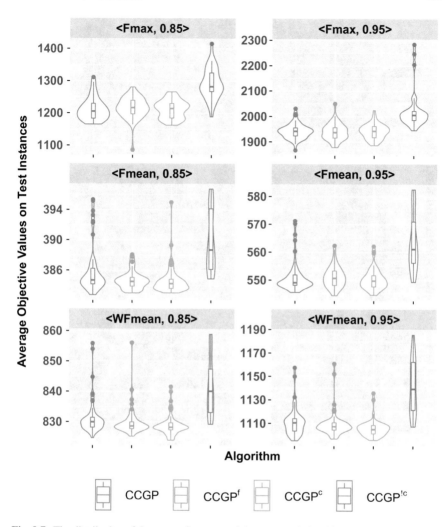

Fig. 9.7 The distribution of the test performance of the compared algorithms

and standard deviation). Furthermore, CCGP^c beat CCGP^f in <WFmean, 0.95>, which has the largest complexity. It is clear that CCGP$^{!c}$ was defeated by all the other compared methods. This is consistent with our expectation since it makes no sense to bias to the unimportant subtrees. In other words, we can see that selecting the subtrees based on the correlation with the entire tree is a promising strategy for crossover.

Figure 9.7 shows the distribution of the test performance of the four compared algorithms. From the figure, we can observe that CCGP^f performed slightly better than CCGP in most scenarios, although the difference is not significant. The distributions obtained by CCGP^c were at a much lower position than that of the other

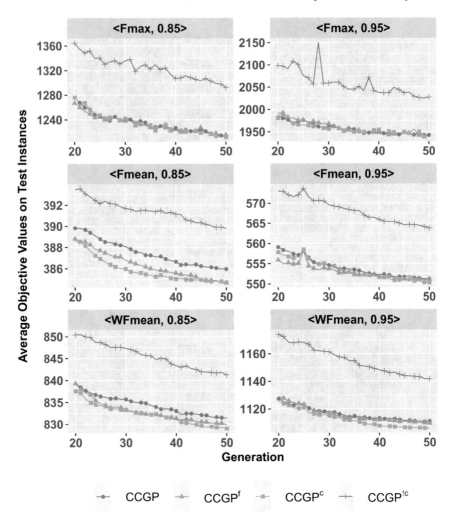

Fig. 9.8 The average test curves of compared algorithms

compared algorithms in most scenarios, showing the advantage of $CCGP^c$. On the contrary, the distributions of $CCGP^{lc}$ were much higher (worse) than that of the other algorithms. This is again consistent with our expectation and the results shown in Table 9.4.

Figure 9.8 shows the test curves (average over 50 independent runs) of the four compared algorithms. From the figure, we can see that $CCGP^c$ can obtain test curves that are almost always the lowest (best) in most scenarios. In <Fmean, 0.85>, <WFmean, 0.85> and <WFmean, 0.95>, $CCGP^c$ showed faster convergence speed than the other two methods. Again, the test curves of $CCGP^{lc}$ are much higher than that of the other compared methods. This enhances our observations that it is promising to consider the importance of a subtree proportional with its correlation coefficient with the entire tree. It is also observed that for the scenarios that minimises maximal flow time, CCGP, CCGP^f and CCGP^c performed very similarly. A possible explanation is that it is not easy to minimise the max-flowtime, which is solely determined by the job with the largest flow time.

9.4.2 Depth Ratios of Selected Subtrees

Both CCGP^c and CCGP^f attempt to smartly select the subtrees during the crossover. The experimental results show that CCGP^c performed much better than CCGP^f. To further analyse how CCGP^c and CCGP^f behave differently during the search process, we investigate the *depth ratio* during the process of the two algorithms. The depth ratio is calculated as the depth of the root node of the subtree (e.g., the root node of the entire tree has a depth of 0) divided by the depth of the tree. If this ratio is smaller, it means that the root node of the selected subtree is closer to the root node of the entire tree. In an extreme case, a ratio of 0 means that the entire tree is selected.

Figures 9.9 and 9.10 show the depth ratios (averaged over 50 runs) of the important and unimportant subtrees selected by CCGP^c and CCGP^f over generations. From the figures, it is clear to see that for both methods, the important subtrees had much smaller depth ratios than the unimportant subtrees. This is easy to understand since it is well known from the GP perspective that the important subtrees are more likely to be closer to the root node. In addition, we can see that the depth ratio of the unimportant subtrees selected by CCGP^c is much smaller than that of CCGP^f, although the depth ratios of the important subtrees are relatively similar for the two algorithms. Furthermore, CCGP^c seems to be able to identify more proper subtrees than CCGP^f early in the search process.

Figure 9.11 illustrates the average depth ratios of the important subtrees selected by the compared methods. From the figure, it is clear that all the three compared methods can obtain very similar depth ratios of the selected important subtrees, which fluctuates from 0.4 to 0.45 after the 10th generation. This suggests that the algorithms do not always select important subtrees that are too close to the root node. At the early stage of the search, however, CCGP^c tends to select the important subtrees with larger depth ratios than the other two algorithms, that is, the subtrees more distant to the root node. On the other hand, CCGP^f shows very similar patterns with CCGP all along the search process, which means that its ability to detect important subtrees may be limited, especially at the beginning of the search process. A possible explanation is that at the early stage, the individuals are not good enough and have a lot of redundant/noisy branches. The number of occurrences can be very inaccurate then, which greatly affects the effectiveness of the detection.

Figure 9.12 gives the average depth ratios of the unimportant subtrees selected by the four compared algorithms. From this figure, one can clearly see that the three compared algorithms selected the unimportant subtrees in very different ways. First, compared with CCGP, the depth ratios obtained by CCGP^f and CCGP^c are larger.

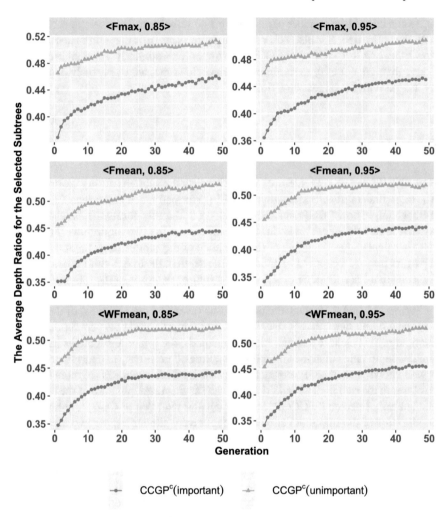

Fig. 9.9 The average depth ratios between the important and unimportant subtrees selected by CCGP^c over generations

That is, the unimportant subtrees selected by CCGP^f and CCGP^c are more distant to the root node and closer to the leaf nodes. Second, between CCGP^c and CCGP^f, CCGP^c has a much larger depth ratio of the selected unimportant subtrees than CCGP^f, especially after the 10th generation. In other words, CCGP^c tends to select unimportant subtrees as those further away from the root node.

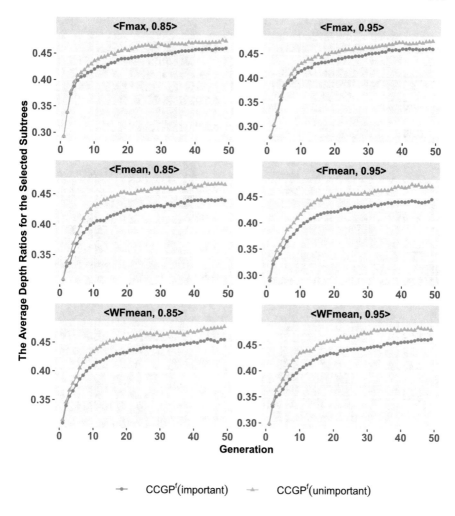

Fig. 9.10 The average depth ratios between the important and unimportant subtrees selected by CCGP^f over generations

9.4.3   Correlations of Selected Subtrees

In CCGP^c, the subtrees to be crossed over are selected based on the correlation coefficient. The probabilities are set according to the correlation. To understand this behaviour, we plot the distribution of the correlation coefficients of the subtrees chosen by CCGP^c at different phases of the search. An example for <WFmean, 0.95> is shown in Fig. 9.13, where "Gen X Large (Small)" means that a subtree will be more likely to be selected if its phenotypic behaviour has a stronger (weaker)

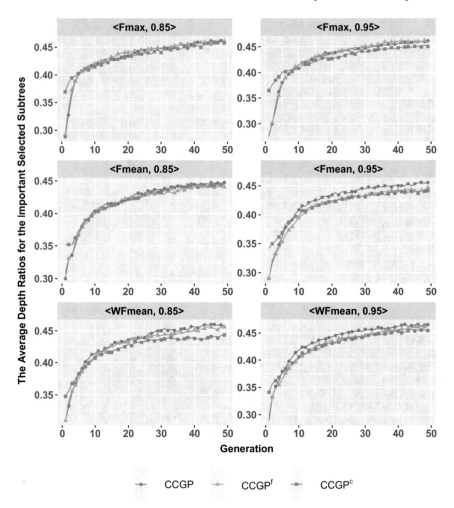

Fig. 9.11 The average depth ratios of the **important subtrees** selected by the compared algorithms

correlation with that of the root node. The ranges of the distributions are $[-1, 1]$, since the correlation coefficient ranges from -1 to 1.

The above row of sub-figures is for selecting important subtrees. These sub-figures show that the algorithm often selects the subtrees with large absolute correlation coefficient values, and rarely selects those with about zero correlation coefficients. This indicates that even from the beginning of the search, $CCGP^c$ is able to identify the subtrees with strong correlation (e.g., the subtrees close to the root node) and select them as important subtrees.

The second row of sub-figures is for selecting unimportant subtrees. An interesting observation is that, even for selecting unimportant subtrees, $CCGP^c$ often selects the

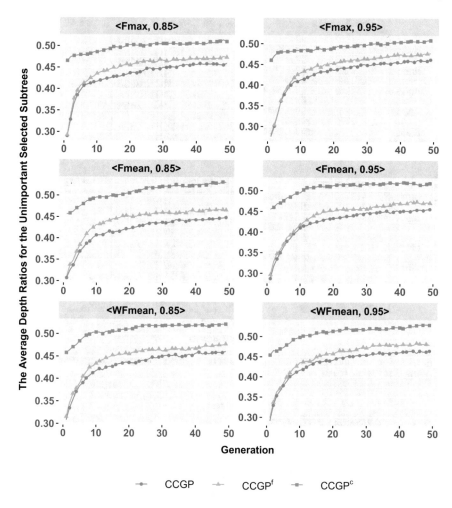

Fig. 9.12 The average depth ratios of the **unimportant subtrees** selected by the compared algorithms

subtrees with very high correlation (close to 1). After further looking into the cases, we observe that, in these cases, the tree is very small and most of its subtrees are highly correlated with the root node. Other than that, it can be seen that more selected unimportant subtrees have weaker correlation coefficients (closer to zero). This is consistent with our expectations.

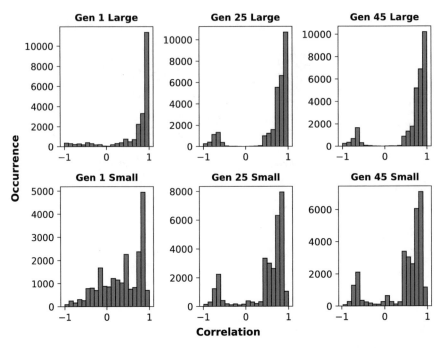

Fig. 9.13 The histogram plot for the **correlations** of the selected subtrees of CCGP^c at different stages in <WFmean, 0.95>

9.4.4   Probability Difference

The core of this chapter is to smartly decide the probability of selecting the crossover points rather than purely random.

To investigate how biased crossover we have achieved by CCGP^c, we plot the *probability difference* of the subtrees selected by CCGP^c at different phases. The probability difference is calculated as the probability set to the selected subtree by CCGP^c minus the pure random probability set by the traditional GP approach. Such difference can take any value between −1 and 1. If the probability difference is greater than zero, it means that the probability of selecting the subtree by CCGP^c is larger than the purely random probability. On the other hand, if the probability difference is negative, then CCGP^c selects a subtree that could have been less likely to be selected by the traditional GP approach. We expect an effective CCGP^c would often have large positive probability differences.

An example plot for <WFmean, 0.95> is shown in Fig. 9.14. From the figure, we can clearly see that cross different phases of the search process of CCGP^c, the probability difference is always larger than zero, meaning that the selection is better than random. In addition, there are a larger number of probability differences whose values are very large (e.g., half of them are larger than 0.5), indicating that there is a

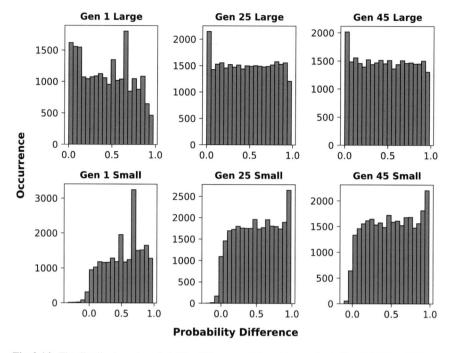

Fig. 9.14 The distribution of **probability difference** of the subtrees selected by CCGP^c at different phases in <WFmean, 0.95>

big gap between the CCGP^c selection and pure random selection. This suggests that CCGP^c can often identify the subtrees with very strong or weak correlation with the root nodes during the search.

9.4.5 Training Time

To see how efficient the algorithms train the heuristics, we show the training time of the compared algorithms in Table 9.5. From the table, we cannot see any obvious difference between the training time of the compared algorithms. This suggests that the extra subtree importance calculations of CCGP^f and CCGP^c are efficient compared with the GP evolution and evaluation, since it does not incur significantly longer training time.

Overall, the experimental results verify the effectiveness of biasing the crossover point selection. The two ways mentioned in this chapter (i.e., using the number of occurrences of features or utilising the correlation coefficient of phenotypic behaviour) to measure the subtree importance are effective, especially the use of correlation coefficient according to phenotypic behaviour.

Table 9.5 The training time (in minutes) of the compared algorithms

Scenario	CCGP	CCGP^f	CCGP^c
<Fmax, 0.85>	73(9)	74(13)($\approx$)	74(11)($\approx$)
<Fmax, 0.95>	87(15)	88(13)($\approx$)	89(12)($\approx$)
<Fmean, 0.85>	71(10)	72(10)($\approx$)	72(9)($\approx$)
<Fmean, 0.95>	80(13)	81(11)($\approx$)	81(12)($\approx$)
<WFmean, 0.85>	73(13)	75(16)($\approx$)	74(15)($\approx$)
<WFmean, 0.95>	82(13)	82(12)($\approx$)	83(13)($\approx$)

9.5   Further Analysis

In this section, we will do some further analysis, including the number of unsuccessful crossover operations, the sizes of learned scheduling heuristics, the insight of the learned scheduling heuristics of CCGP^c, and the occurrences of features of CCGP^f.

9.5.1   Number of Invalid Crossover Operations

From the previous section, we have seen that the important subtrees are likely to be close to the root node, and the unimportant subtrees tend to be closer to the leaf nodes. If we replace an unimportant subtree (closer to the leaf nodes) with an important subtree (closer to the root node, which can usually be large), the resultant offspring can have a much larger tree depth and the maximum depth limit is likely to be violated. In this case, the crossover becomes invalid and the generated offspring is abandoned.

We record the number of "invalid" crossover which generates an offspring whose depth is larger than eight. We name the "invalid" crossover as *potential invalid replacements* since the produced offspring are ignored, and the crossover actually does not happen. The number of potential invalid replacements can be used to investigate how the presented algorithm influences the process of generating offspring. Figure 9.15 shows the average potential invalid replacements for the crossover of CCGP, CCGP^f and CCGP^c at each generation over 50 independent runs in the six DFJSS scenarios. In all the scenarios, CCGP^c leads to more potential invalid replacements than CCGP^f over the generations. It is consistent with the analyses in Sect. 5.4.4. In CCGP^c, the unimportant subtrees with larger depth ratios are more likely to be replaced by the important subtrees with smaller depth ratios, which leads to more potential invalid replacements. Fortunately, it does not have a significant impact on the effectiveness of the presented algorithm.

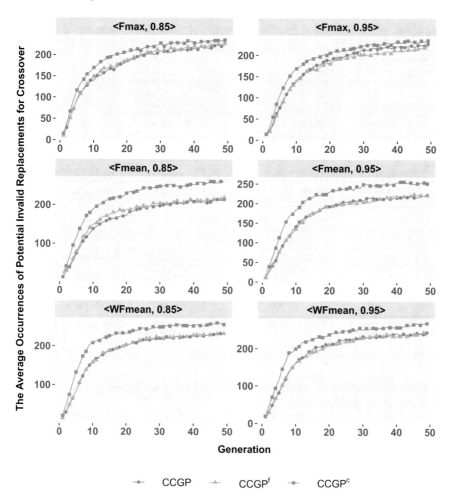

Fig. 9.15 The number of **invalid crossover** of CCGP^f and CCGP^c

9.5.2 Sizes of Learned Scheduling Heuristics

The size (i.e., the number of nodes) can be a measure for the "interpretability" [75] of the learned rules. A smaller rule can be more easily interpreted than a larger rule. In this subsection, we investigate how the presented algorithm influences the sizes of the learned rules in terms of the sizes of the learned best rules. Table 9.6 shows the mean and standard deviation of the sizes of the learned best routing and sequencing rules in six DFJSS scenarios. Compared with CCGP, there is no statistically significant difference between the sizes of learned routing and sequencing rules obtained by CCGP^f and CCGP^c. We can conclude that the presented CCGP^c with recombinative

Table 9.6 The mean (standard deviation) of the sizes of learned the best routing and sequencing rules of CCGP, CCGP^f, and CCGP^c over 50 independent runs in six dynamic flexible job shop scheduling scenarios

Scenario	Routing rule		
	CCGP	CCGP^f	CCGP^c
\<Fmax, 0.85\>	61.48(18.30)	68.68(18.68)($\approx$)	62.32(19.07)($\approx$)
\<Fmax, 0.95\>	59.28(19.20)	66.28(20.30)($\approx$)	60.68(18.87)($\approx$)
\<Fmean, 0.85\>	59.84(15.05)	61.52(16.21)($\approx$)	59.40(18.09)($\approx$)
\<Fmean, 0.95\>	64.16(19.42)	65.28(16.45)($\approx$)	59.60(16.52)($\approx$)
\<WFmean, 0.85\>	59.00(17.35)	63.12(17.99)($\approx$)	64.52(17.20)($\approx$)
\<WFmean, 0.95\>	63.88(15.95)	61.44(15.30)($\approx$)	65.20(18.11)($\approx$)
Scenario	Sequencing rule		
	CCGP	CCGP^f	CCGP^c
\<Fmax, 0.85\>	54.40(18.12)	51.36(15.01)($\approx$)	53.72(18.23)($\approx$)
\<Fmax, 0.95\>	51.32(16.34)	53.92(16.77)($\approx$)	50.08(19.32)($\approx$)
\<Fmean, 0.85\>	46.64(19.98)	47.32(18.43)($\approx$)	45.32(15.22)($\approx$)
\<Fmean, 0.95\>	44.92(16.04)	44.80(15.47)($\approx$)	42.12(19.94)($\approx$)
\<WFmean, 0.85\>	46.44(18.77)	50.92(13.93)($\approx$)	46.32(18.32)($\approx$)
\<WFmean, 0.95\>	47.04(18.45)	52.92(20.33)($\approx$)	51.68(19.11)($\approx$)

guidance achieves better performance without having an impact on the sizes of the learned scheduling heuristics.

9.5.3 Insight of Learned Scheduling Heuristics

To study the behaviours of the learned rules obtained by CCGP^c, this section conducts structural analyses on the learned sequencing rules. Specifically, the best sequencing rules obtained by CCGP^c for minimising max-flowtime and mean-weighted-flowtime with the utilisation level of 0.95 are further investigated, respectively.

Learned Rule for Max-flowtime: Figure 9.16 shows one of the best learned sequencing rules by CCGP^c in the scenario \<Fmax, 0.95\>. It is observed that the rule is a combination of six simple terminals (TIS, WKR, PT, NIQ, WIQ, and NOR), and TIS and WKR are the most frequently used terminals for building this rule. In addition, "TIS + WKR" might be an effective constructed building block for this sequencing rule, since it appears five times in this rule.

To make analysis easy, the rule in Fig. 9.16 is further simplified, as shown in Eq. (9.1).

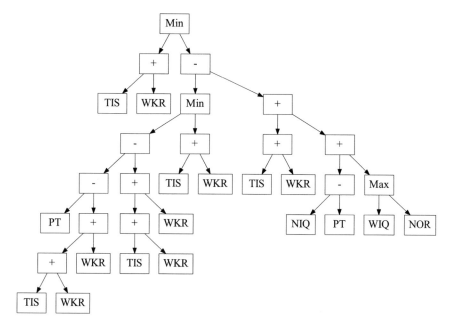

Fig. 9.16 One of the best learned sequencing rules learned by CCGP^c in scenario $<$Fmax, 0.95$>$

$$
\begin{aligned}
S_1 =\ & Min\{TIS + WKR, \\
& Min\{PT - 2TIS - 4WKR, TIS + WKR\}- \\
& (TIS + WKR + NIQ - PT + Max\{WIQ, NOR\})\} \\
\approx\ & Min\{TIS + WKR, \\
& 2PT - 3TIS - 5WKR - NIQ - WIQ\} \\
\approx\ & 2PT - 3TIS - 5WKR - NIQ - WIQ \\
=\ & 2PT - 3TIS - 5WKR
\end{aligned}
\tag{9.1}
$$

From step 1 to step 2, "Min{PT − 2TIS − 4WKR, TIS + WKR}" is simplified as "PT − 2TIS − 4WKR", since "PT − 2TIS − 4WKR" is almost always smaller than "TIS + WKR". In addition, "Max{WIQ, NOR}" is represented as WIQ, since WIQ (time) tends to be larger than NOR (between 1 and 10). Similarly, the rule in step 2 can be mostly replaced by the rule in step 3. Finally, NIQ and WIQ are the same for all operations in the same queue, and can be safely ignored, since they do not affect the final decision of choosing an operation. This rule suggests that, when a machine is idle, the machine prefers to process operations with smaller processing times. In addition, the jobs that arrive at the shop floor earlier or have more remaining work should be processed earlier. Otherwise, if they are completed too late, the max-flowtime will be increased. It is consistent with our intuition for minimising max-flowtime due to its sensitivity to the worst case. The weights of the terminals may require domain knowledge and many rounds of trial-and-error if manually designed.

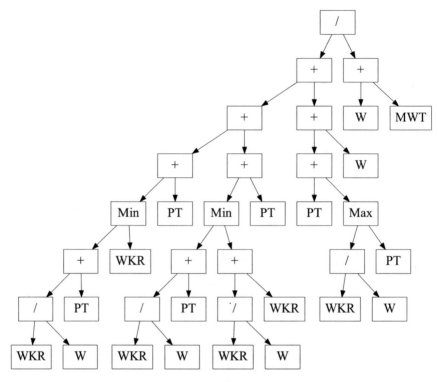

Fig. 9.17 One of the best learned sequencing rules learned by CCGP^c in the scenario <WFmean, 0.95>

Learned Rule for Mean-weighted-flowtime: Fig. 9.17 shows one of the best learned sequencing rules obtained by CCGP^c in the scenario <WFmean, 0.95>. This rule consists of four simple terminals (WKR, W, PT, MWT), and four functions (+, /, Max, Min). "WKR / W" is an important learned component in this rule, and it appears four times. W tends to play its role as a denominator. PT is also an important terminal which mainly plays its role as a component for addition.

The simplification of the sequencing rule in Fig. 9.17 is shown in Eq. (9.2).

$$\begin{aligned}
S_2 &= (Min\{WKR/W + PT, WKR\} + PT \\
&\quad + Min\{WKR/W + PT, WKR/W + WKR\} + PT \\
&\quad + PT + Max\{WKR/W, PT\} + W)/(W + MWT) \\
&= (Min\{WKR/W + PT, WKR\} + 3PT \\
&\quad + Min\{PT, WKR\} \\
&\quad + Max\{WKR/W, PT\} + W)/(W + MWT)
\end{aligned} \quad (9.2)$$

This rule suggests to process the important operation with a large W earlier. In addition, the operations with short processing time and the jobs with small remaining work are preferred to be processed as soon as possible. Otherwise, the weighted-flowtime will increase.

In summary, this section shows the advantage of evolving scheduling heuristics with the presented algorithm. The learned scheduling heuristics consist of simple heuristics but are combined in an effective way, which is not easy to be designed manually. In addition, the learned scheduling heuristics have good interpretability, which is important for real-world applications.

9.5.4 Occurrences of Features

Figure 9.18 shows the curves of the occurrence of features in routing rules during the evolutionary process of CCGP^c. The MWT (i.e., machine waiting time) is the most important feature for the routing rules in all the scenarios. The importance of MWT is much higher than the other features. In the scenarios whose utilisation levels are 0.85, WIQ (i.e., the workload in the queue) plays a second important role. In the scenarios with a high utilisation level (i.e., 0.95), NIQ (i.e., the number of operations in the queue) plays a significant role. Intuitively, both WIQ and NIQ are important indicators for measuring the workload for machines, they might have the same functionality, and one might take over the other. However, we do not know how they work in different scenarios. It is interesting to see that the role of NIQ is significantly higher than that of WIQ in the scenarios that have higher utilisation level. This indicates that NIQ is an important factor in busy scenarios.

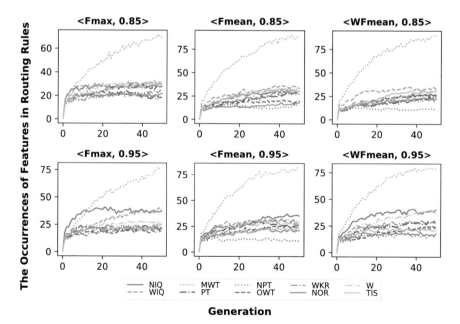

Fig. 9.18 The curves of the occurrence of features in **routing rules** during the evolutionary process of CCGP^c

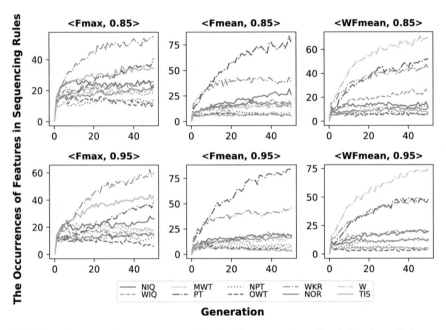

Fig. 9.19 The curves of the occurrence of features in **sequencing rules** during the evolutionary process of CCGP[c]

Figure 9.19 shows the curves of the occurrence of terminals in sequencing rules during the evolutionary process of CCGP[c]. Different from routing rules, three terminals (i.e., WKR, TIS, and PT) play a vital role in minimising max-flowtime. PT and WKR are also the two important terminals in minimising mean-flowtime and weighted mean-flowtime. Except for them, W plays a dominant role in minimising the weighted mean-flowtime, which is consistent with our intuition. In addition, W plays its role mainly in sequencing rules instead of routing rules.

9.6   Chapter Summary

This chapter introduces an effective recombinative guidance strategy for GP to automatically learn effective scheduling heuristics by improving the quality of produced offspring for DFJSS. This chapter presents novel techniques measure the subtree importance. The presented algorithm is based on the idea that the subtrees containing the important features are more important for the whole tree. In addition, this chapter describes an effective way to measure the importance of subtrees of an individual based on the characteristics of DFJSS with the correlation coefficient technique. A properly designed recombinative guidance mechanism is then developed for the crossover operator of GP to generate offspring by replacing the unimportant subtrees of one parent with the important subtrees from the other parent.

The results show that the learned scheduling heuristics, by the presented algorithm with the correlation coefficient based recombinative guidance, have better performance in most scenarios while no worse in all the other scenarios due to its effectiveness for producing offspring. This is also verified by the analyses in terms of the depth ratios, the correlations, and the probability difference of selected subtrees during the evolutionary process. In addition, the presented algorithm does not need extra computational time compared with its counterparts. This verifies the advantages of utilising the information produced by GP during the evolutionary process and the efficient information calculation techniques such as correlation coefficient.

In this part, we have already studied different machine learning techniques (i.e., surrogate, feature selection, and specialised genetic operators) for different chapter goals. However, this part only works on single objective optimisation. In the next part, we will learn more about multi-objective optimisation for GP in dynamic JSS.

Part IV
Genetic Programming for Multi-objective Production Scheduling Problems

Apart from challenges due to dynamic changes and uncertainty in the production environments, production scheduling is a complex task because of the needs to optimise multiple conflicting objectives. For example, delivery reliability, e.g., measured by tardiness, and inventory levels, e.g., measured by mean flow times, are conflicting objectives as attempts to reduce tardiness may require more jobs entering the shops, which will increase the inventory levels. Also, different departments can have different preferences, quite often conflicting, which need to be taken into consideration when developing a schedule. Handling multiple conflicting objectives has been an important topic in production scheduling research. Multi-objective optimisation techniques are usually categorised into two main groups: (1) aggregating multiple objectives into a single objective (e.g., weighted summation) and (2) Pareto-based optimisation which tries to find the Pareto front of non-dominated solutions.

As GP is an evolutionary computation technique, multi-objective optimisation techniques developed in evolutionary computation communities can be easily adopted. However, different from other traditional optimisation problems, the goal of GP in this case is to find a Pareto front of non-dominated heuristics rather than solutions. The complexity of the heuristic search space and generalisation of evolved heuristics also influence the effectiveness of multi-objective GP (MOGP) algorithms.

Part IV will discuss the challenges of MOGP for automatic design of production scheduling heuristics and some solutions to address these challenges. This part will include three chapters:

- Chapter 10 shows a straightforward adoption of the multi-objective optimisation technique to help MOGP explore the Pareto front of heuristics for a dynamic job shop scheduling problems with five objectives. This chapter also compares evolved heuristics to those developed in the literature and examines the robustness of evolved heuristics.

- Chapter 11 investigates how cooperative coevolution is used within MOGP to improve the search efficiency when due-date assignment and job sequencing are considered simultaneously.

- Chapter 12 further explores the applications of MOGP to dynamic flexible job shop scheduling problems when machine assignment and job sequencing are considered simultaneously.

Chapter 10
Learning Heuristics for Multi-objective Dynamic Production Scheduling Problems

Production scheduling is one of the first application domains in which multi-objective optimisation is applied and demonstrates practical values. In real-world production environments, production scheduling usually has to take into consideration multiple aspects or performance metrics such as resource utilisation, throughput, and delivery performance, which are formally formulated as popular scheduling objectives in the literature such as mean flowtime, makespan, maximum tardiness, and total weighted tardiness. It is difficult, if not impossible, to find a solution that can simultaneously optimise all of these objectives because they are usually conflicting. The most straightforward approach to deal with multiple objectives is to formulate an aggregate objective, e.g., by linearly combining original objectives. This approach works well if the objective search space is well understood and the preferences of decision-makers are clear. An arbitrary aggregate objective may mislead the search to less desirable solutions. Another approach that less relies on prior knowledge is to identify Pareto or non-dominated solutions. Instead of aiming to find a single best solution, Pareto-based optimisation or multi-objective algorithms are designed to efficiently find a set of non-dominated solutions, which can be further analysed by decision-makers. This approach usually reveals an interesting trade-off solutions and helps decision-makers gain a better understanding of potential solutions before final decisions are made. Evolutionary multi-objective optimisation (EMO) has been investigated extensively in the scheduling literature, especially for static scheduling problems. Past studies have shown that the multiple-objective optimisation approach is more attractive than the single objective approach as optimal solutions for primary (if available) can be still identified while other trade-off solutions are also available for comparison and selection.

F. Zhang et al., *Genetic Programming for Production Scheduling*, Machine Learning: Foundations, Methodologies, and Applications, https://doi.org/10.1007/978-981-16-4859-5_10

10.1 Challenges and Motivations

While EMO has been very popular in solving multi-objective static scheduling problems, its applications in dynamic scheduling are much less explored. Heuristics manually designed by experts are limited and usually only focus on a single objective. For the automated design of production scheduling heuristics, there are only a handful number of studies investigating multiple conflicting objectives simultaneously. In this chapter, we will show how multi-objective optimisation techniques can be applied to learn non-dominated heuristics for dynamic job shop scheduling (MO-DJSS). In this chapter, we are interested in the following questions:

- What are the advantages of multi-objective genetic programming (MOGP) algorithm for scheduling heuristics?
- What is the generalisation of evolved non-dominated scheduling heuristics?
- Can MOGP help us explore good and unexplored non-dominated scheduling heuristics?

Experiments with different simulation scenarios also demonstrate the advantages of the multi-objective genetic programming (MOGP) algorithm. In addition, systematic comparisons between learned heuristics and existing heuristics designed by human experts show the effectiveness and novelties of learned heuristics.

In Sect. 10.2, a detailed description of the proposed MOGP is presented. Simulation models for DJSS, benchmark heuristics in the literature, and statistic analysis procedures to compare heuristics are also provided in this section. Section 10.3 examines the performance of the learned heuristics through extensive experiments. More insights regarding the learned Pareto fronts and the robustness of learned heuristics are investigated in Sect. 10.4. Section 10.5 concludes this chapter.

10.2 Algorithm Design and Details

This section shows how MOGP is used to solve DJSS problems. We first show how heuristics are represented in MOGP. Then, the MOGP algorithm is presented. Finally, we describe the simulation model of DJSS used for training/test purposes and the statistical procedure to analyse the results.

10.2.1 Representation

Similar to previous applications of GP for JSS problems introduced in the earlier chapter, the heuristics or priority rules (as priority function is the only evolvable part of the heuristic for this DJSS problem) are represented by GP trees [115]. A GP tree in this case will play the role of a priority function which will determine the priorities

Table 10.1 Terminal and function sets for dispatching rules

Symbol	Description
rJ	Job release time (arrival time)
RJ	Operation ready time
RO	Number of remaining operation within the job
RT	Work remaining of the job
PR	Operation processing time
DD	Due date of the job
RM	Machine ready time
SL	Slack of the job $= DD - (t + RT)$
WT	Is the current waiting time of the job $= \max(0, t - RJ)$
#	Random number from 0 to 1
NPR	Processing time of the next operation
WINQ	Work in the next queue
APR	Average operation processing time of jobs in the queue
Function set	$+, -, \times, \%$, min, max, abs, and If

of jobs waiting in the queue. The terminal and function sets of learned heuristics are presented in Table 10.1. In this table, the upper part shows a number of terms that usually appear in the heuristics in the literature. The next part in this table shows three terms (i.e., NPR, WINQ, and APR) that reflect the status of the current and downstream machines.

10.2.2 Multi-objective Genetic Programming Algorithm

In this chapter, we aim to evolve scheduling heuristics to minimise five popular objectives in the DJSS literature, which are the mean-flowtime (F), maximum flowtime (F_{max}), percentage of tardy jobs (%T), mean tardiness (T), and maximum tardiness (T_{max}) [96, 197, 201]. The HaD-MOEA algorithm [223] is applied here to explore the Pareto front of non-dominated heuristics regarding the five objectives mentioned above. HaD-MOEA is an extension of NSGA-II [58] which works well on problems with many objectives. Figure 10.1 shows how the proposed MOGP works. At first, a number of training simulation scenarios (more details will be shown in the next section) are loaded and the initial archive $\mathcal{P}^e$ (parent population) is empty. These scenarios will be used to evaluate the performance of a learned heuristic.

The initial GP population is created using the ramped-half-and-half method [115]. In each generation of MOGP, all individuals in the population will be evaluated by applying them to each simulation scenario. The quality of each individual in the pop-

load training simulation scenarios $\mathbb{S} \leftarrow \{S_1, S_2, \ldots, S_T\}$
randomly initialise the population $P \leftarrow \{\mathcal{H}_1, \mathcal{H}_2, \ldots, \mathcal{H}_{popsize}\}$
$\mathcal{P}^e \leftarrow \{\}$ and $generation \leftarrow 0$
1: **while** $generation \leq maxGeneration$ **do**
2:     **for all** $\mathcal{H}_i \in P$ **do**
3:         $\mathcal{H}_i.objectives \leftarrow$ apply $\mathcal{H}_i$ to each scenario $S_k \in \mathbb{S}$
4:     **end for**
5:     calculate the Harmonic distance [227] and the ranks for individuals in $P \bigcup \mathcal{P}^e$
6:     $\mathcal{P}^e \leftarrow$ select($P \bigcup \mathcal{P}^e$)
7:     $P \leftarrow$ apply crossover, mutation to $\mathcal{P}^e$
8:     $generation \leftarrow generation + 1$
9: **end while**
10: return $\mathcal{P}^e$

Fig. 10.1 MOGP to evolve heuristics for DJSS problems

ulation will be measured by the average value of the objectives across all simulation scenarios. After all individuals have been evaluated, we calculate the Harmonic distance [223] for each individual. Then, individuals in both archive $\mathcal{P}^e$ and population P are selected to update the archive $\mathcal{P}^e$ based on the Harmonic distance and the non-dominated rank [58]. The new population will be generated by applying subtree crossover and subtree mutation to the current population. Binary tournament selection [58] is used to select the parents for the two genetic operations. The crossover rate and mutation rate used in this method are 90% and 10%, respectively. The maximum depth of GP trees is eight. A population size of 200 is used in this study and the results will be obtained after the proposed method runs for 200 generations. These parameters are selected based on our preliminary experiments to balance between the effectiveness and the (Pareto) diversity of learned heuristics.

10.2.3   Simulation Models for Dynamic Job Shop Scheduling Problems

The scenarios for training and test of heuristics are shown in Table 10.2. The simulation experiments have been conducted in a job shop with 10 machines. The triplet $\langle m, u, c \rangle$ represents the simulation scenario in which the average processing time is m (m is 25 or 50 when processing times follow discrete uniform distribution [1, 49] or [1, 99], respectively), the utilisation is $u\%$ and the allowance factor is c (due date = release time + $c \times$ total processing time). In the training stage, two simulation scenarios (corresponding to the two utilisation levels) will be selected and five replications will be performed for each scenario. The average value for each objective from $2 \times 5 = 10$ replications will be used to measure the quality of the learned heuristics (as described in the previous section). We use the shop with different characteristics here in order to evolve heuristics with good generality. The

Table 10.2 Scenarios for training and test

Factor	Training	Test
$F1$	Discrete uniform $[1, 49]$	Discrete uniform $[1, 49]$ and $[1, 99]$
$F2$	70%, 80%	85%, 95%
$F3$	c is randomly selected from $(3, 5, 7)$	$c = 4, c = 6, c = 8$
Summary	$\langle 25, 70, (3, 5, 7)\rangle$, $\langle 25, 80, (3, 5, 7)\rangle$	$\langle 25, 85, 4\rangle, \langle 25, 85, 6\rangle,$ $\langle 25, 85, 8\rangle, \langle 25, 95, 4\rangle,$ $\langle 25, 95, 6\rangle, \langle 25, 95, 8\rangle,$ $\langle 50, 85, 4\rangle, \langle 50, 85, 6\rangle,$ $\langle 50, 85, 8\rangle, \langle 50, 95, 4\rangle,$ $\langle 50, 95, 6\rangle, \langle 50, 95, 8\rangle$

allowance factors, which decide the due date tightness, are selected randomly from the three values 3, 5, and 7 instead of a fixed allowance factor (for each scenario) in common simulation experiments for DJSS problems. If we train on scenarios with fixed allowance factors, the learned heuristics will tend to focus more on the scenarios with small allowance factors to improve the due date performance (mean tardiness, maximum tardiness, etc.) because the values of the due date based objectives are higher in these cases. This may cause an overfitting problem for the learned heuristics. Moreover, training on different scenarios with different fixed allowance factors will also increase the training time of our MOGP method. Simulating multiple utilisation levels in a simulation scenario can also be used to reduce the number of training scenarios but will increase the running time of a replication significantly to obtain the steady-state performance of the heuristics, and indirectly increase the training time of the MOGP method.

In the test stage, 12 simulation scenarios with 50 replications for each scenario (or shop condition) resulting in $12 \times 50 = 600$ replications will be used to have a comprehensive assessment of the learned heuristics. In each replication of a simulation scenario, we start with an empty shop and the interval from the beginning of the simulation until the arrival of the 500th job is considered as the warm-up time and the statistics from the next completed 2000 jobs [96] will be used to calculate the five objective values. The number of operations for each new job is randomly generated from the discrete uniform distribution $[2, 14]$ and the routing for each job is randomly generated, with each machine having equal probability to be selected (re-entry is allowed here but consecutive operations are not processed on the same machine). The arrival of jobs will follow a Poisson process with the arrival rate adjusted based on the utilisation level [197].

10.2.4   Benchmark Heuristics

Table 10.3 shows 31 priority heuristics that will be used to compare with the learned heuristics in our work. The upper part of this table shows some original heuristics and the lower part shows some extensions of the original heuristics that have been proposed in the literature. The parameters of ATC and COVERT are the same as those used in the literature ($k = 3$ for ATC, $k = 2$ for COVERT, and the leadtime estimation parameter $b = 2$). More detailed discussion on these heuristics can be found in [96, 101, 171, 197].

10.2.5   Statistical Analysis

Since DJSS is a stochastic problem, statistical analysis is required to compare the performance of heuristics obtained from simulation. In this work, we use the one-way ANOVA and Duncan's multiple range tests [153] to compare the performance of heuristics. It is noted that the *common random number* technique is used in our experiments for variance reduction.

While Pareto dominance is easily determined in most multi-objective optimisation problems in the literature, it is not well-defined for stochastic problems such as DJSS. For such a problem, statistical analysis is necessary to examine the Pareto dominance. In this section, we describe two procedures to check for the *statistical Pareto dominance* between two different heuristics, which can be used to validate each other.

10.2.5.1   Objective-Wise Procedure

In multi-objective optimisation, solution (or heuristic in this work) a is said to *Pareto dominate* solution b if and only if $\forall i \in \{1, 2, \ldots, n\} : f_i(a) \leq f_i(b) \land \exists j \in \{1, 2, \ldots, n\} : f_j(a) < f_j(b)$ where n is the number of objectives to be minimised. However, if $f_j(a)$ and $f_j(b)$ are random variables (i.e. solutions a and b produce different outputs in different runs/replications), we cannot use the above definitions to check for the Pareto dominance. Therefore, we need to redefine the Pareto dominance for this context. For the objective-wise procedure, solution a *statistically Pareto dominates* solution b if and only if $\forall i \in \{1, 2, \ldots, n\} : f_i(a) \leq_{\mathcal{T}} f_i(b) \land \exists j \in \{1, 2, \ldots, n\} : f_j(a) <_{\mathcal{T}} f_j(b)$, where $f_i(a) \leq_{\mathcal{T}} f_i(b)$ means that a is significantly smaller (better) than or not significantly different from b based on the statistical test $\mathcal{T}$ (e.g., z-test); similarly, $f_j(a) <_{\mathcal{T}} f_j(b)$ means that a is significantly smaller than b based on $\mathcal{T}$. It should be noted that since multiple comparisons (n comparisons for n objectives) need to be done here, we have to adjust the value of the pre-set probability α of a type-1 error [153] in order to control the false positive

Table 10.3 Benchmark heuristics

Heuristics	Descriptions
SPT	Shortest processing time
EDD	Earliest due date
FIFO	First in first out
LWKR	Least work remaining
NPT	Next processing time
CR	Critical ratio
MOD	Modified due date
SL	Negative slack
PW	Process waiting time
ATC	Apparent tardiness cost
LPT	Longest processing time
FDD	Earliest flow due date
LIFO	Last in first out
MWKR	Most work remaining
WINQ	Work in next queue
AVPRO	Average processing time/operation
MOPNR	Most operations remaining
Slack	Slack
RR	Raghu and Rajendran
COVERT	Cost over time
OPFSLK/PT	Operational flow slack per processing time
LWKR+SPT	Least work remaining plus processing time
CR+SPT	Critical ratio plus processing time
SPT+PW	Processing time plus processing waiting time
SPT+PW+FDD	SPT+PW plus earliest flow due date
Slack/OPN	Slack per remaining operations
Slack/RPT+SPT	Slack per remaining processing time plus operation processing time
PT+WINQ	Processing time plus work in next queue
2PT+WINQ+NPT	Double processing time plus WINQ and NPT
PT+WINQ+SL	Processing time plus WINQ and slack
PT+WINQ+NPT+WSL	PT+WINQ plus next processing time and waiting slack

rate. Many methods have been proposed for this problem such as the Bonferroni method and Scheffe method [194].

10.2.5.2   Replication-Wise Procedure

Different from the above method that examines the Pareto dominance of two solutions based on the relative performance of each objective, the replication-wise procedure focuses on the Pareto dominance in each replication/observation to detect the difference between two solutions. This procedure is adapted from the method proposed by Bhowan et al. [23] to compare the performance of different multi-objective GP methods on a run-by-run basis and determine whether a method significantly dominates another overall runs.

In this procedure, the traditional Pareto dominance is used to examine the dominance relation between two solutions in each replication. For instance, $f_j(a) = \{f_j^1(a), f_j^2(a), \ldots, f_j^N(a)\}$ and $f_j(b) = \{f_j^1(b), f_j^2(b), \ldots, f_j^N(b)\}$ are the values for objective j obtained by solutions a and b from N replications. In replication k, $\{f_1^k(a), f_2^k(a), \ldots, f_n^k(a)\}$ is compared to $\{f_1^k(b), f_2^k(b), \ldots, f_n^k(b)\}$ to determine the Pareto dominance between a and b in this replication. Three possible outcomes from the comparison are (1) win for a if a dominates b, (2) loss for a if b dominates a, or (3) draw otherwise. The proportions of win (p_w), lose (p_l), and draw (p_d) over N replications is then recorded. Figure 10.2 gives an example to show how p_w, p_l and p_d are calculated in the case with two objectives and $N = 5$. The outcomes here form a *multinominal* distribution since the proportions or probabilities for all outcomes always sum to one. In a multinominal distribution, the $(1 - \alpha)\%$ confidence interval of the difference in the probability of win and lose ($p_w - p_l$) can be calculated as follows:

$$(p_w - p_l) \pm z_{\alpha/2}\sqrt{var(p_w - p_l)} \tag{10.1}$$

where

$$var(p_w - p_l) = var(p_w) + var(p_l) - (var(p_w + p_l) - var(p_w) - var(p_l))$$
$$= 2var(p_w) + 2var(p_l) - var(p_w + p_l)$$
$$var(p_w) = \frac{p_w(1 - p_w)}{N}$$
$$var(p_l) = \frac{p_l(1 - p_l)}{N}$$
$$var(p_w + p_l) = \frac{(p_w + p_l)(1 - p_w - p_l)}{N}$$

The confidence interval obtained by the Eq. (10.1) can be used to determine whether one solution significantly dominates the other. If the lower bound of the confidence interval is positive, solution a significantly dominates solution b. If the

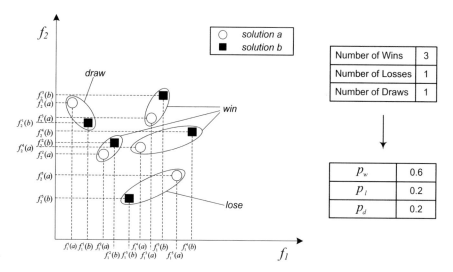

Fig. 10.2 Wins, losses and draws in replication-wise procedure

upper bound of the confidence interval is negative, solution b significantly dominates solution a. Otherwise, there is no significant dominance between the two solutions.

There are some key differences between these two procedures. While the objective-wise procedure focuses more on the magnitude of the difference between average objectives obtained by the two methods, the replication-wise procedure only cares about the Pareto dominance regardless of the difference between the obtained objective values in each replication. If the variances of the objectives obtained from the simulation are high, the replication-wise procedure may not accurately determine the statistical Pareto dominance between two solutions. For example, when p_w and p_l are very close, it is very likely the replication-wise procedure will conclude that there is no dominance between the two solutions. However, it is intuitively not true if there are some "*big*" wins (there are large difference between pairs of objective values $f_j^k(a)$ and $f_j^k(b)$ for some $j \in \{1, 2, \ldots, n\}$) for a solution in some replications. The advantage of the replication-wise procedure is that one statistical significance test needs to be performed as compared to multiple tests (which make the procedure more complicated) in the objective-wise procedure. In Sect. 10.3.2, we will apply both procedures to determine the *statistical Pareto dominance* between learned heuristics and the heuristics reported in the literature.

10.3 Empirical Study

Thirty independent runs of the proposed MOGP method are performed and the non-dominated heuristics from the learned Pareto front $\mathcal{P}^e$ are recorded. We perform a

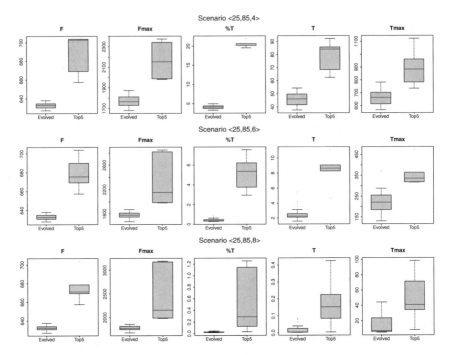

Fig. 10.3 Performance of learned heuristics (processing times from [1, 49] and utilisation of 85%)

post-processing step to extract the Pareto front $\mathcal{P}$ from $\mathcal{P}^e$ for each testing scenario based on the average values of five objectives in that scenario. The performance of the learned heuristics in $\mathcal{P}$ will be presented in this section.

10.3.1 MOGP Performance with Single Objective

Even though our target is to solve the MO-DJSS problems, it is important to know whether the learned heuristics can provide satisfactory results for every single objective. Figures 10.3, 10.4, 10.5, and 10.6 show the performance of the learned heuristics for each objective under different shop conditions. For each GP run, the learned heuristic within $\mathcal{P}$ that performs best on the objective O (O can be F, F_{max}, %T, T, or T_{max}) is denoted as $\mathcal{H}_O^*$. The left box-plot in each plot in Figs. 10.3, 10.4, 10.5, and 10.6 represents the average values of the objective O obtained by $\mathcal{H}_O^*$ from the 30 GP runs. The right box-plot shows the corresponding values obtained by the top five heuristics for the objective O among the 31 existing heuristics shown in Table 10.3.

A quick observation of Figs. 10.3, 10.4, 10.5, and 10.6 shows that the proposed MOGP method can effectively find heuristics that are better than, or as competitive as, the best existing heuristics for each objective under different shop conditions.

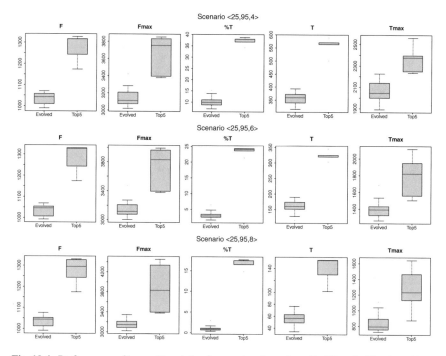

Fig. 10.4 Performance of learned heuristics (processing times from [1, 49] and utilisation of 95%)

The learned heuristic $\mathcal{H}_O^*$ can dominate the existing heuristics regarding F, F_{max}, %T, and T. For T_{max}, the proposed MOGP can find the heuristics that dominate the majority of the existing heuristics and the obtained $\mathcal{H}_O^*$ from some GP runs can also dominate the best existing heuristic. This suggests that it is totally possible to evolve a superior heuristic for every single objective by the proposed MOGP. However, there are objectives that are more difficult to minimise, e.g., T_{max} in this case. Given that we try to evolve heuristics to minimise five objectives simultaneously in the general case, the results obtained here for single objective are very competitive.

Further, statistical tests are also performed here to confirm the quality of the learned heuristics. For a specific objective O and shop condition $\langle m, u, c \rangle$, we perform statistical analysis of the $\mathcal{H}_O^*$ heuristic from each GP run and the best five heuristics in the literature (based on the average values of the corresponding objective) using the one-way ANOVA and Duncan's multiple range tests [153] with $\alpha = 0.01$. The summary of all statistical tests is shown in Table 10.4. For each shop condition, the first row shows the number of times the proposed MOGP method is able to find the $\mathcal{H}_O^*$ that is significantly better than the best existing heuristic for minimising O, which is shown in the second row. In general, the results here are similar to those shown in Figs. 10.3, 10.4, 10.5, and 10.6. It is clear that the MOGP method can almost always find a superior heuristic for minimising F while 2PT+WINQ+NPT is the best existing heuristic. These observations are consistent with those in [93, 96]. Similar to

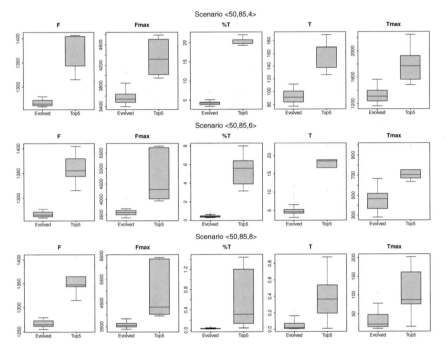

Fig. 10.5 Performance of learned heuristics (processing times from [1, 99] and utilisation of 85%)

[93], when using GP to evolve heuristics for minimising F, the learned heuristics can easily beat 2PT+WINQ+NPT across different simulation scenarios. For F_{max}, the learned heuristics also dominate the best heuristic, i.e., SPT+PW+FDD in this case, in the majority of GP runs. It is interesting to note that the 2PT+WINQ+NPT heuristic and SPT+PW+FDD heuristic are always the best existing heuristics for the two objectives (F and F_{max}) under all shop conditions. This suggests that the shop condition does not really have a large impact on the performance of the heuristics. However, the complexity of the objective may make the design of an effective heuristic more difficult.

The due date-based performance measures such as %T, T, and T_{max} are more sensitive to the shop condition since the best existing heuristics are different under different shop conditions. %T is an easy objective as it does not take into account the magnitude in which the job misses the due date. Therefore, MOGP is able to find superior heuristics for this objective in almost all the scenarios. The number of superior learned heuristics is not large only in the scenarios with large allowance factor ($c = 8$) and low utilisations (85%). The reason is that the number of tardy jobs is very low (near zero as seen in Figs. 10.3, 10.4, 10.5, and 10.6) when due dates are too loose and the shop is not very busy. It is noted that many other existing heuristics (besides Slack/OPN) can also achieve near zero %T in this case. Therefore, it is very difficult to detect superior learned heuristics here. In other cases, the differences

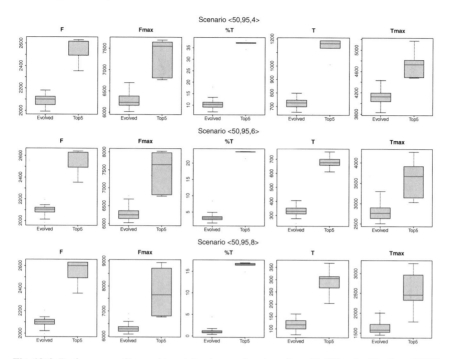

Fig. 10.6 Performance of learned heuristics (processing times from [1, 99] and utilisation of 95%)

between the learned heuristics and existing heuristics for minimising %T are very clear. A similar conclusion can also be applied to T. Perhaps, T_{max} is the most difficult objective among the five objectives that we consider in this study since it is hard to minimise and also quite sensitive to the shop condition. Even though our MOGP method can find superior heuristics in most runs overall, the number of superior heuristics is usually lower than those for other objectives.

In general, the experimental results show that the proposed MOGP can effectively find the good heuristics for each specific objective we consider in this work. It is obvious that the existing heuristics that are supposed to be the best for an objective can also be outperformed by the learned heuristics. Since we learned the Pareto front of non-dominated heuristics for five objectives with a modest population of 200 individuals, the method may not always find the superior heuristics for some hard objectives. However, as shown in Table 10.4, because the shop condition can affect the performance of priority heuristics and their relative performance, the heuristics that are superior under one shop condition may not be the superior one under other shop conditions. Therefore, evolving a set of non-dominated heuristics in our method is actually more beneficial than evolving a single heuristic (either for a single objective in [93] or aggregate objective of multiple objectives in [213]) since it can provide potential heuristics to deal with different shop conditions.

Table 10.4 Performance of learned heuristics under different shop conditions

		F	F_{max}	%T	T	T_{max}
(25, 85, 4)	*	30/30	30/30	30/30	30/30	26/30
	**	2PT+WINQ+NPT	SPT+PW+FDD	MOD	COVERT	PT+WINQ+NPT+WSL
(25, 85, 6)	*	30/30	30/30	30/30	29/30	29/30
	**	2PT+WINQ+NPT	SPT+PW+FDD	RR	RR	Slack/OPN
(25, 85, 8)	*	30/30	30/30	25/30	17/30	18/30
	**	2PT+WINQ+NPT	SPT+PW+FDD	Slack/OPN	Slack/OPN	Slack/OPN
(25, 95, 4)	*	30/30	29/30	30/30	30/30	30/30
	**	2PT+WINQ+NPT	SPT+PW+FDD	LWKR+SPT	2PT+WINQ+NPT	PT+WINQ+SL
(25, 95, 6)	*	30/30	29/30	30/30	30/30	26/30
	**	2PT+WINQ+NPT	SPT+PW+FDD	LWKR+SPT	RR	PT+WINQ+NPT+WSL
(25, 95, 8)	*	30/30	30/30	30/30	30/30	22/30
	**	2PT+WINQ+NPT	SPT+PW+FDD	LWKR+SPT	RR	PT+WINQ+NPT+WSL
(50, 85, 4)	*	30/30	29/30	30/30	30/30	28/30
	**	2PT+WINQ+NPT	SPT+PW+FDD	LWKR+SPT	COVERT	PT+WINQ+NPT+WSL
(50, 85, 6)	*	30/30	29/30	30/30	29/30	29/30
	**	2PT+WINQ+NPT	SPT+PW+FDD	MOD	RR	PT+WINQ+NPT+WSL
(50, 85, 8)	*	30/30	30/30	24/30	10/30	8/30
	**	2PT+WINQ+NPT	SPT+PW+FDD	RR	Slack/OPN	Slack/OPN
(50, 95, 4)	*	30/30	30/30	30/30	30/30	30/30
	**	2PT+WINQ+NPT	SPT+PW+FDD	Slack/OPN	2PT+WINQ+NPT	PT+WINQ+NPT+WSL
(50, 95, 6)	*	30/30	29/30	30/30	30/30	25/30
	**	2PT+WINQ+NPT	SPT+PW+FDD	LWKR+SPT	2PT+WINQ+NPT	PT+WINQ+NPT+WSL
(50, 95, 8)	*	30/30	29/30	30/30	30/30	24/30
	**	2PT+WINQ+NPT	SPT+PW+FDD	LWKR+SPT	RR	PT+WINQ+NPT+WSL

10.3.2 MOGP Performance with Multiple Objectives

The comparison above has shown that the proposed MOGP method can simultaneously evolve superior heuristics for each specific objective. However, these superior performances come with some trade-offs on other objectives. Previous studies have shown that there is no priority heuristic that can minimise all objectives. Therefore, priority heuristics in the literature are designed for minimising a specific objective only. Although it is true that these heuristics can effectively minimise the objective that it focuses on, it usually deteriorates other objectives significantly. For example, the 2PT+WINQ+NPT heuristic can successfully reduce the average flowtime but it performs badly on almost all other objectives. Since the existence of multiple conflicting objectives is a natural requirement in real-world scheduling applications, it is crucial to include this issue into the design process of priority heuristics as well. In this part, we will examine the *Pareto dominance* of the learned heuristics against other priority heuristics in the literature.

For each MOGP run, the learned heuristics in the Pareto front $\mathcal{P}$ are compared to the set $\mathcal{D}$ of 31 benchmark heuristics in Table 10.3. For each shop condition, we employ the objective-wise (OBJW) and replications-wise (REPW) procedures discussed in Sect. 10.2.5 to determine the statistical Pareto dominance between each pair $(\mathcal{H}_i, \mathcal{B}_j)$ for all $\mathcal{H}_i \in \mathcal{P}$ and $\mathcal{B}_j \in \mathcal{D}$. Therefore, there are $|\mathcal{P}| \times |\mathcal{D}|$ comparisons in total for each MOGP run and each statistical procedure. Both OBJW and REPW procedures are performed with $\alpha = 0.01$. In the OBJW procedure, we use the Bonferroni method [153] to adjust the value of $\alpha^t = \alpha/n$ in each z-test (for each objective). From this point forward in this chapter, we use *dominate* or *dominance* when mentioning about the statistical Pareto dominance, unless otherwise indicated. After all the comparisons in each MOGP run were done, each learned heuristic $\mathcal{H}_i$ is classified into three categories:

1. *Non-dominated* if there is no dominance between $\mathcal{H}_i$ and $\mathcal{B}_j$ for
   $\forall \mathcal{B}_j \in \mathcal{D}$
2. *Dominating* if $\mathcal{H}_i$ is not dominated by any $\mathcal{B}_j \in \mathcal{D}$ and $\exists \mathcal{B}_j \in \mathcal{D}$ such that $\mathcal{H}_i$
   dominates $\mathcal{B}_j$
3. *Dominated* if $\exists \mathcal{B}_j \in \mathcal{D}$ such that $\mathcal{H}_i$ is dominated by $\mathcal{B}_j$.

The proportions of learned heuristics in the three categories for each $\mathcal{P}$ are determined and the average proportions from 30 MOGP runs are shown in Fig. 10.7. The triplets in the figure indicate the shop conditions as explained in the previous section. It is clear that the proposed MOGP method can always find heuristics that can dominate heuristics reported in the literature across all objectives. In the worst cases $\langle 25, 95, 4 \rangle$ and $\langle 50, 95, 4 \rangle$, there are still about 20% of the learned heuristics that are dominating heuristics. The number of dominated learned heuristics is also very low and the highest proportions (about 10%) of dominated heuristics are in $\langle 25, 95, 8 \rangle$ and $\langle 50, 95, 8 \rangle$. There are also some interesting patterns in Fig. 10.7. Different from our comparison for a single objective where there are fewer superior heuristics found when the allowance factor increases, it is easy to see that the num-

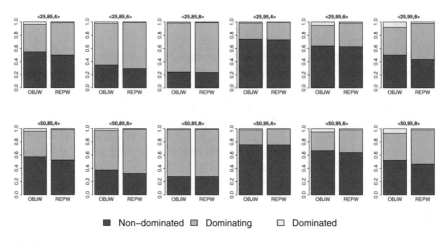

Fig. 10.7 Average Pareto dominance proportion of learned heuristics

ber of non-dominated heuristics decreases and the number of dominating heuristics increases when the allowance factor increases from 4 to 8. This suggests that even when the MOGP method cannot find a superior heuristic for a specific objective, it can easily find heuristics that can perform as good as the best existing heuristic on that objective while significantly improving other objectives. Another interesting pattern in Fig. 10.7 is that the number of dominated heuristics decreases when the allowance factor increases with the shop utilisation of 85%. However, a reverse trend is found with the utilisation of 95% when the number of dominated heuristics increases when the allowance factor increases. For the cases with the utilisation of 85%, the higher allowance made the DJSS problems easier, at least for the due date-based performance measures. Therefore, it is difficult for existing heuristics to dominate the learned priority heuristics. In the case with the utilisation of 95% and low allowance factor, it is very difficult to make a good sequencing decision to satisfy multiple objectives and to find a heuristic that is superior on all objectives. For that reason, the number of dominating and dominated heuristics are relatively small compared to the number of non-dominated heuristics. When the utilisation is 95% and the allowance factor is high, the number of dominated heuristics increases because these shop conditions (very busy shop and loose due dates) are quite different from the shop conditions used in the training stage.

It is also noted that the results from OBJW and REPW in Fig. 10.7 are very consistent. The REPW procedure results in slightly more dominating heuristics (fewer non-dominated heuristics) as compared to the OBJW procedure. Perhaps, this is because the OBJW procedure with the Bonferroni adjustment method is quite conservative, which makes the OBJW procedure more difficult to detect significant differences between two heuristics. However, the differences between the two procedures in our application are very small. Therefore, both OBJW and REPW are suitable procedures to analyse the results from our experiments. A more detailed Pareto dominance of

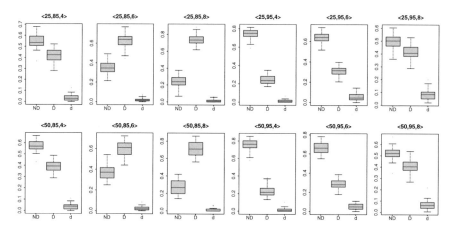

Fig. 10.8 Pareto dominance proportions of learned heuristics

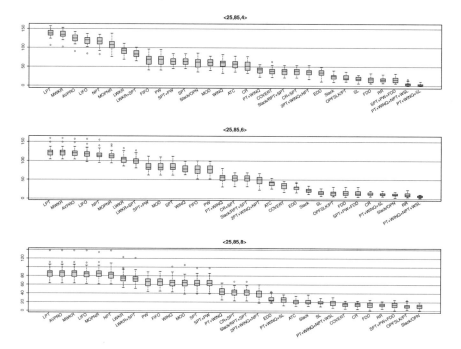

Fig. 10.9 NDLH for each existing heuristics (processing times from [1, 49] and utilisation of 85%)

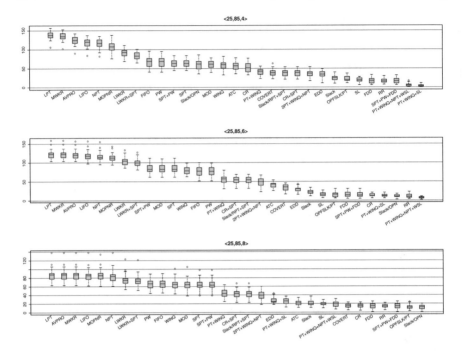

Fig. 10.10 NDLH for each existing heuristics (processing times from [1, 49] and utilisation of 95%)

learned heuristics is shown in Fig. 10.8. In this figure, the box plots represent the proportions from the OBJW procedure of non-dominated (ND), dominating (D), and dominated (d) from each run of MOGP. This figure shows that the proposed MOGP is quite stable since the obtained dominance proportions have low variances. Moreover, the proportions of non-dominated and dominating heuristics are always larger than that of dominated heuristics. In general, these results suggest that the learned heuristics are significantly better or at least very competitive when compared to the existing heuristics.

Through all the comparisons, we also count the number of dominating learned heuristics (NDLH) in each MOGP run that dominate a specific heuristic $\mathcal{B}_j$. These values can be used as an indicator for the competitiveness of the existing heuristics when multiple objectives are considered. The values of NDLH for each heuristic $\mathcal{B}_j$ shown in Table 10.3 under different shop conditions from 30 MOGP runs are shown in Figs. 10.9, 10.10, 10.11, and 10.12. In these figures, the heuristics are arranged from left to right in the order of decreasing values of the average NDLH. It is easy for the MOGP method to evolve heuristics that dominate the simple heuristics such as LPT, MWKR, FIFO, etc. It is noted that most heuristics with low values of NDLH are the ones that are designed for minimising due date-based performance measures and the ones that achieve the best performance for each objective as shown back in Table 10.4. Since the MOGP method can almost always find superior heuristics

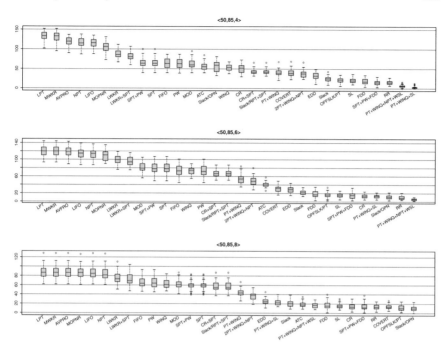

Fig. 10.11 NDLH for each existing heuristics (processing times from [1, 99] and utilisation of 85%)

for minimising F and F_{max}, the best existing heuristics for these two objectives, i.e., 2PT+WINQ+NPT and SPT+PW+FDD, are also easily dominated by the learned heuristics (dominating learned heuristic for these two heuristics can be found in all MOGP runs). The most competitive existing heuristics are actually the ones that give reasonably good performance across all objectives such as OPFSLK/PT, which is not the best heuristic for any particular objective. PT+WINQ+NPT+WSL and PT+WINQ+SL are the most competitive heuristics overall (with low NDLH in most simulation scenarios) and the MOGP method cannot find heuristics that dominate these two heuristics in some runs.

Although a lot of effort have been made in the literature to improve the competitiveness of heuristics, it is clear that the search space of potential heuristics is very large and there are still many highly competitive heuristics that have not been explored, especially when different multiple conflicting objectives are simultaneously considered. Manually exploring this search space seems to be an impossible task. For that reason, there is a need for automatic design methods such as the MOGP proposed in this work. The extensive experimental results shown here have convincingly confirmed the effectiveness of the proposed MOGP method for evolving heuristics for DJSS problems. It is totally possible for the proposed method to evolve heuristics that are significantly better than heuristics reported in the literature, not only on a specific objective, but also on different objectives of interest.

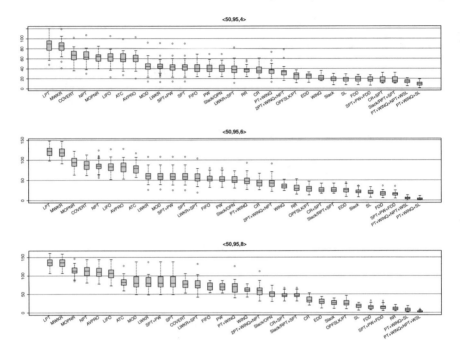

Fig. 10.12 NDLH for each existing heuristics (processing times from [1, 99] and utilisation of 95%)

10.4   Further Analyses

The previous section has shown the performance of the learned heuristics when a single objective and multiple objectives are considered. In this section, we will provide more insights on the distribution and robustness of the learned heuristics on the obtained Pareto front. Some examples of learned heuristics are also shown here to demonstrate their robustness as well as how the learned heuristics are more effective as compared to the existing heuristics.

10.4.1   Learned Pareto Front

The comparison results have shown that the proposed MOGP method can evolve very competitive heuristics. However, we have not fully assessed the advantages of the proposed MOGP methods, more specifically the advantages of the learned Pareto front of non-dominated learned heuristics. In Fig. 10.13, we show the *aggregate Pareto front* including the non-dominated learned heuristics extracted from Pareto fronts generated by all MOGP runs (based on the traditional Pareto dominance concept) in

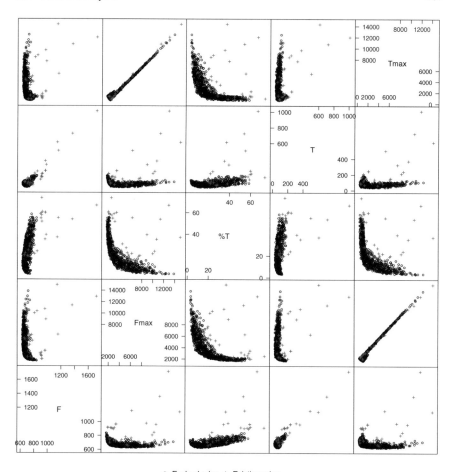

o Evolved rules + Existing rules

Fig. 10.13 Distribution of heuristics on the learned Pareto front for the scenario ⟨25, 85, 4⟩

the scenario with the shop condition ⟨25, 85, 4⟩. This figure is a scatter plot matrix that contains all the pairwise scatter plots of the five objectives (the two scatter plots that are symmetric with respect to the diagonal are similar except that the two axes are interchanged). The objective values obtained by 31 existing heuristics are also plotted in this figure (as +).

The first observation is that the Pareto front can cover a much wider range of potential non-dominated heuristics compared to heuristics that have been discovered in the literature. The figure not only shows that the learned heuristics can dominate the existing heuristics, but the Pareto front of learned heuristics also helps to understand better the possible trade-offs in this scenario. For example, it can be seen that the percentage of tardy jobs %T can be substantially reduced with only minor deterioration on other objectives. Obviously, this insight cannot be obtained with the available priority heuristics since these heuristics only suggest that other objectives will deteriorate significantly when we try to reduce %T below 20%. However, we can see from the Pareto front that it is possible to reduce %T further to 10% without major deteriorations in other objectives. In fact, F and T will not be affected when we try to reduce %T to a level above 10%. When we try to reduce %T below 10%, F_{max} and T_{max} will be greatly deteriorated. In this scenario, we also see that there is a strong correlation between F_{max} and T_{max} when the values are high and the trade-offs between these two objectives are only obvious when they reach their lowest values. This makes sense since high values of F_{max} and T_{max} are caused by some extreme cases. Thus, as long as these extreme cases are handled well, both F_{max} and T_{max} can also be reduced. This observation also suggests that focussing on one of them should be enough if these two objectives are not very important.

This visualisation shows that decision-makers can benefit greatly from the Pareto front found by the proposed MOGP method. For DJSS problems, the ability to understand all possible trade-offs is very important since many aspects need to be considered when a decision needs to be made. Without the knowledge from these trade-offs, the decisions will be too extreme (only focus on a specific objective) and they can be practically unreasonable sometimes (e.g., double the maximum tardiness just for reducing %T by 1%). Moreover, the decision-makers do not need to decide their preferences on the objectives before the design process, which could be quite subjective in most cases.

10.4.2   Robustness of Learned Heuristics

It has been shown that the learned Pareto fronts contain very competitive heuristics. In this section, we will investigate the robustness of the learned heuristics, which is their ability to maintain their performance across different simulation scenarios. In the single objective problem, the robustness of the learned heuristics can be easily examined by measuring and comparing the performance of the heuristics on different scenarios. However, the assessment of the robustness of the learned heuristics is not trivial in the case of multi-objective problems because the robustness of heuristics depends not only on the values of all the objectives, but also on the Pareto dominance relations of the heuristics. Unfortunately, there has been no standard method to measure the robustness of the learned heuristics for the multi-objective problems. Therefore, we develop a method to help roughly estimate the robustness of the learned heuristics. In this work, the robustness of a heuristic $\mathcal{H}_i$ will be calculated as follows:

$$robustness_i = 1 - \frac{\sum_{s \in \mathbb{S}} Hamming\_Distance(dom_{is}, dom_{is}^*)}{|\mathbb{S}| \times |\mathbb{B}|} \qquad (10.2)$$

where $dom_{is} = \{d_{is1}, \ldots, d_{isj}, \ldots, d_{is|\mathbb{B}|}\}$ is a binary array which stores the Pareto dominance between $\mathcal{H}_i$ and each heuristic $\mathcal{B}_j$ in the set $\mathbb{B}$ of reference heuristics. In a simulation scenario $s \in \mathbb{S}$ (12 test scenarios in our work), d_{isj} is assigned 1 when

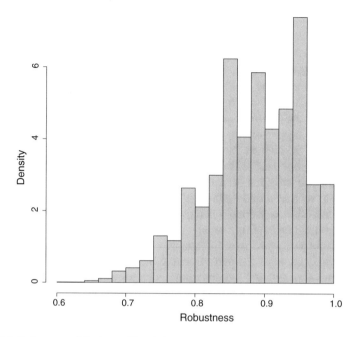

Fig. 10.14 Robustness of the learned heuristics

$\mathcal{H}_i$ statistically dominates $\mathcal{B}_j$, and 0 otherwise. Here, we include in $\mathbb{B}$ ten benchmark heuristics that are most competitive in Figs. 10.9 and 10.10 (FDD, Slack/OPN, OPFSLK/PT, SPT+PW+FDD, PT+WINQ+NPT+WSL, PT+WINQ+SL, RR, 2PT+WINQ+NPT, COVERT, and SL). Meanwhile, dom^*_{is} is also a binary array which contains the most frequent value of d_{isj} across all $s \in \mathbb{S}$. The second term in Eq. (10.2) measures the average Hamming distance per dimension between dom_{is} and dom^*_{is}. From this calculation, if the Pareto dominance relations between $\mathcal{H}_i$ and each heuristic $\mathcal{B}_j$ are consistent across all $s \in \mathbb{S}$, this term will be zero and the robustness is one. In the worst case, when the Pareto dominance relations are greatly different for each scenario s, the second term in Eq. (10.2) will approach 1 and the robustness will be near zero.

A histogram of robustness values of all learned heuristics obtained by 30 MOGP runs is shown in Fig. 10.14. The density in this figure is the number of heuristics with robustness within the corresponding range (bin) of the histogram. It is clear that the distribution of robustness values is skewed to the right, which indicates that the learned heuristics are reasonably robust. The majority of the heuristics have robustness values from 0.8 to 0.95 and there is only a small proportion of learned heuristics with small robustness. This result is consistent with our observation in Sect. 10.3.2 that a small number of learned heuristics that do not perform well on unseen scenarios can be dominated by the benchmark heuristics, in which case their Pareto dominance relations are changed.

10.5   Chapter Summary

Most of the heuristics for DJSS problems proposed in the literature are designed for minimising a specific objective. However, the choice of suitable heuristics depends on the performance of the heuristic across multiple conflicting objectives. In this chapter, we show how GP can handle this challenge. The proposed MOGP method aims at exploring the Pareto front of learned heuristics which can be used to support the decision-making process. Extensive experiments have been performed and the results show that the learned Pareto front contains superior heuristics as compared with heuristics reported in the literature when both single and multiple objectives are considered. Moreover, it has been shown that the obtained Pareto front can provide valuable insights on how trade-offs should be made.

We have also discussed and implemented different analyses on the experimental results, which help us confirm the effectiveness of the learned heuristics. In these analyses, we focus on two issues. First, we try to define a standard procedure in order to properly compare the performance of heuristics within the multi-objective stochastic environments. Second, we need to find a way to assess the robustness of the learned heuristics under different simulation scenarios. Even though there are still some limitations with these approaches, they can nevertheless be used as a good way to assess the performance of heuristics in such a complicated problem.

It is noted that conflicting objectives may come from different scheduling decisions made by different parts of the organisation and potential heuristics can be found without taking these decisions into account. The next chapter will extend the work in this chapter to investigate the dynamic multi-objective job shop scheduling problems in which due date assignment and job sequencing must be handled simultaneously.

Chapter 11
Cooperative Coevolution for Multi-objective Production Scheduling Problems

Real-world production scheduling usually involves a wide range of decisions. However, most studies on production scheduling only consider one of the many decisions and fix the others to ensure that the problems are solvable. This approach is valid when these decisions are independent, which is often not the case for real-world applications. Although production scheduling has been an active research topic for decades and investigation of the interactions among various decisions is essential for the development of effective and comprehensive scheduling systems, studies on the interactions among different scheduling decisions are limited. The studies [11, 47, 150, 195] mainly examined the performance of simple combinations of different existing dispatching rules (DRs) and due-date assignment rules (DDARs). One of the reasons for the lack of research in this direction is that dealing with each scheduling decision is already difficult, and thus considering multiple scheduling decisions simultaneously will be even more complicated.

11.1 Challenges and Motivations

In this chapter, we will explore how GP is used for the automatic design of scheduling heuristics which include sequencing rules and due date assignment rules for dynamic JSS (DJSS) problems. GP is a suitable method for designing such heuristics because of its flexibility to encode different scheduling rules in the representation. Moreover, as an evolutionary approach, GP can be applied to handle the multiple conflicting objectives of JSS problems, as shown in the previous chapter. Another advantage of GP is that learned scheduling heuristics are potentially interpretable, which is important and useful for understanding how the problem is solved by the learned heuristics and how the trade-offs among the different objectives of JSS can be obtained.

© The Author(s), under exclusive license to Springer Nature Singapore Pte Ltd. 2021
F. Zhang et al., *Genetic Programming for Production Scheduling*, Machine Learning:
Foundations, Methodologies, and Applications,
https://doi.org/10.1007/978-981-16-4859-5_11

Designing scheduling heuristics as described above is challenging, even with an automated approach such as GP because the search space is much larger than those explored in the previous chapter. This chapter will discuss the complexity of this problem and introduce a multi-objective cooperative coevolution algorithm to overcome the challenges of handling two complex decisions simultaneously. Specifically, this chapter will address the following questions:

- How to represent scheduling heuristics with multiple scheduling decisions?
- How to evaluate evolved scheduling heuristics with multiple scheduling decisions?
- How is cooperative coevolution used to improve the efficiency of GP for this complex design task?

Section 11.2 describes the proposed MOGP methods and settings for our experiments. The experimental results and comparison of the learned scheduling heuristics and the existing scheduling heuristics are presented in Sect. 11.3. The analysis of the proposed methods and the learned scheduling heuristics are shown in Sect. 11.3.4. Conclusions are drawn in Sect. 11.4.

11.2   Algorithm Design and Details

This section describes the new GP algorithm for evolving scheduling heuristics, which include rules for due date assignment and sequencing decisions in dynamic job shop environments. We will first show how DDARs and DRs are represented by individuals in GP. Then, new MOGP methods based on NSGA-II, SPEA2, HaD-MOEA and a proposed cooperative coevolution method are applied to deal with the dynamic JSS problems. These MOGP methods are different in the way that the Pareto fronts of non-dominated scheduling heuristics are explored. Lastly, the job shop simulation model used for the training and test will be described.

11.2.1   Representations

As investigated in previous chapters, there are several ways to represent scheduling rules (either DDARs or DRs). For due date assignment, we employ an operation-based estimation approach which iteratively estimates the operation flow times of operations belonging to a job. The estimator is represented by an expression tree by GP, which estimates the job flowtime by iteratively predicting the operation flow times of operations that belong to that job. The terminals used to construct DDARs are shown in Table 11.1. Figure 11.1 shows when DDARs are activated and illustrates how the job flowtime is estimated with learned DDARs. It is noted that estimating job flow time is similar to symbolic regression, in which the goal is to discover a function that can predict a target with minimal errors.

Table 11.1 Terminal sets for due date assignment rules (σ is the considered operation, and δ is the machine that process σ)

Terminal	Description
N	Number of jobs in the shop
SAR	Sampled arrival rate
APR	Average processing times of jobs in queue of the machine that processes σ
OT	Processing time of σ
LOT	Time for δ to finish the leftover job
OTR	Percentage of jobs in queues of δ require less processing time less than OT
SOTR	Percentage of sampled jobs processed at δ that require less processing time less than OT
QWL	Total processing time of jobs in queue of δ
SAPR	Sampled average processing time of jobs processed at δ
RWL	Total processing time of jobs that need to be processed at δ
W	Weight of the considered job
PEF	Partial estimated flowtime
#	Random number from 0 to 1

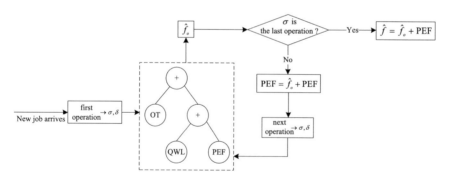

Fig. 11.1 Operation-based DDAR

Regarding DRs, we use the simple arithmetic representation to construct composite dispatching rules (CDRs), similar to those introduced in the earlier chapter. The terminal set used to evolve CDRs in this chapter is shown in Table 11.2. Figure 11.2 shows the moment at which DRs are activated and illustrates how decisions are made with these learned rules.

Table 11.2 Terminal set for composite dispatching rules

Terminal	Description
rJ	Job release time (arrival time)
RJ	Operation ready time
RO	Number of remaining operation of the job
RT	Work remaining of the job
PR	Operation processing time
W	Weight of the job
DD	Due date of the job
RM	Machine ready time
SJ	Slack of the job $= DD - (t + RT)$
#	Random number from 0 to 1
WR	Workload ratio $= \frac{\sum_{\sigma \in \Omega} p(\sigma)}{\sum_{\sigma \in I} p(\sigma)}$
MP	Machine progress $= \frac{\sum_{\sigma \in K} p(\sigma)}{\sum_{\sigma \in \Lambda} p(\sigma)}$
DJ	Deviation of jobs in queue $= \frac{\min_{\sigma \in \Omega}\{p(\sigma)\}}{\max_{\sigma \in \Omega}\{p(\sigma)\}}$
CWR	Critical workload ratio $= \frac{\sum_{\sigma \in \Omega^c} p(\sigma)}{\sum_{\sigma \in \Omega} p(\sigma)}$
CWI	Critical machine idleness, WR of the critical machine
BWR	Bottleneck workload ratio $= \frac{\sum_{\sigma \in \Omega^b} p(\sigma)}{\sum_{\sigma \in \Omega} p(\sigma)}$

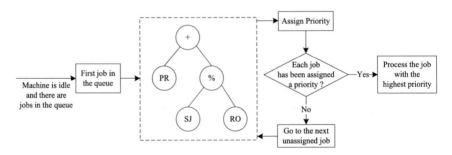

Fig. 11.2 Example of an evolved dispatching rule

11.2.2 Cooperative Coevolution MOGP for Dynamic Job Shop Scheduling

As discussed earlier, this work aims to evolve scheduling heuristics that include two key components, i.e., due date assignment rules and dispatching rules. While the representation of the rules has been discussed previously, we need to specify how these rules are evolved in our proposed GP algorithms. In this work, two approaches are examined. In the first approach, a GP individual contains two GP trees for the two rules as presented above. In this case, each individual is equivalent to a scheduling heuristic. The scheduling heuristic is evaluated by applying the first tree as a DDAR when a new job arrives at the job shop to assign a due date to that job. Meanwhile, the second tree is applied when a machine becomes idle, and there are jobs in the queue of that machine to find the next job to be processed. For this approach, we apply NSGA-II [58], SPEA2 [254], and HaD-MOEA [223] to explore the Pareto front of non-dominated scheduling heuristics.

The second approach to evolving scheduling heuristics is to employ cooperative coevolution [76, 77, 211] to evolve two decision rules in two subpopulations. This approach is similar to the cooperative coevolution framework proposed by Potter and de Jong [193], in which the scheduling heuristic is the combination of an individual in a subpopulation with a *representative* from the other subpopulation, and some specialised operations are also employed here to help explore the Pareto front of the scheduling heuristics. We introduce a new diversified multi-objective cooperative coevolution (DMOCC) method based on the second approach. An overview of the proposed DMOCC is shown in Fig. 11.3. Here each subpopulation (P_1 for DDARs and P_2 for DRs) represents one rule of the complete scheduling heuristic. For each individual $p_i^r \in P_r$, the objective values which determine the quality (fitness) of p_i^r are obtained by combining that individual with a representative from the other population to form a complete scheduling heuristic $\mathcal{H}$. When a complete scheduling heuristic is applied to the job shop, the quality of that heuristic is characterised by the expected values of three performance measures: (1) makespan (C_{max}) [185]; (2) total weighted tardiness (TWT) [185]; and (3) mean absolute percentage error ($MPEA$) [19] (see Table 11.5). C_{max} and TWT are two popular performance measures for evaluating dispatching rules or scheduling methods while $MPEA$ is used to indicate the accuracy of the due date assignment rules. In this work, the scheduling heuristics are evolved such that these three performance measures are minimised.

Within DMOCC, we use the *crowding distance* (individuals with higher crowding distances are located in less crowded areas of the objective space) and *non-dominated rank* [58] (individuals with the same *rank* are not dominated by each other and dominated by at least one individual with a smaller *rank*) to select GP individuals for genetic operations and for collaboration between the two subpopulations. *Representatives* for collaboration are selected based on a *binary tournament selection* method [58], which randomly selects two individuals and the one with a lower non-dominated rank will be chosen. In the case that two individuals have the same rank, the individual with a higher crowding distance will be selected. The binary tournament selection is employed in DMOCC because it takes into account both the quality of the non-dominated individuals and their spread/distribution.

An external archive A is employed in this method to store the non-dominated scheduling heuristics. After all individuals have been evaluated, a set of non-dominated scheduling heuristics are extracted from individuals in the two subpopulations and the current archive to form a new archive. Besides storing the non-dominated scheduling heuristics, the archive in DMOCC is also used for two other purposes. First, it is used to evaluate the quality (rank and crowding distance) of the learned scheduling heuristics in the two subpopulations. Instead of evaluating the quality of the learned rules independently in each subpopulation, it is better to assess

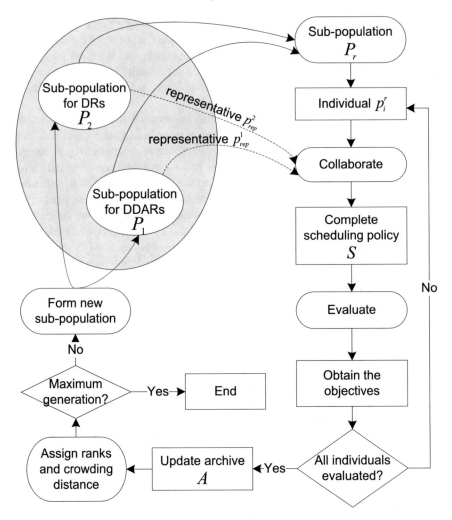

Fig. 11.3 Overview of the proposed algorithm

their quality based on comparisons with those in the archive and the other subpopulation to identify other potential non-dominated scheduling heuristics. Secondly, the archive can provide genetic materials which are needed for the crossover operation (more details are provided in Sect. 11.2.3).

Different from NSGA-II, SPEA2, and HaD-MOEA, the size of the archive in DMOCC is not fixed, although the number of complete scheduling heuristics stored in the archive cannot exceed a predefined maximum size. When the number of non-dominated scheduling heuristics extracted from a generation is more than the maximum size, only individuals with the highest crowding distance will be preserved in the archive. Since new individuals will be created from parents in the archive through

```
1: initialise each sub-population P_r with r = {1, 2} P_r ← {p_1^r, p_2^r, ..., p_N^r}
2: A ← {}
3: while maxGeneration is not reached do
4:     for r = 1 → 2 do
5:         for i = 1 → N do
6:             H ← collaborate(p_i^r, p_rep^r') where r' ≠ r
7:             p_i^r.objectives ← evaluate(H)
8:         end for
9:     end for
10:    A ← update(A, P_1, P_2)
11:    assign ranks and crowding distance
12:    for r = 1 → 2 do
13:        P_r ← genetic_operations(P_r, A)
14:    end for
15: end while
16: return A
```

Fig. 11.4 DMOCC algorithm

crossover, such a dynamic archive will help focus the search towards non-dominated scheduling heuristics at the early stage of the evolution. When the number of individuals in the archive increases, the shape of the Pareto front will be characterised and the method will focus on distributing the individuals uniformly.

The pseudo-code of DMOCC is shown in Fig. 11.4. The algorithm starts by populating the two subpopulations P_1 and P_2 with randomly generated individuals. In each generation, each individual p_i^r of the two populations collaborates with the representative $p_{rep}^{r'}$ from the other subpopulation to create a complete scheduling heuristic H. Then, the objective values of p_i^r are obtained by applying S to the simulated job shop. When all individuals have been evaluated, the archive A will be updated. Ranks and crowding distances are then assigned to individuals in A, P_1, and P_2. Here, new subpopulations are generated by genetic operations and the algorithm starts a new generation if the maximum generation is not reached.

11.2.3 Genetic Operators

Traditional genetic operators are employed by the proposed MOGP methods. For crossover, GP uses the subtree crossover [115], which creates new individuals for the next generation by randomly recombining subtrees from two selected parents. SPEA2 uses tournament selection to select parents in the population with the highest fitness value in the tournament. NSGA-II and HaD-MOEA use binary tournament selection based on rank and crowding distance as explained in the previous section. For DMOCC, binary tournament selection is used to select one parent from a subpopulation and one parent from the archive. Since individuals in the archive have a rank of zero, the selection is made only based on the crowding distance in order to

Table 11.3 Parameter settings for MOGP algorithms

Parameter	Value
Initialisation	Ramped-half-and-half
Crossover rate	90%
Mutation rate	10%
Maximum depth	8
Number of generations	100
Population size	200 for NSGA-II, SPEA2, and HaD-MOEA, and 100 for each subpopulation of DMOCC

direct the search to less crowded areas. Here, mutation is performed by the subtree mutation [115], which randomly selects a node of a chosen individual in the population and replaces the subtree rooted at that node by a new randomly generated subtree. For NSGA-II, SPEA2, and HaD-MOEA, the genetic operations will first randomly choose which tree (either DR or DDAR) of the parents to perform the operations on since each individual includes two trees for the two scheduling rules. If the crossover is applied, only genetic materials from the selected tree of the same type will be exchanged (e.g., a tree representing DR in one parent will only replaced with a tree representing DR of the other parent).

11.2.4 Parameters

Table 11.3 shows the parameters used by the four proposed MOGP methods. SPEA2 applied tournament selection with a tournament size of 5 to select individuals for the genetic operations. NSGA-II, SPEA2, and HaD-MOEA used a population size of 200 while DMOCC used a population size of 100 for each subpopulation to ensure that the number of program evaluations remains the same for all methods. The archive size of SPEA2 and *maximum-size* of DMOCC are set to 200. These settings are used so that the proposed methods will give the same number of non-dominated scheduling heuristics at the end of each run.

11.2.5 Job Shop Simulation Model

The same simulation model of DJSS in the previous chapter is employed here for training and test purposes. Table 11.4 presents all the training and test simulation scenarios in our experiments. In each replication of a simulation scenario, we start with an empty shop and the interval from the beginning of the simulation until the arrival of the 1000th job is considered as the warm-up time and the statistics from the

Table 11.4   Scenarios for training and test

Factor	Training	Test
Number of machines	4, 6	5, 10, 20
Utilisation	80%, 90%	70%, 80%, 90%, 95%
Distribution of processing time	Exponential	Exponential, uniform
Distribution of # of operations	Missing	Missing, full

Table 11.5   Performance measures of scheduling heuristics

Metric	Calculation
Makespan	$C_{\max} = \max_{j \in \mathbb{C}}\{f_j\}$
Normalised total weighted tardiness	$TWT = \dfrac{\sum_{j \in \mathbb{C}} w_j T_j}{|\mathbb{C}| \times M \times \frac{1}{\mu} \times \bar{w}}$
Mean absolute percentage error	$MAPE = \frac{1}{|\mathbb{C}|} \sum_{j \in \mathbb{C}} \frac{|e_j|}{f_j}$

next completed 5000 jobs (set $\mathbb{C}$) are recorded to evaluate the performance measures of the scheduling heuristics as shown in Table 11.5. In this table, M is the number of machines in the shop, $\bar{w}$ is the average weight, and $\frac{1}{\mu}$ is the average processing time of an operation. The average values of these performance measures across different simulation scenarios/replications are the objectives to be minimised by the proposed MOGP methods.

In the training stage, due to the heavy computation time, we only perform one replication for each scenario. In Table 11.4, there are $(2 \times 2 \times 1 \times 1) = 4$ simulation scenarios used for evaluating the performance of the learned scheduling heuristics. It should be noted that the performance measures are obtained for each scenario by applying the learned scheduling heuristics thousands of times since there are thousands of due date assignment and sequencing decisions needed to be made during a simulation replication of that scenario. During the test stage, each of the non-dominated scheduling heuristics from a GP run is applied to $(3 \times 4 \times 2 \times 2) = 48$ simulation scenarios (see Table 11.4) and 5 simulation replications are performed for each scenario; therefore, we perform $48 \times 5 = 240$ simulation replications for test the performance of the obtained non-dominated scheduling heuristics. The test scenarios are selected to help us assess the performance of learned rules on unseen cases.

11.2.6   Performance Measures for MOGP Methods

Similar to other multi-objective optimisation applications, we are interested in the quality of the obtained Pareto fronts in terms of (1) *convergence* to the trade-off solutions and (2) the *spread* or *distribution* of the solutions on the obtained Pareto front. Three popular performance metrics for multi-objective optimisation are used here: hypervolume ratio (HVR) [219, 255]; SPREAD [58]; and generational distance (GD) [8].

11.2.6.1 Hypervolume (HV) and Hypervolume Ratio (HVR)

Hypervolume is used to measure the "volume" in the objective space covered by the obtained non-dominated solutions for minimisation problems,

$$HV = volume(\bigcup_{i=1}^{n_{PF}} v_i) \tag{11.1}$$

where n_{PF} is the number of members in the obtained Pareto front PF_{known}, v_i is the hypercube constructed with a reference point and the member i as the diagonal of the hypercube [255]. van Veldhuizen and Lamont [219] normalised HV by using the hypervolume ratio which is the ratio of the hypervolume of PF_{known} and the hypervolume of the reference Pareto front PF_{ref},

$$HVR = \frac{HV(PF_{known})}{HV(PF_{ref})} \tag{11.2}$$

The Pareto fronts found with larger HVRs are preferred as they contain quality and widespread solutions.

11.2.6.2 SPREAD

This metric measures the non-uniformity of PF_{known} [58]. A widely and uniformly spread out set of non-dominated solutions in the PF_{known} will result in a small $SPREAD$.

$$SPREAD = \frac{d_f + d_l + \sum_{i=1}^{n_{PF}-1} |d_i' - \bar{d}|}{d_f + d_l + (n_{PF} - 1)\bar{d}} \tag{11.3}$$

where d_i' is the Euclidean distance between member i and its nearest member in PF_{known}, $\bar{d}$ is the average of all d_i', and d_f and d_l are the Euclidean distances between the extreme solutions and the boundary solutions of PF_{known}.

11.2.6.3 Generational Distance (GD)

This metric is used to measure the distance between the obtained Pareto front (PF_{known}) and the reference Pareto front (PF_{ref}) [8],

$$GD = \left(\frac{1}{n_{PF}} \sum_{i=1}^{n_{PF}} d_i^2 \right)^{1/2} \tag{11.4}$$

where d_i is the Euclidean distance between the member i in PF_{known} and its nearest member in PF_{ref}.

PF_{ref} is normally the true Pareto front, which is unknown in the simulation here. Therefore, we need to adopt a reference Pareto front PF_{ref} in the calculation of these performance metrics. In this work, PF_{ref} includes the non-dominated scheduling heuristics extracted from all scheduling heuristics found by the four MOGP methods (NSGA-II, SPEA2, HaD-MOEA, and DMOCC) in all independent runs. Basically, the learned scheduling heuristics from 4 (methods) × 30 (runs) = 120 Pareto fronts are combined into a common pool, and the non-dominated sorting technique (as employed in NSGA-II) is used to find the non-dominated scheduling heuristics from this pool.

11.3 Empirical Study

In order to evaluate the effectiveness of the proposed methods, 30 independent runs of each MOGP method are performed and the learned Pareto fronts obtained from each run are recorded. The learned non-dominated scheduling heuristics are then compared with the existing heuristics based on combinations of well-known dispatching rules with dynamic and regression-based due date assignment rules.

11.3.1 Pareto Front of Learned Scheduling Heuristics

Figure 11.5a, b show the *aggregate Pareto fronts* extracted from all the learned scheduling heuristics, which were obtained by the four proposed MOGP methods in 30 independent runs for both the training and test scenarios. It can be observed that the three objectives C_{max}, TWT, and $MAPE$ are conflicting objectives. When tracing along the Pareto front to find heuristics that are able to minimise C_{max} and TWT, it can be seen that the value of $MAPE$ tends to be increased. This suggests that scheduling heuristics that provide better shop performance (small C_{max} and TWT) will result in flow times that are hard to predict accurately (large $MAPE$). Given a similar value of $MAPE$, the trade-off between C_{max} and TWT can also be observed, which suggests that there is no learned dispatching rule that can simultaneously optimise these objectives. Such an observation is consistent with those discussed in the literature [101, 213].

In both the test and training scenarios, it is observed that C_{max} and TWT can be significantly reduced by using heuristics with $MAPE$ smaller than 0.5. The use of more sophisticated heuristics can provide slightly better C_{max} and TWT but they

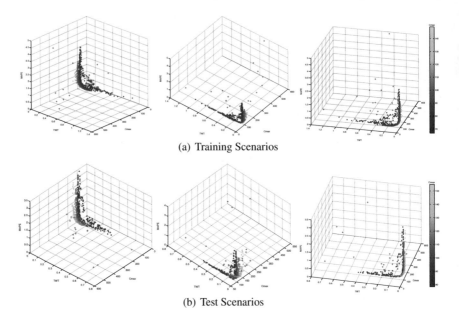

(a) Training Scenarios

(b) Test Scenarios

Fig. 11.5 Pareto front of non-dominated scheduling heuristics. In these figures, • and +, respectively, represent learned and existing scheduling heuristics as shown in Sect. 11.3.2

also make the job flow times much more difficult to be estimated. These results show that there are many trade-offs to be considered when selecting an appropriate heuristic for a scheduling system and the knowledge about these trade-offs is useful in making a better decision. For example, the obtained Pareto fronts suggest that much better delivery reliability (a small $MAPE$) can be achieved with a reasonable sacrifice in C_{max} or TWT. However, if a single objective such as C_{max} or TWT is to be minimised in this case, the learned heuristics will lead to very poor delivery reliability (a high $MAPE$) and thus reduce the customer satisfaction. Also, given the shape of the Pareto fronts in Fig. 11.5, it would be difficult to apply a traditional linear combination of objective values for fitness assessment to find desirable rules due to the difficulty in identifying suitable weights for each objective. These observations show that handling multiple objectives with knowledge about their Pareto front is crucial for the design of effective heuristics.

11.3.2   Comparison to Heuristics Combining Existing DRs and Dynamic DDARs

The combination of six popular DRs and three dynamic DDARs is evaluated on both the training and test scenarios. The results (red +) are compared with the learned non-

Table 11.6 Performance of existing scheduling heuristics for *training* scenarios ($C_{\max}$, TWT, $MAPE$)

	DTWK	DPPW	ADRES
FIFO	(101.5, 1.25, 0.81)	(101.5, 0.71, 2.00)	(101.5, 0.36, 1.05)
CR	(174.6, 0.53, 0.73)	(127.4, 0.52, 2.47)	(178.0, 0.49, 1.33)
S/OPN	(156.9, 0.42, 0.57)	(114.0, 0.42, 1.63)	(123.9, 0.14, 1.68)
SPT	(575.8, 0.60, 0.67)	(575.8, 0.69, 1.45)	(575.8, 0.46, 4.96)
ATC	(476.9, 0.32, 0.58	(504.3, 0.36, 1.32)	(173.8, 0.06, 2.17)
COVERT	(301.6, 0.25, 0.40)	(362.9, 0.23, 1.00)	(145.4, 0.09, 0.96)

Table 11.7 Performance of existing scheduling heuristics for *test* scenarios ($C_{\max}$, TWT, $MAPE$)

	DTWK	DPPW	ADRES
FIFO	(149.7, 0.73, 0.35)	(149.7, 0.47, 0.80)	(149.7, 0.21, 0.60)
CR	(160.5, 0.19, 0.18)	(115.2, 0.18, 0.71)	(206.1, 0.03, 0.60)
S/OPN	(159.9, 0.20, 0.20)	(114.2, 0.19, 0.58)	(184.6, 0.02, 0.74)
SPT	(510.2, 0.68, 0.56)	(510.2, 0.74, 0.86)	(510.2, 0.45, 2.59)
ATC	(302.8, 0.19, 0.29)	(306.4, 0.19, 0.53)	(220.4, 0.01, 1.05)
COVERT	(242.8, 0.15, 0.21)	(237.9, 0.14, 0.46)	(152.2, 0.02, 0.55)

dominated heuristics as shown in Fig. 11.5. The six DRs used in this comparison are first-in-first-out (FIFO), critical ratio (CR), slack-per-operation (S/OPN), shortest processing time (SPT), weighted apparent tardiness cost (ATC), and weighted cost over time (COVERT). The parameters of ATC and COVERT are the same as those used in the literature ($k = 3$ for ATC, $k = 2$ for COVER, and the leadtime estimation parameter $b = 2$). The three dynamic DDARs are DTWK, DPPW [48], and ADRES [19]. These DDARs are selected for comparison because they are well-known in the scheduling literature and the application of these rules does not require predetermination of any parameter or coefficient for each simulation scenario. The objective values obtained by these 18 combinations for the training and test scenarios are shown in Tables 11.6 and 11.7 and are visualised as crosses in Fig. 11.5a, b.

Among these existing scheduling heuristics, the ones given with FIFO provide the best $C_{\max}$. The scheduling heuristics with DTWK provide the best $MAPE$, and the combination of ATC and ADRES achieves the best TWT. However, these existing scheduling heuristics are easily dominated by the learned scheduling heuristics from the *aggregate Pareto fronts* as shown in Fig. 11.5. Moreover, when compared with the non-dominated scheduling heuristics obtained by *each independent run* of NSGA-II, SPEA2, HaD-MOEA, and DMOCC, it can be observed from our experiments that these scheduling heuristics are dominated by at least one of the learned scheduling heuristics using the proposed methods in both the training and test scenarios. These results show that the non-dominated scheduling heuristics learned by the proposed

MOGP methods not only show good performance on the training scenarios but can also be effectively reused for unseen scenarios.

11.3.3   Comparison to Heuristics Combining Existing DRs and Regression-Based DDARs

We further examine the effectiveness of the learned heuristics by comparing them with existing DRs and regression-based DDARs. The four due date assignment models used here are TWK, NOP, JIQ, and JIS in combination with the six dispatching rules reported in the previous section.

Different from the dynamic DDARs, the coefficients of the employed models have to be determined by regression methods for each job shop setting. Figure 11.6 and Table 11.8 show the performance of these $(6 \times 4) = 24$ combinations and the *aggregate Pareto front* of the non-dominated scheduling heuristics for the case with the utilisation of 90%, 5 machines, *full* setting and processing times following an exponential distribution. In this case, the coefficients of the due date assignment models TWK, NOP, JIQ, and JIS were determined by using iterative multiple regression (IMR) [73]. The values shown in the figure are the average values of the three objectives obtained from 30 independent simulation replications.

Since this work deals with a dynamic JSS environment with stochastic factors (such as arrival process, processing time), we also examine the Pareto dominance of heuristics under uncertainty. We will utilise the concept of *statistical Pareto dominance* based on OBJW introduced in the previous chapter to help determine the Pareto dominance relation between two scheduling heuristics in this case.

The results show that each of the 24 existing heuristics considered here is statistically dominated (with a significant level $\alpha = 0.05$ and using the Bonferroni method [153] used to adjust the value of each individual statistical test) by at least one learned heuristic in the aggregate Pareto front, which is indicated as a *dominating learned*

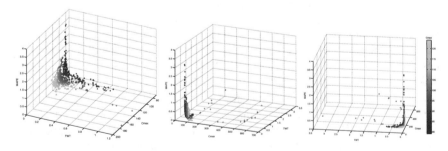

Fig. 11.6 DRs and Regression-based DDARs versus learned scheduling heuristics. (• and ▲ are, respectively, the non-dominated and dominating learned scheduling heuristics; and + represents existing scheduling heuristics)

Table 11.8 Performance of heuristics combining simple DRs and regression-based DDARs

Heuristic	$C_{\max}$	TWT	$MAPE$
FIFO+TWK	129.381 ± 27.26	3.231 ± 1.38	0.495 ± 0.05
FIFO+NOP	129.381 ± 27.26	1.933 ± 1.31	0.451 ± 0.09
FIFO+JIQ	129.381 ± 27.26	0.924 ± 0.09	0.178 ± 0.01
FIFO+JIS	129.381 ± 27.26	0.871 ± 0.13	0.178 ± 0.01
CR+TWK	183.169 ± 44.14	1.239 ± 1.06	0.291 ± 0.07
CR+NOP	139.275 ± 40.01	1.183 ± 0.94	0.310 ± 0.09
CR+JIQ	103.784 ± 15.15	0.406 ± 0.04	0.084 ± 0.01
CR+JIS	102.935 ± 16.73	0.384 ± 0.04	0.085 ± 0.01
S/OPN+TWK	156.279 ± 27.73	1.650 ± 1.36	0.440 ± 0.10
S/OPN+NOP	126.500 ± 29.73	1.698 ± 1.44	0.375 ± 0.10
S/OPN+JIQ	102.718 ± 15.44	0.383 ± 0.05	0.088 ± 0.08
S/OPN+JIS	101.404 ± 15.74	0.390 ± 0.07	0.089 ± 0.09
SPT+TWK	603.661 ± 235.69	0.996 ± 0.30	0.608 ± 0.03
SPT+NOP	603.661 ± 235.69	1.370 ± 0.33	0.844 ± 0.03
SPT+JIQ	603.661 ± 235.69	0.979 ± 0.29	0.632 ± 0.02
SPT+JIS	603.661 ± 235.69	0.990 ± 0.30	0.616 ± 0.02
ATC+TWK	600.539 ± 213.47	0.507 ± 0.19	0.481 ± 0.04
ATC+NOP	609.517 ± 200.21	0.702 ± 0.21	0.441 ± 0.03
ATC+JIQ	570.168 ± 201.18	0.409 ± 0.10	0.395 ± 0.02
ATC+JIS	548.237 ± 190.89	0.381 ± 0.08	0.362 ± 0.02
COVERT+TWK	511.948 ± 204.55	0.424 ± 0.18	0.225 ± 0.03
COVERT+NOP	540.604 ± 210.92	0.483 ± 0.19	0.233 ± 0.04
COVERT+JIQ	379.399 ± 141.83	0.258 ± 0.03	0.141 ± 0.02
COVERT+JIS	336.875 ± 116.26	0.237 ± 0.02	0.135 ± 0.02

heuristic in Fig. 11.6 and Table 11.8. This further shows the high quality of the learned heuristics even when they are compared with customised heuristics. Figure 11.6 also reveals that the combinations of existing DRs and DDARs do not cover all promising regions in the objective space. This observation suggests that automatic design methods like the proposed MOGP methods are essential in order to provide informed knowledge about any potential heuristics. Moreover, these results suggest that the learned heuristics are robust to uncertain JSS environments even though they are trained/evolved based on the mean values of the objectives across different simulation scenarios.

11.3.4   Further Analysis

The comparison results in the previous section have shown the effectiveness of the proposed MOGP methods for evolving efficient heuristics. In this section, we will compare the ability of the proposed MOGP methods in exploring the Pareto front of non-dominated heuristics.

11.3.4.1   Performance of MOGP Methods

The performance indicators of the four MOGP methods are shown in Figs. 11.7 and 11.8 (better methods have higher HVR and smaller SPREAD and GD). In the training scenarios, *Wilcoxon signed-rank tests* (with a significance level of 0.05) show that the HVRs obtained by DMOCC, NSGA-II, and HaD-MOEA are significantly better (higher) than that of SPEA2. This means that the heuristics obtained by these meth-

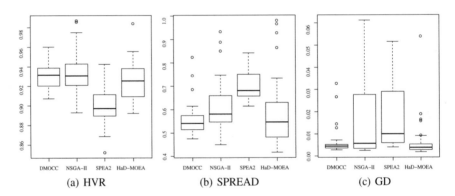

| (a) HVR | (b) SPREAD | (c) GD |

Fig. 11.7 Performance of MOGP methods on the *training* scenarios. (HVR to be maximised, and SPREAD and GD to be minimised)

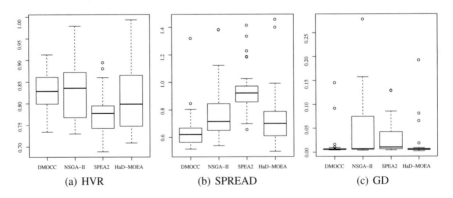

| (a) HVR | (b) SPREAD | (c) GD |

Fig. 11.8 Performance of MOGP methods on the *test* scenarios

ods can significantly dominate those obtained by SPEA2. In terms of HVR, there is no significant difference between DMOCC, NSGA-II, and HaD-MOEA but the standard deviations of HVRs obtained by DMOCC and HaD-MOEA are slightly smaller than those obtained by NSGA-II. For the distribution of the obtained heuristics on the Pareto fronts, the SPREAD values obtained by DMOCC and HaD-MOEA are significantly better than those obtained by NSGA-II and SPEA2. Although DMOCC uses *crowding distance* (like NSGA-II) as the indicator for individuals in less crowded areas, the selection method for choosing representative individuals as well as individuals for crossover has significantly improved the uniformity of the Pareto fronts obtained by DMOCC. Given a better distribution of scheduling heuristics, GD of DMOCC is significantly smaller than NSGA-II although there is no significant difference in HVR. Overall, DMOCC and HaD-MOEA are the two most competitive methods for the problems studied in this chapter. It should be noted that performances of the obtained non-dominated heuristics on the test scenarios are rather consistent with those obtained in the training scenarios. However, the SPREAD of DMOCC is significantly better than all the other methods. These experimental results show that the proposed DMOCC is a very promising approach for evolving highly efficient heuristics.

11.3.4.2 Complexity of the Proposed Algorithm

The complexity of DMOCC depends on the operations performed at each generation. Similar to NSGA-II, the three basic operations of DMOCC are (1) non-dominated sorting, (2) crowding-distance assignment, and (3) sorting for genetic and representative selection. For non-dominated sorting, we adopt the procedure proposed by Deb et al. [58], which results in the worst-case complexity of $O(MR^2)$ where M is the number of objective functions to be minimised and R is the size of the joined population. Assuming that the size of each subpopulation N is the same and the maximum size of the archive is A, the size of the joined population is $R = 2N + A$. The worst-case complexity of the crowding-distance assignment and sorting for genetic operators and representative selection are $O(MR \log(R))$ and $O(R \log(R))$, respectively. It is obvious that the complexity of the algorithm is $O(MR^2)$, governed by the non-dominated sorting procedure. Therefore, both the subpopulation size N and archive size A will influence the complexity of DMOCC. For complex problems where GP needs a large population size in order to maintain the diversity of the population, the complexity of NSGA-II ($O(MN^2)$) will increase since its complexity depends mainly on the population size. Because the number of final non-dominated solutions is not necessarily as large as the population, the complexity of DMOCC can be smaller than that of NSGA-II by maintaining a small archive.

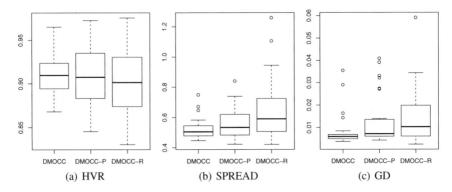

Fig. 11.9 Influence of representative selection methods on the *training* scenarios

11.3.4.3 Representative Selection

As mentioned earlier, representative selection is an important factor in the proposed cooperative coevolution method. Here, we will examine the influence of representative selection methods on the performance of the proposed DMOCC. Apart from the representative selection method discussed above, two other methods are also examined here. The first is a problem-based method that applies two different representative selection strategies for each subpopulation. In this method, the representatives of the subpopulation of DRs are selected by using a similar method to that in DMOCC (based on the non-dominated rank and crowding distance). On the other hand, the representatives of DDARs are selected based on the values of $MAPE$. This method assumes that good DDARs (with small values of $MAPE$) are able to cope with a wide range of DRs, and thus it will only focus on $MAPE$ when selecting representatives to form complete heuristics with the learned DRs. The second method simply selects random representatives from each subpopulation. The performances of the three representative selection methods are shown in Figs. 11.9 and 11.10. The DMOCC-P and DMOCC-R are similar to DMOCC, except that they employ problem-based and random representative selection methods, respectively.

The results from these figures show that the HVR values of the three selection methods are not significantly different. However, DMOCC gives significantly better SPREAD and GD performances as compared to DMOCC-P and DMOCC-R. Also, the DMOCC-P is better than DMOCC-R according to these two performance metrics. The results show that it is important to include the representative selection method based on the non-dominated rank and crowding distance. Although individuals selected for genetic operations for each subpopulation also employ the non-dominated rank and crowding distance, the features of non-dominated rank and crowding distance still have a strong impact on the performance of the representative selection method. Also, the learned DDARs with good $MAPE$ are not necessarily suitable for a wide range of DRs, since DMOCC-P does not produce Pareto fronts with the performance of SPREAD as good as DMOCC.

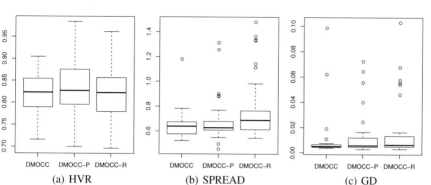

Fig. 11.10 Influence of representative selection methods on the *test* scenarios

11.3.4.4 Influence of Training Scenarios

Like other machine learning methods, it would be interesting to examine how the choices of training sets or training scenarios may influence the ability of the proposed MOGP in exploring effective scheduling heuristics. Previously, we have trained the proposed MOGP on scenarios with the *missing* setting of arriving jobs. This section will further examine the cases where *full* and *missing+full* settings are used. The first case used 4 simulation scenarios and the second case used 8 simulation scenarios for training. Figure 11.11 shows the performance of DMOCC on the test scenarios when different training scenarios have been used.

The results show that there is no significant difference between the cases where either *missing* or *full* setting is used. When both *missing* and *full* settings are used for training, the obtained HVRs are significantly better than those obtained in the cases where either *missing* or *full* setting is used and there are no significant differences

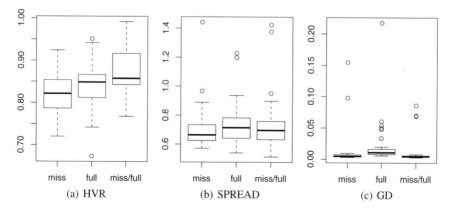

Fig. 11.11 Influence of training scenarios on test performance of DMOCC

in SPREAD and GD. This indicates that more general training scenarios are necessary in order to improve the quality of the learned scheduling heuristics. Although the simulation scenarios with jobs following the *missing* setting also include jobs following the *full* setting, it is still unable to cover all situations that happened in the simulation scenarios with jobs following the *full* setting. The major problem is that the use of a large number of simulation scenarios will increase the computation cost of the proposed methods. Thus, there is a trade-off between the computational effort and the reusability of the learned scheduling heuristics. Depending on the available computational resources and the environments where the learned scheduling heuristics will be applied, the training simulation scenarios should be logically selected.

11.3.4.5   Examples of Learned Scheduling Heuristics

This section investigates how the learned scheduling heuristics can effectively solve the problem and how trade-offs can be made among different objectives. Since many heuristics have been obtained from our experiments, we will present some examples of the heuristics. Figure 11.12 is the same as Fig. 11.5a with a different view and the points surrounded by rectangles are the example heuristics, which are also presented in Table 11.9.

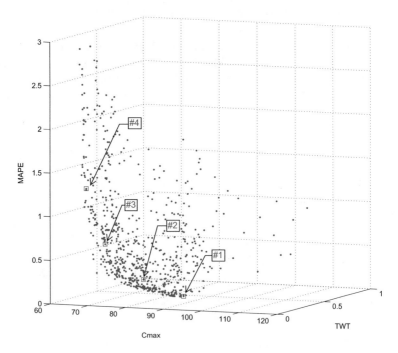

Fig. 11.12 Pareto front and selected learned scheduling heuristics

Table 11.9 Examples of the learned scheduling heuristics

Scheduling heuristic #1 ($C_{\max} = 90.774$, $TWT = 0.170$, $MAPE = 0.098$)	
DR:	$\frac{2RO^2MP}{(0.9087407 - RJ)} + IF(\max(BWR, WR) - 2SJ, -PR, -RJ)$
$DDAR$:	$OT + LOT + QWL$

Scheduling heuristic #2 ($C_{\max} = 83.597$, $TWT = 0.048$, $MAPE = 0.312$)	
DR:	$\max(-0.043989424, \frac{1}{PR}(\frac{2PR+RT}{PR} + \max(RT, PR) - SJ))$
$DDAR$:	$OT + LOT + QWL$

Scheduling heuristic #3 ($C_{\max} = 66.844$, $TWT = 0.311$, $MAPE = 0.608$)	
DR:	$RT\frac{RO}{PR} + IF(-CWR\frac{SJ}{PR}, DJ, RT(\frac{RO}{PR})^2) - 2rJ + W$
$DDAR$:	$OT + 2LOT + QWL$

Scheduling heuristic #4 ($C_{\max} = 68.162$, $TWT = 0.059$, $MAPE = 1.321$)	
DR:	$RT(2\frac{RO}{PR} - CWR) - 3rJ + W$
$DDAR$:	$OT + 2LOT + 2QWL + 2SOTR$

* These rules have been simplified for better presentation but still ensure to achieve the same objective values obtained by the original learned rules
$^* IF(a, b, c)$ will return b if $a \geq 0$; otherwise it will return c

In general, the learned rules are not very complicated and are in forms that are explainable, especially for the DDARs, which are simply linear combinations of different terms. Scheduling heuristic #1 is the one that achieves the best $MAPE$ among the four heuristics. Since scheduling heuristic #2 also employs the same DDAR, the better $MAPE$ obtained by the first SP is strongly influenced by its DR. The first component of the DR in Scheduling heuristic #1 will be negligible at the latter stage of the simulation since it has RJ in its denominator, which increases with the time. Therefore, this component has little impact on the performance of the scheduling heuristic and the performance of the rules will be governed by the second component. At first glance, the second component is a combination of both SPT and FIFO because the priorities of jobs are either $-PR$ (higher priority for jobs with shorter processing time) or $-RJ$ (higher priority for jobs arriving at the machine earlier). The switch between FIFO and SPT is controlled by $\max(BWR, WR) - 2SJ$. In the case that the slack of jobs SJ is positive (not late) and larger than $\frac{1}{2}\max(BWR, WR)$, FIFO will be applied; otherwise SPT will be used. The purpose of this rule is to maintain a more predictable flow (by FIFO) of jobs when the jobs are not late and to finish jobs with a shorter processing time first so as to reduce the number of jobs in the shop. These features make the flowtime prediction by $OT + LOT + QWL$ more accurate because it is important for a shop to have a smooth flow of jobs.

Scheduling heuristic #2 provides a better $C_{\max}$ and TWT as compared to the first SP. Different from the DR in the first SP, the DR in this SP emphasises more on reducing the duration that jobs stay in the shop as well as on reducing the lateness of jobs. In this case, the rule will give a higher priority to jobs with a higher RT in order to reduce its flowtime (and the makespan in general). In the case that jobs have large negative slacks, the rule will give a higher priority to jobs with larger negative slacks and reduce the lateness of jobs. However, because this rule will disturb the flow of jobs, the prediction will become less reliable.

Scheduling heuristics #3 and #4 are the two heuristics that provide the smallest makespans among the four. In general, the DRs in these two heuristics give a higher priority for jobs with a larger RT and RO and a shorter release time rJ in order to reduce the makespan. However, when these values are similar for the considered jobs, W is used to break the ties and gives a higher priority to jobs with higher weights. The focus on makespan has made the delivery performance of Scheduling heuristic #3 worse as compared to the first and the second heuristics. Scheduling heuristic #4 tries to minimise TWT by over-estimating the flow times for reducing the tardiness of jobs. The consequence is that the reliability of the due date deteriorates significantly.

From Scheduling heuristics #1 to #4, $MAPE$ tends to be increased and either C_{max} or TWT is reduced, especially for C_{max}. Tracking the learned heuristics along this direction helps explain how the trade-offs can be achieved, mainly between C_{max} and $MAPE$. When observing the learned heuristics on the Pareto fronts along with other directions, we are also able to explore other types of trade-offs, e.g., accepting a higher makespan for a better TWT. Since the learned heuristics are constructed based on the genetic materials of other heuristics or individuals, we can easily examine the connection among these heuristics and understand what creates the trade-offs. In other words, the learned DRs and DDARs are interpretable in this case.

11.4   Chapter Summary

Designing an effective scheduling heuristic is challenging and time-consuming because it needs to take into account multiple scheduling decisions and conflicting objectives in a production system. This chapter introduces a genetic programming approach for the automatic design of scheduling heuristics with multiple scheduling decisions. The empirical results show that the learned scheduling heuristics out-perform the existing scheduling heuristics created from combinations of popular dispatching and dynamic or regression-based due-date assignment rules on both the training and test scenarios. Moreover, the Pareto fronts obtained also provide much better knowledge about the space of potential scheduling rules, which cannot be achieved by simple combinations of existing scheduling rules or by methods using an aggregate objective function to handle multiple objectives. Another advantage of the proposed methods is that they perform well in unseen situations, which makes the learned scheduling heuristics more robust when they are employed in stochastic and dynamic job shops. Analysis of the learned scheduling heuristics also shows that the proposed methods can evolve not only effective but also very meaningful scheduling heuristics. In addition, it is easy to apply the proposed method to track the learned scheduling heuristics along the Pareto front for a better understanding of the trade-off among different objectives.

This chapter also shows that cooperative coevolution is a promising search strategy to discover effective and complex heuristics. The experimental results show that the proposed DMOCC method can evolve Pareto fronts that are better than NSGA-II and SPEA2. It is also very competitive on all performance metrics for the training scenarios and provides a better spread of the learned scheduling heuristics for the test scenarios. Further analysis also shows that the representative selection strategies based on non-dominated rank and crowding distance play a very important role in the proposed cooperative coevolution for evolving well-distributed Pareto fronts. Another advantage of the coevolution approach is that multiple scheduling decisions can be evolved in different subpopulations in order to reduce the complexity of evolving the sophisticated scheduling heuristics. This method can be modified to take advantage of parallelisation techniques so as to reduce the computational time.

The next chapter will further explore the application of GP in dynamic multi-objective flexible job shop scheduling, in which machine assignment and job sequencing need to be considered simultaneously.

Chapter 12
Learning Scheduling Heuristics for Multi-objective Dynamic Flexible Job Shop Scheduling

The studies in the last two chapters aim to learn scheduling heuristics for multi-objective DJSS. The investigations related to multi-objective DFJSS with machine assignment and operation sequencing are important for expanding the multi-objective GP approach. One may prefer to achieve a trade-off between different objectives with multiple decisions, i.e., routing and sequencing. Non-dominated sorting genetic algorithm II (NSGA-II) [58, 206] and strength Pareto evolutionary algorithm 2 (SPEA2) [255] are two common candidates for multi-objective methods for this purpose. However, the studies that use the strategies of NSGA-II and SPEA2 with GP for multi-objective DFJSS are not fully explored. This chapter will investigate how GP can learn scheduling heuristics for multi-objective DFJSS.

12.1 Challenges and Motivations

To the best of our knowledge, little is yet known about using multi-objective GP in DFJSS. Zhang et al. proposed to apply the strategies of NSGA-II and SPEA2 in GP to learn scheduling heuristics for DFJSS [242], which is the first piece of work related to multi-objective GP in DFJSS. The decision-making in DFJSS is more complex than that in DJSS, since the routing decision and the sequencing decision need to be made simultaneously. Since DFJSS involves multiple decisions and the behaviour of individuals considers both the routing rule and the sequencing rule, there are a number of challenges for multi-objective GP in DFJSS. First, the multi-objective GP algorithm with the strategies of NSGA-II and SPEA2 may not be effective for DFJSS, since each GP individual makes two decisions and the quality of individuals is affected by both the decisions. Second, the learned scheduling heuristics from the training instances may have different behaviour on unseen instances, which is more likely to happen in hyper-heuristic approaches, such as multi-objective GP. The goal of this chapter is to explore how the strategies of NSGA-II or SPEA2 can be used in multi-objective DFJSS. Specially, we are interested in the following questions:

- Is it possible to incorporate the strategies of NSGA-II or SPEA2 into multi-objective GP to learn scheduling heuristics for DFJSS?
- Which strategy is better for multi-objective GP in DFJSS, the strategy of NSGA-II or SPEA2?
- How is the consistency of rule behaviour between training and test?

Detailed descriptions of the proposed algorithm are given in Sect. 12.2. The experiment design is shown in Sect. 12.3, followed by results and discussions in Sect. 12.4. Finally, Sect. 12.5 concludes this chapter.

12.2   Algorithm Design and Details

This section will show how the strategies of NSGA-II and SPEA2 can be applied to GP to learn scheduling heuristics for DFJSS. The overall framework of the presented algorithms will be introduced firstly. Then, the details of the components of the proposed algorithm will be described. In this chapter, we will incorporate the strategies of NSGA-II and SPEA2 into GP with multi-tree representation. It is worth mentioning that multi-tree representation is used to optimise the routing rule and the sequencing rule simultaneously by assigning two rules (trees) to a GP individual.

12.2.1   Genetic Programming Multi-objective with NSGA-II Strategy

Figure 12.1 shows the flowchart of GP mutli-objective algorithm incorporated with the dominance relation strategies from NSGA-II. Each individual consists of a routing rule and a sequencing rule. All the individuals are evaluated at the beginning, and the individuals are then ranked based on the dominance relation and the sparsity value. Each individual will be assigned a rank and a sparsity value. A good individual is the individual with a small rank and a large sparsity value. A new population which combines the parent population and the newly generated offspring population is formed, and all the individuals in the new population are sorted according to the dominance relation and sparsity value. It is noted that the parent population from the previous generation needs to be evaluated again at the current generation, since we rotate the simulation at each generation in DFJSS. Only the original population size of individuals will be moved to the next generation. Finally, if the termination condition is met, the algorithm will output the Pareto front as the outcomes of the algorithm. Otherwise, the algorithm will continue to the next iteration of the evolutionary loop.

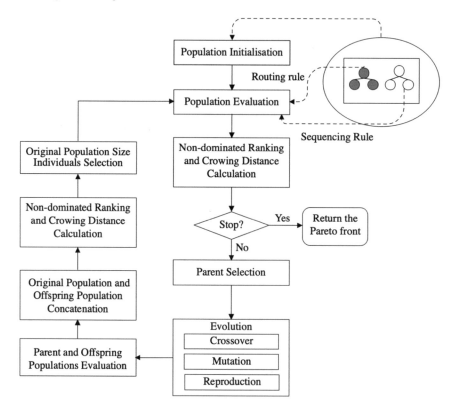

Fig. 12.1 The flowchart of multi-objective GP algorithm with NSGA-II strategy

12.2.2 Genetic Programming Multi-objective with SPEA2 Strategy

Figure 12.2 shows the flowchart of GP mutli-objective algorithm incorporated with the strategies from SPEA2. The main feature of SPEA2 is to use archive to keep good individuals. At the beginning, the population is initialised and the archive is empty. During evaluation, the individuals in both the population and the archive are evaluated. The non-dominated solutions in the population are copied to the archive. If the size of archive is smaller than the predefined size S, the non-dominated ones in the population will be added to the archive to fill the empty slots. If the size of the archive is larger than the predefined size, the individuals in the archive will be removed to keep the size S. If the stop criteria is met, the solutions in the archive will be the final output of the algorithm. Otherwise, the individuals in the archive will be used to generate offspring for the next population. Thus, the algorithm continues to the next iteration of the loop.

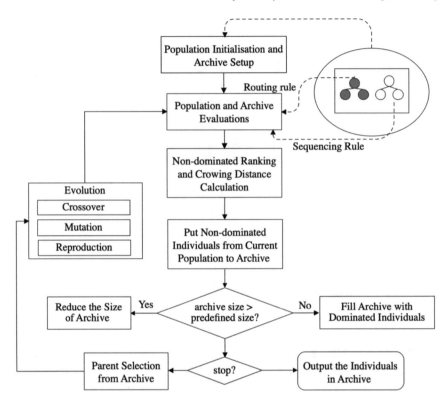

Fig. 12.2 The flowchart of multi-objective GP algorithm with SPGP2 strategy

12.3   Experiment Design

12.3.1   *Comparison Design*

The multi-objective GP algorithm with the strategy of NSGA-II is named NSMTGP-II, while the one with the strategy of SPEA2 is named SPMTGP2. To verify the performance of the presented algorithms NSMTGP-II and SPMTGP2, six different scenarios are designed based on three objectives (i.e., max-flowtime, mean-flowtime, and mean-weighted-flowtime) and two utilisation levels (i.e., 0.85 and 0.95). The examined multi-objective optimisation scenarios are shown in Table 12.1. For a more comprehensive comparison, we also compare NSMTGP-II and SPMTGP2 with the weighted sum based GP (i.e., named WMTGP). WMTGP simply combines the two objectives into a single one during the optimisation. In the experiment, we examine six weight vectors, i.e., (0, 1), (0.2, 0.8), (0.4, 0.6), (0.6, 0.4), (0.8, 0.2), and (1, 0), which correspond to different preferences between the objectives.

Table 12.1 The details of six multi-objective dynamic flexible job shop scheduling scenarios

Scenario	Objective 1	Objective 2	Utilisation level
1	Fmax	Fmean	0.85
2	Fmax	Fmean	0.95
3	Fmax	WFmean	0.85
4	Fmax	WFmean	0.95
5	Fmean	WFmean	0.85
6	Fmean	WFmean	0.95

12.3.2 Specialised Parameter Settings of Genetic Programming

The general parameter settings are the same as shown in Chap. 6. The number of individual evaluations at each generation are set to 1024 for NSMTGP-II, SPMTGP2, and WMTGP. The archive size of SPMTGP2 is set as 1024 (i.e., the same as the population size of SPMTGP2).

12.4 Results and Discussions

12.4.1 Quality of Learned Scheduling Heuristics

For multi-objective problems, two widely used performance indicators are hyper-volume (HV) [255] and inverted generational distance (IGD) [164]. HV is used to measure the Pareto front by taking the number of optimal solutions of Pareto front and its uniformity. IGD is applied to indicate the optimal degree (i.e., how close is the optimal solution to the true Pareto front) and coverage of the optimal solutions. Since the true Pareto front is unknown in this research, we will use a reference Pareto front instead. The reference Pareto front is generated by getting all the non-dominated dispatching rule found by three approaches (e.g., NSMTGP-II, SPMTGP2, and WMTGP with all weight vectors) in 50 independent runs. It is worth mentioning that the non-dominated dispatching rules of WMTGP are generated by combining the results of WMTGP with all weight vectors together. A larger HV value indicates a better performance, while a smaller IGD value indicates a better performance.

Table 12.2 shows the mean and standard deviation of HV of WMTGP, SPMTGP2, and NSMTGP-II in the training and test instances. For training, NSMTGP-II shows its superiority over SPMTGP2 and WMTGP in most of the scenarios, i.e., sceanrios 1–4 and 6. NSMTGP-II performs significantly better than both SPMTGP2 and WMTGP

Table 12.2 The mean and standard deviation of the **HV** of WMTGP, SPMTGP2, and NSMTGP-II over 50 independent runs in **training and test instances** in six dynamic flexible job shop scheduling scenarios

Sce.	Training			Test		
	WMTGP	SPMTGP2	NSMTGP-II	WMTGP	SPMTGP2	NSMTGP-II
1	0.787(0.061)	0.889(0.046)(−)	**0.923(0.041)***	0.833(0.032)	0.954(0.056)(−)	**0.991(0.050)***
2	0.505(0.050)	0.908(0.022)(−)	**0.928(0.019)***	0.602(0.049)	0.895(0.036)(−)	**0.919(0.028)***
3	0.723(0.079)	0.877(0.056)(−)	**0.912(0.042)***	0.350(0.049)	0.439(0.031)(−)	**0.458(0.021)***
4	0.494(0.062)	0.857(0.033)(−)	**0.879(0.020)***	0.536(0.045)	0.746(0.040)(−)	**0.772(0.042)***
5	0.904(0.054)(−)*	0.660(0.263)	**0.815(0.184)**	0.120(0.007)	0.109(0.143)(−)	**0.114(0.029)**
6	0.737(0.059)	0.696(0.144)	**0.761(0.109)***	0.189(0.014)	0.169(0.052)	0.204(0.121)

bold The significantly better approach between NSMTGP-II and SPMTGP2
* The significantly better approach between NSMTGP-II and WMTGP
− The significantly better approach between SPMTGP2 and WMTGP

Table 12.3 The mean and standard deviation of the **IGD** of WMTGP, SPMTGP2, and NSMTGP-II over 50 independent runs in **training and test instances** in six dynamic flexible job shop scheduling scenarios

Sce.	Training			Test		
	WMTGP	SPMTGP2	NSMTGP-II	WMTGP	SPMTGP2	NSMTGP-II
1	0.234(0.043)	0.057(0.027)(−)	**0.037(0.023)***	0.153(0.017)	0.088(0.027)(−)	**0.074(0.030)***
2	0.324(0.029)	0.044(0.013)(−)	**0.032(0.012)***	0.351(0.045)	0.055(0.022)(−)	**0.039(0.018)***
3	0.225(0.058)	0.060(0.031)(−)	**0.037(0.023)***	0.214(0.015)	0.053(0.025)(−)	**0.036(0.014)***
4	0.354(0.044)	0.060(0.025)(−)	**0.041(0.015)***	0.066(0.013)	0.035(0.009)(−)	0.032(0.08)*
5	0.130(0.035)(−)	0.305(0.233)	**0.177(0.146)**	0.999(0.015)(−)	1.053(0.185)	**1.013(0.072)**
6	0.144(0.055)	0.151(0.111)	**0.104(0.083)***	0.853(0.027)	0.902(0.092)	0.850(0.139)

bold The significantly better approach between NSMTGP-II and SPMTGP2
* The significantly better approach between NSMTGP-II and WMTGP
− The significantly better approach between SPMTGP2 and WMTGP

in most of the scenarios (i.e., from scenario 1 to scenario 4). In scenario 5, NSMTGP-II is significantly better than SPMTGP2, and SPMTGP2 is better than WMTGP.

Table 12.3 shows the mean and standard deviation of IGD of WMTGP, SPMTGP2, and NSMTGP-II in training and test instances. For training, we can see that NSNTGP-II achieve smaller IGD values than that in WMTGP and SPMTGP2 in all most of the scenarios, i.e., scenarios 1–4 and 6. In addition, NSMTGP-II get better performance than SPMTGP2 in scenario 5. For the test, we can see that NSMTGP-II and SPMTGP2 are significantly better than WMTGP in four scenarios (i.e., scenario 1–4). In scenario 5, NSMTGP-II is significantly better than SPMTGP2, and SPMTGP2 is significantly worse than WMTGP.

Overall, from the perspective of HV and IGD, NSMTGP-II is the most promising among the three approaches in most scenarios, which verifies its effectiveness.

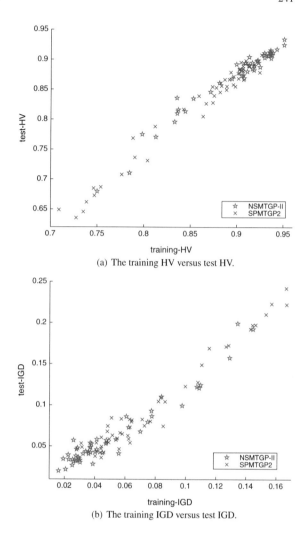

Fig. 12.3 The training HV and IGD versus test HV and IGD based on the 50 final results in the scenario 3 of NSMTGP-II and SPMTGP2

(a) The training HV versus test HV.

(b) The training IGD versus test IGD.

12.4.2   *Generalisation of Learned Scheduling Heuristics*

To show the generalisation of NSMTGP-II and SPMTGP2, Fig. 12.3 shows the scatter plot of the training HV (IGD) versus test HV (IGD) based on 50 final results of NSMTGP-II and SPMTGP2 in scenario 3. Figure 12.3 shows that both in the training and test, NSMTGP-II is much better than SPMTGP2 in terms of the HV and IGD. In addition, the generalisations of both algorithms are quite good in terms of the correlation between training and test HV and IGD.

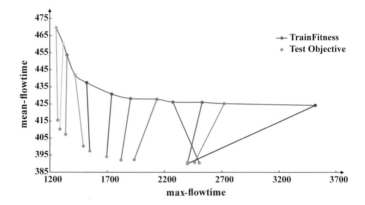

Fig. 12.4 An example of training fitness versus test objective of NSMTGP-II of scenario 1 in one of the independent runs

12.4.3   Consistency of Rule Behaviour

It is interesting to investigate the behaviour consistency of the learned scheduling heuristics from training to test. This is a very important issue when using GP as a hyper-heuristic to learn scheduling heuristics because the output of GP is a heuristic (dispatching rule) rather than a solution. The scheduling heuristics learned from the training may have different behaviour on test. The learned scheduling heuristics from training are expected to show consistent test performance in terms of the trade-off between the conflicting objectives.

Figure 12.4 shows the fitness of the Pareto front learned by NSMTGP-II versus their corresponding test objectives in one of the independent runs in the first scenario. We can see that most of the learned scheduling heuristics from the training have consistent behaviour on test. It indicates that the user can always expect consistent preference between the objectives on the unseen test data. We also find the same pattern of the scheduling heuristics learned by SPMTGP2.

In order to measure the behaviour consistency, we propose to use the *ratio difference* of training points and test points as a measure. For example, if the training fitness of a rule is (230, 100) and its test objective is (150, 50), the ratio difference is 0.7 ($|230/100 - 150/50|$). The ratio here is designed to show the different degrees (behaviours) of concentration of the evolved rules on different objectives. Thus, the ratio difference can be used to measure the behaviour difference between different rules. The minimum value of ratio difference is zero, which indicates the rules have the same behaviour. Given a set of rules, the mean value of the ratio differences is recorded. A smaller ratio difference indicates a better consistency.

Table 12.4 shows the mean and standard deviation of ratio difference obtained by NSMTGP-II and SPMTGP2 over 50 independent runs in six scenarios. We can see that there is no significant difference between NSMTGP-II and SPMTGP2. However, the mean and standard deviation values obtained by NSMTGP-II are smaller than

Table 12.4 The mean and standard deviation value of **ratio difference** between training fitness and test objective of NSMTGP-II, SPMTGP2 over 50 independent runs in six dynamic flexible job shop scheduling scenarios

Scenario	SPMTGP2	NSMTGP-II
1	0.317(0.143)	**0.303(0.088)**
2	2.461(1.350)	**2.082(0.710)**
3	0.171(0.110)	**0.151(0.071)**
4	1.858(0.985)	1.952(0.636)
5	0.008(0.001)	0.009(0.001)
6	0.035(0.006)	0.035(0.006)

that of SPMTGP2 in half scenarios. It means that the proposed methods can learn scheduling heuristics with good behaviour consistency, especially NSMTGP-II.

12.4.4 Insight of the Distribution of Learned Scheduling Heuristics

The learned Pareto fronts (i.e., non-dominated solutions generated by the optimal solutions of 50 independent rules) by SPMTGP2, NSMTGP-II, and WMTGP from scenario 1 to scenario 6 are shown in Fig. 12.5. In Fig. 12.5, for weights setting, the first (second) value is the weight for the objective indicated on the x-axis (y-axis). It is obvious that the learned Pareto front by NSMTGP-II is significantly better than SPMTGP2 and WMTGP (i.e., with different set of weights). It is interesting that the weighted sum method gets a good scheduling heuristics with the weight (0.6, 0.4) in scenario 6. This indicates in some cases, the weighted sum method can get better performance, however, it is not easy to define appropriate weights in advance. In addition, it is more complex than NSMTGP-II and SPMTGP2 method because different independent runs should be conducted independently.

12.5 Chapter Summary

In this chapter, the presented algorithms incorporate the strategies of NSGA-II and SPEA2 into GP with multi-tree representation to solve the DFJSS problems. To the best of our knowledge, the presented algorithms try the first time to incorporate the strategies of NSGA-II and SPEA2 into GP with multi-tree representation to solve the DFJSS problem by evolving routing and sequencing rules simultaneously. In addition, the presented algorithms propose to use ratio difference as a metric to measure the consistency of rule behaviour, which can promote the studies of

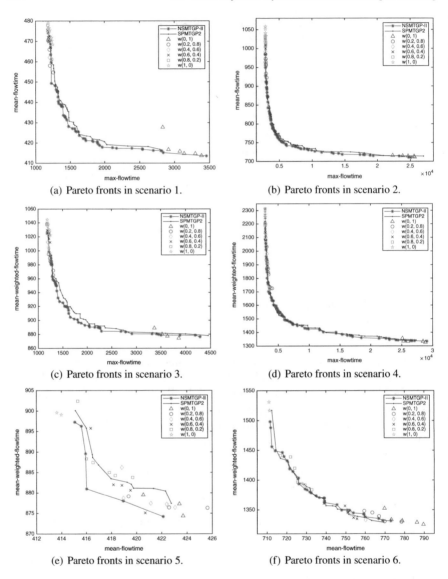

(a) Pareto fronts in scenario 1.   (b) Pareto fronts in scenario 2.

(c) Pareto fronts in scenario 3.   (d) Pareto fronts in scenario 4.

(e) Pareto fronts in scenario 5.   (f) Pareto fronts in scenario 6.

Fig. 12.5 Pareto fronts in different scenarios in training process

consistency measure between training and test. The experimental results show that both the presented algorithms can work on DFJSS well. In addition, NSMTGP-II performs better than SPMTGP2 and the scheduling heuristics learned by NSMTGP-II have promising behaviour consistency.

In the previous parts, all the studies aim to solve one task at a time. In the next part, multitask GP will be studied to solve multiple DFJSS tasks simultaneously with the investigated machine learning techniques in the previous chapters such as the surrogate and phenotypic characterisation.

Part V
Multitask Genetic Programming for Production Scheduling Problems

Multitask learning aims at solving multiple related tasks simultaneously [1], which is an important type of transfer learning [2]. Multitask learning encourages multiple self-contained tasks to be unified and searched concurrently. Although evolutionary multitask learning [1, 3] is a relatively new paradigm, it has recently received much research interest. Its success relies on the knowledge sharing mechanism between tasks during the evolutionary process of evolutionary algorithms. Evolutionary multitask learning has been successfully applied to different problem domains such as continuous numeric optimisation [1, 4, 5], feature selection [6], symbolic regression [7], and job shop scheduling [8, 9].

In the previous parts, we mainly focus on solving one task at a time with different machine learning techniques. In this part, we will focus on developing multitask GP to solve multiple JSS tasks simultaneously. The literature has identified a number of gaps that lead us to explore multitask GP in DFJSS from different aspects. First, the multitask learning algorithm in the *hyper-heuristics domain*, especially heuristic generation, for solving multiple tasks simultaneously is still an unexplored area. Second, *how to measure the relatedness between tasks is still an open question*, especially in combinatorial optimisation [10]. Last, although *surrogate techniques* have been incorporated into multitask learning [11, 12, 13, 14], they are used to improve the performance of a single task rather than *enhancing the core mechanism of multitask learning*, such as the knowledge sharing between tasks.

Part V is organised into three chapters:

- Chapter 13 proposes a multitask GP algorithm by adapting the traditional multitask algorithm according to the characteristics of GP in dynamic scheduling. In addition, this chapter develops an origin-based offspring reservation strategy to further improve the proposed multitask GP algorithm. The proposed multitask GP algorithm with an origin-based offspring reservation strategy can achieve good performance in homogeneous and heterogeneous multitask scenarios.

- Chapter 14 develops an adaptive multitask GP algorithm that aims to realise positive knowledge sharing between tasks by measuring the relatedness between tasks. This chapter shows how the relatedness between combinatorial optimisation problems can be measured.

- Chapter 15 proposes a surrogate-assisted multitask GP algorithm. The surrogate is used to not only evaluate the individuals efficiently but also help share knowledge between tasks. This chapter also shows the details of the individual allocations and the diversity of the individuals for each task.

Chapter 13
Multitask Learning in Hyper-Heuristic Domain with Dynamic Production Scheduling

Researchers have developed a wide range of evolutionary multitask methods so far [83, 85]. For solving multiple related combinatorial optimisation problems simultaneously, multitask GP algorithms have also been developed for team orienteering [107], dynamic JSS [173, 234], etc. In [107], the training simulations are divided into groups (islands), and individuals that have good quality are transferred between islands to realise knowledge sharing. Nevertheless, this method is not directly applicable to dynamic production scheduling, e.g., DFJSS, since it is difficult to collect the training dynamic simulation in advance and group them. In [173], a new niching-based GP was developed for solving dynamic JSS, which is still not efficient due to the extra requirement of individual evaluations. In [234], a new GP method was developed, which can obtain comparable scheduling heuristics for multiple related DFJSS problems in a much shorter time. This chapter will focus on presenting an effective multitask GP algorithm by adapting the traditional evolutionary multitask learning algorithm to GP for DFJSS with a focus on knowledge sharing.

13.1 Traditional Evolutionary Multitask Learning Algorithms

The paradigm of multifactorial optimisation toward evolutionary multitask (MFEA) was given in [83, 85] for solving multiple tasks simultaneously in evolutionary algorithms. Figure 13.1 shows the MFEA for solving k related tasks [83]. First, the initial population is randomly generated. The evaluation of each individual is done with all the tasks to be solved, and the individuals are assigned the task they are good at solving. The core of MFEA is the transmission of cultural genetic materials from parents to their offspring, which enables the search process to exchange useful knowledge among the tasks. Specifically, assortative mating and vertical cultural transmission

© The Author(s), under exclusive license to Springer Nature Singapore Pte Ltd. 2021   249
F. Zhang et al., *Genetic Programming for Production Scheduling*, Machine Learning: Foundations, Methodologies, and Applications,
https://doi.org/10.1007/978-981-16-4859-5_13

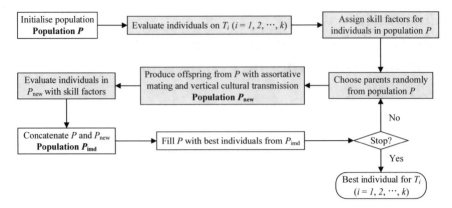

Fig. 13.1 The MFEA to solve k related tasks

work together to generate offspring based on two randomly selected parents. Assortative mating means that individuals are more likely to exchange genetic materials with other individuals coming from the same cultural environment. This is reflected in MFEA by the so-called skill factor. In general, vertical cultural transmission implies that the behaviour of a parent individuals can be inherited by its offspring. In MFEA, this is realised by setting the skill factor of an offspring the same as that of its parents. After generating all the offspring, the current population P and the offspring population P_{new} are merged together to obtain an intermediate population P_{imd}. Then, the best *popsize* individuals from P_{imd} survive in the next generation. MFEA finally returns k individuals, each for solving a specific task. The main features of MFEA are highlighted in blue in Fig. 13.1.

13.2 Challenges and Motivations

Evolutionary multitask learning has been successfully used in continuous numeric optimisation with benchmarks [61, 83, 127, 128], and regression problems [248]. However, the studies on discrete, combinatorial problems with complex situations are still very limited. This limits its applications in practice, since there are lots of combinatorial problems in the real-world. For example, in the clothing industry, the number of orders for the down jacket in winter is usually larger than in summer. For the down jacket job shop, the frequencies of orders for producing down jacket vary between seasons. The processing plan of the same commodity has similarities and differences in peak season and off-season. Intuitively, giving multiple solutions simultaneously to a company is an effective way to improve problem-solving capability. It is thus beneficial to have various kinds of production scheduling heuristics for a company to handle different cases.

Most existing multitask learning approaches aim at improving the qualities of solutions for all the tasks directly [61, 132], thus ignoring the hyper-heuristic research area. There are a number of related dynamic tasks in real-world applications such as cloud computing [87, 210] and JSS [159] with a preference for hyper-heuristic approaches to find a generalised heuristic for handling a number of problem scenarios. A unified framework of graph-based evolutionary multitasking hyper-heuristic approach was proposed and examined on timetabling and graph-colouring problems [90]. However, the proposed approach was a heuristic selection approach rather than a heuristic generation approach. In addition, it was only compared with the simple single-tasking hyper-heuristics, and there is no further analysis of the heuristic structure.

GP and hyper-heuristics have their own characteristics, so directly applying MFEA [83] to GP and hyper-heuristics may not achieve satisfactory effectiveness and efficiency. The main concerns are shown as follows. First, the selection pressure in MFEA may be too greedy for multitask GP. In GP, the selection pressure is typically implemented by the parent selection. However, MFEA selects the best individuals from the combined parent and offspring population to the next generation following a common genetic algorithm framework. Using both the parent selection in GP and offspring selection in MFEA simultaneously will make GP too greedy and lose the population diversity. Second, to improve generalisation of the learned scheduling heuristics, a commonly used strategy in GP hyper-heuristic for dynamic JSS is to rotate the training instances at each generation [27]. Selecting the best individuals from both the parent population and the offspring population requires re-evaluating the individuals in the parent population on the same training instances with the offspring in the current generation. This can make the training process less efficient. Last, MFEA allocates individuals to different tasks by calculating their fitness on all the tasks at the first generation, and then assigns the individuals to the task on which they do best. It is time-consuming due to the requirement of extra individual evaluations. This is especially true for GP in dynamic JSS. The goal of this chapter is to adapt multitask learning to GP as a hyper-heuristic approach for learning scheduling heuristics in dynamic scheduling. Specially, we are interested in the following questions:

- How to build tasks in dynamic scheduling?
- How to design the algorithm framework of GP to handle multiple tasks?
- What is an effective way to share knowledge between dynamic scheduling tasks?
- How can the learned scheduling heuristics from different tasks show the learning processing between tasks?

Detailed descriptions of the proposed algorithm are given in Sect. 13.3. The experiment design is shown in Sect. 13.4, followed by results and discussions in Sect. 13.5. Finally, Sect. 13.6 concludes this chapter.

13.3   Algorithm Design and Details

This section will introduce the framework of the proposed algorithm followed by the details of the developed knowledge sharing mechanism. The advantages of the proposed algorithm are also discussed.

13.3.1   Framework of the Algorithm

For handling multiple tasks $(T_1, T_2, \ldots, T_k)$ simultaneously, we divide the GP individuals in the population into k subpopulations $(Subpop_1, Subpop_2, \ldots, Subpop_k)$, each with equal subpopulation size. The individuals in each subpopulation can be regarded as having the same skill factor. Tasks assist each other by sharing their knowledge with others, which is realised by using the individuals from different subpopulations to generate offspring via crossover.

Algorithm 13.1 shows the outline of the presented multitask GP algorithm. Given k related tasks to be tackled, the algorithm returns a set of learned specialised rules $(h_1^*, h_2^*, \ldots, h_k^*)$, where ($h_i^*$ is to tackle task i). We use $fitness_{h_i}$ to denote the fitness of a heuristic h_i. The presented multitask GP hyper-heuristic differs with the GP methods that tackles only a single task in three main aspects. First, *at the initialisation stage*, the algorithm generates and maintains multiple subpopulations, each of which is specifically for tackling one task (line 1). On the contrary, the single-task GP methods normally contains only one population. Second, *during the evaluation process* of the proposed algorithm, the individuals in each subpopulation are evaluated independently on the corresponding tasks and do not affect other subpopulations (from line 6 to line 11). For the traditional GP, all the individuals in the population are used for optimising one task. Last, *during the evolution stage* of the newly developed algorithm, the crossover operator is used to exchange useful building blocks among the tasks for solving them more effectively. We use a transfer ratio rmp (i.e., within [0, 1]) to control the frequency of knowledge sharing among the tasks. If we randomly sample a value $rand$ between 0 and 1, and $rand < rmp$, then the crossover is conducted between the parents from this subpopulation and some other subpopulation (from line 25 to line 28). Otherwise, both parents will be selected from the current subpopulation, and exchange genetic materials with each other to generate two offspring (from line 29 to line 31). In other words, the knowledge (building blocks) of different tasks is exchanged via doing crossover between the parents from different subpopulations.

Algorithm 13.1 The Framework of the New Multitask GP Algorithm

---

Input: k tasks $T_1, T_2, \ldots, T_k$
Output: The best learned scheduling heuristics: $h_1^*, h_2^*, \ldots, h_k^*$
1: **Initialisation**: Randomly initialise the population with k subpopulations
2: $h_1^*, h_2^*, \ldots, h_k^* \leftarrow null$
3: $fitness_{h_1^*}, fitness_{h_2^*}, \ldots, fitness_{h_k^*} \leftarrow +\infty$
4: $gen \leftarrow 0$
5: **while** $gen < maxGen$ **do**
6:     // **Evaluate** each individual in each subpopulation
7:     **for** i = 1 to k **do**
8:         **for** j = 1 to $subpopsize_i$ **do**
9:             Compute $fitness_{h_j}$ by the simulation of task T_i
10:         **end for**
11:     **end for**
12:     **for** i = 1 to k **do**
13:         **for** j = 1 to $subpopsize_i$ **do**
14:             **if** $fitness_{h_j} < fitness_{h_i^*}$ **then**
15:                 $h_i^* \leftarrow h_j$
16:             **end if**
17:         **end for**
18:     **end for**
19:     **if** $gen < maxGen - 1$ **then**
20:         Change the random seed of the training simulations
21:         // **Evolution**: Generate offspring for each subpopulation
22:         **for** i = 1 to k **do**
23:             **if** $newsubpopsize_i \leq subpopsize_i$ **then**
24:                 **if** Crossover operator is chosen **then**
25:                     **if** $rand \leq rmp$ **then**
26:                         Select the first parent from $Subpop_i$ with tournament selection
27:                         Select the second parent from $Subpop_{\neg i}$ with tournament selection
28:                         Apply the **_origin-based offspring reservation_** crossover on the parents to generate one offspring
29:                     **else**
30:                         Select both parents from $Subpop_i$ with tournament selection method, and apply the sub-tree crossover to produce two offspring
31:                   **end if**
32:                 **else**
33:                     Select a parent from $Subpop_i$ by tournament selection
34:                     Do mutation or reproduction accordingly
35:                 **end if**
36:             **end if**
37:         **end for**
38:     **end if**
39:     $gen \leftarrow gen + 1$
40: **end while**
41: **return** $h_1^*, h_2^*, \ldots, h_k^*$

Fig. 13.2 The process of the
presented multitask GP

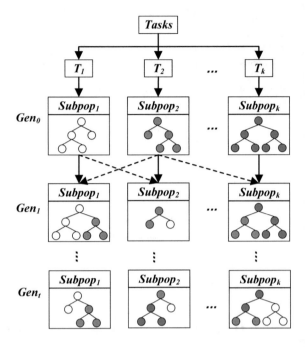

13.3.2   Knowledge Sharing Among Tasks

13.3.2.1   Overview of Evolutionary Process

Figure 13.2 shows the evolutionary process of the presented multitask GP based generative hyper-heuristic with the proposed knowledge sharing mechanism. At generation 0, the individuals are initialised for each task separately. The individuals for different tasks (i.e., T_1, T_2, and T_3) are distinguished with white, grey, and blue colours in Fig. 13.2. We can see that the individuals in different subpopulations consist of the genetic materials of individuals that originally belong to other tasks in the following generations. This is due to the knowledge sharing between tasks via the crossover operator.

13.3.2.2   Knowledge Sharing Ratio

To balance the exploration and exploitation of multitask GP, we use a parameter rmp to control the likelihood of knowledge exchange among tasks. For crossover, two parents are needed. The first parent is always selected from the current subpopulation (by tournament selection). The second parent, however, can be selected either from the current subpopulation or a different subpopulation. Specifically, if we randomly sample a value $rand$ (i.e., between 0 and 1) and it is smaller than rmp, then the

Fig. 13.3 An example of crossover with knowledge exchange with $Subpop_1$ and $Subpop_2$

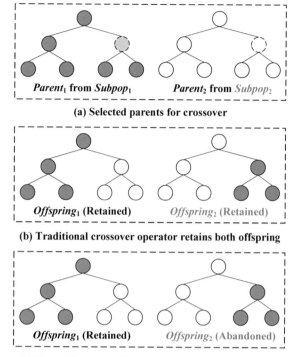

(a) Selected parents for crossover

(b) Traditional crossover operator retains both offspring

(c) The origin-based offspring reservation retains only the offspring derived from $Subpop_1$

second parent $parent_2$ will be selected from another subpopulation (by tournament selection). If there are multiple other subpopulations, we randomly select one of them to select the second parent. Taking $Subpop_1$ as an example, assume that $Subpop_2$ is selected to share knowledge to $Subpop_1$ as shown in Fig. 13.3a. The generated two offspring with knowledge sharing for $Subpop_1$ consists of white (from $Subpop_2$) and grey (from $Subpop_1$) genetic materials as shown in Fig. 13.3b, c. Thus, the knowledge sharing between the tasks is realised. Otherwise, if a random sampled value $rand$ is larger than rmp, two parents will be selected from the current subpopulation (i.e., the same subpopulation) to produce two offspring for the next generation.

13.3.2.3 Origin-Based Offspring Reservation

An effective knowledge sharing mechanism should let the individuals learn useful information from other individuals for other tasks, and maintain the characteristics of the individuals themselves for their own task. To achieve this, this chapter presents an origin-based offspring reservation strategy to generate offspring for optimising one task in multitask learning. Specifically, for generating an offspring for a task, we keep the main structure of the parent selected from the current subpopulation, and swap a random sub-tree with another random sub-tree of the other parent (either from the same or a different subpopulation). An example of the origin-based offspring reservation is shown in Fig. 13.3c, in which only the left offspring is retained. On the other hand, Fig. 13.3b shows that the existing crossover operator will keep both the generated offspring by the sub-tree swapping.

13.3.3 Algorithm Summary

Inspired from MFEA [85], the presented algorithm adapts the multitask optimisation process to fit into the special characteristics of GP as a hyper-heuristic approach in dynamic scheduling. The presented algorithm is generic, and has potential to be used to solve other multitask problems such as evolving routing rules for multiple arc routing problems [6], perhaps with slight modifications. A number of advantages of the presented algorithm are shown as follows.

- First, the presented multitask GP algorithm has a more efficient initialisation than MFEA. The initialised individuals are randomly allocated to the subpopulations (i.e., randomly assigned skill factors). However, MFEA has to evaluate each initial individual on each task, which requires $popsize * k$ more fitness evaluations.
- Second, the presented multitask GP algorithm can save $popsize * maxGen$ evaluations compared with the traditional multitask learning algorithm [85], since the offspring population is not needed to be merged with the parent population. This can greatly help with solving dynamic problems, in which the simulation is typically changed at each generation. Without merging parents and offspring, it is not necessary to compare between parents and offspring, which needs to re-evaluate the parents.
- Third, the individuals for each task are fixed, which is easy to manage. This reduces the number of parameters in the algorithm, such as the skill factor. In addition, it is not necessary to update the skill factor with extra calculations.

13.4 Experiment Design

13.4.1 Multitask Dynamic Flexible Job Shop Scheduling Task Definition

Most of existing multitask studies are on benchmark problems [54]. For DFJSS problems, it is still unclear which tasks are related and can be solved together in a multitask optimisation framework. In this section, we define some related tasks that have similar problem properties.

13.4.1.1 Same Objective but Different Utilisation Levels

In practice, the requests for a specific product in the market vary over time [71]. For example, a clothing factory is more likely to get much more orders for T-shirts in the summer season than in the winter season. A larger number of orders can result in a more complicated scheduling scenario that is harder to solve. However, the job shops with different complexities still have common properties. For example, they have the same objective, such as minimising the total flow time or tardiness. Inspired by this example, we define related tasks to have the same objective but different utilisation levels. A multitask scenario with such kind of related DFJSS tasks is called *homogeneous multitask* scenario.

13.4.1.2 Different Objectives but Same Utilisation Level

Different factories may have different requirements/expectations [185] on their production scheduling. One may want to minimise the flowtime for producing products to reduce the total cost. Others may prefer to minimise the tardiness to hand out products to the customers on time. Although the objectives can be different (i.e., either minimise flowtime or tardiness), they are in common to reduce the idle time of the machines in the shop floor (i.e., keep machines as busy as possible). The knowledge learned by optimising one objective might be also helpful for optimising other objectives. To simplify the problems, we use the same utilisation level for all tasks. Therefore, we define the tasks with different objectives but with the same utilisation level as related tasks to be considered in a multitask scenario, named *heterogeneous multitask* scenario.

13.4.2 Comparison Design

Table 13.1 shows three homogeneous multitask scenarios used in the experiments. The tasks in the same scenario have the same objective, while the tasks in different scenarios have different objectives (i.e., mean-flowtime, mean-tardiness, and mean-weighted-tardiness). The tasks in the same scenario have different utilisation levels (i.e., 0.75, 0.85, and 0.95), which are commonly considered configurations in DFJSS studies [236, 243].

Table 13.2 shows three heterogeneous multitask scenarios used in the experiments. We choose the most complex scenario with a utilisation level of 0.95 for the studies in this chapter [160]. For each heterogeneous multitask scenario, the tasks have different objectives (i.e., max-flowtime and max-tardiness, mean-flowtime and mean-tardiness, and mean-weighted-flowtime and mean-weighted-tardiness) but the same utilisation level (i.e., 0.95).

The GP algorithm with k subpopulations to solve k tasks independently is regarded as the baseline algorithm. The second compared algorithm adapts MFEA [83] to GP

Table 13.1 The **homogeneous** multitask scenarios used in the experiments

Scenario	Task 1	Task 2	Task 3
Scenario 1	<Fmean, 0.75>	<Fmean, 0.85>	<Fmean, 0.95>
Scenario 2	<Tmean, 0.75>	<Tmean, 0.85>	<Tmean, 0.95>
Scenario 3	<WTmean, 0.75>	<WTmean, 0.85>	<WTmean, 0.95>

Table 13.2 The **heterogeneous** multitask scenarios used in the experiments

Scenario	Task 1	Task 2
Scenario 1	<Fmax, 0.95>	<Tmax, 0.95>
Scenario 2	<Fmean, 0.95>	<Tmean, 0.95>
Scenario 3	<WFmean, 0.95>	<WTmean, 0.95>

Table 13.3 The availability of instance rotation and re-evaluation of GP, MFGP, MFGP^{r-}, MFGP^{r+}, M^{2}GP, and M^{2}GP^f

	GP	MFGP	MFGP^{r-}	MFGP^{r+}	M^{2}GP	M^{2}GP^f
Instance Rotation	✓		✓	✓	✓	✓
Re-evaluation				✓		

without rotating training instances (i.e., no need to do re-evaluation), which is named MFGP. In addition, the algorithm adapts MFEA to GP with *rotating* training instances but without re-evaluation is named MFGP^{r-}. While the algorithm adapts MFEA to GP with both rotating training instances and re-evaluation is named MFGP^{r+}. It is noted that MFGP, MFGP^{r-}, and MFGP^{r+} contain one population. The presented multitask GP-based generative hyper-heuristic approach without the presented offspring reservation strategy is named M^{2}GP, since it involves both multitask and multi-population. M^{2}GP with the presented offspring reservation strategy named M^{2}GP^f. Table 13.3 summaries the characteristics of GP, MFGP, MFGP^{r-}, MFGP^{r+}, M^{2}GP, and M^{2}GP^f according to whether they involve instance rotation or re-evaluation or not.

To verify the adaptability of MFEA to GP in dynamic scheduling, MFGP, MFGP^{r-}, and MFGP^{r+} are compared. To verify the effectiveness of the presented M^{2}GP and the origin-based offspring reservation strategy, GP, MFGP, M^{2}GP, and M^{2}GP^f are compared. In addition, the effectiveness of the presented M^{2}GP^f on the common tasks between homogeneous and heterogeneous scenarios is further compared with MFGP. The effectiveness of knowledge sharing in multitask GP is also examined by analysing the learned scheduling heuristics for each task in a multitask scenario.

Table 13.4 The specialised parameter settings of genetic programming

Parameter	Value
*Number of subpopulations	k
*Subpopulation size	400
**Number of tasks	k
**Population size with re-evaluation	$200 * k$
**Population size without re-evaluation	$400 * k$
The transfer ratio	0.6

*: for the algorithms (i.e., GP, M^2GP, and M^2GP^f) with multiple subpopulations only
**: for the algorithms (i.e., MFGP, $MFGP^{r-}$, and $MFGP^{r+}$) with one population only

13.4.3 Specialised Parameter Settings of Genetic Programming

Table 13.4 shows the specialised parameter settings of GP in this chapter. For the algorithms (i.e., GP, M^2GP, and M^2GP^f) with k subpopulations for solving k tasks, each subpopulation of the algorithms contains 400 individuals. MFGP, $MFGP^{r-}$, and $MFGP^{r+}$ involve one population. To keep the number of individual evaluations the same between different algorithms for fair comparison, population size of the algorithm with re-evaluation mechanism (i.e., $MFGP^{r+}$) are set to $200 * k$, and the population sizes of the algorithms without re-evaluation mechanism (i.e., MFGP and $MFGP^{r-}$) are set to $400 * k$. The transfer ratio for knowledge transfer is set to 0.6.

13.5 Results and Discussions

13.5.1 Adaptation of MFEA to Genetic Programming in Dynamic Scheduling

Table 13.5 shows the mean and standard deviation of the test performance of $MFGP^{r-}$, $MFGP^{r+}$, and MFGP on the unseen instances in the three homogeneous multitask scenarios. From the table, we can see that compared with $MFGP^{r-}$, the performance of $MFGP^{r+}$ is significantly better for all the tasks. This suggests that changing the random seed of training simulations requires re-evaluating the parent individuals to reach better performance. In addition, the MFGP with no seed change in the simulation manages to show similar (better) performance with $MFGP^{r+}$ in six (three) scenarios. Therefore, MFGP will be used for the comparisons in the subsequent experiments due to its best performance among the compared counterpart algorithms. It is noted that changing simulation seeds is still useful for solving DFJSS, if the parent and offspring populations are merged together for selecting the next generation.

Table 13.5 The mean (standard deviation) of the test performance of MFGP^{r-}, MFGP^{r+}, and MFGP

Scenario	Task	MFGP^{r-}	MFGP^{r+}	MFGP
1	<Fmean, 0.75>	339.97(1.11)	337.01(1.39)(−)	336.60(1.21)(−)(≈)
	<Fmean, 0.85>	396.21(2.90)	387.91(3.72)(−)	386.67(2.93)(−)(≈)
	<Fmean, 0.95>	586.77(6.50)	560.04(8.64)(−)	556.55(5.83)(−)(≈)
2	<Tmean, 0.75>	16.08(0.95)	13.90(0.66)(−)	13.60(0.25)(−)(≈)
	<Tmean, 0.85>	46.32(2.74)	41.51(1.92)(−)	40.54(0.66)(−)(≈)
	<Tmean, 0.95>	202.54(3.40)	182.88(5.60)(−)	180.39(4.46)(−)(≈)
3	<WTmean, 0.75>	33.56(2.55)	28.44(1.87)(−)	27.26(0.61)(−)(−)
	<WTmean, 0.85>	97.12(5.22)	79.90(4.79)(−)	76.95(2.19)(−)(−)
	<WTmean, 0.95>	381.03(29.32)	310.91(15.34)(−)	303.05(8.84)(−)(−)

13.5.2 Quality of Learned Scheduling Heuristics

13.5.2.1 Homogeneous Multitask Scenarios

Table 13.6 shows the mean and standard deviation of the objective values on the test instances of GP, MFGP, M^{2}GP, and M^{2}GP^f over 30 independent runs in the three homogeneous multitask scenarios. MFGP does not show any significant difference from GP in most of the tasks (i.e., <Fmean, 0.75>, <Fmean, 0.85>, <Fmean, 0.95>, <Tmean, 0.75>, <Tmean, 0.85>, <Tmean, 0.95>, <WTmean, 0.75>, and <WTmean, 0.85>), which indicates that directly applying the idea of MFEA is not effective in the context of GP hyper-heuristic. This is because the characteristics of GP are quite different from the genetic algorithm that is used in MFEA. M^{2}GP showed significantly better performance than GP and MFGP in all the scenarios. This verifies the effectiveness of the presented M^{2}GP in the homogeneous multitask scenarios. M^{2}GP^f also shows its superiority compared with GP and MFGP in terms of the mean objective values in six out of the nine scenarios (i.e., <Fmean, 0.95>, <Tmean, 0.75>, <Tmean, 0.95>, <WTmean, 0.75>, <WTmean, 0.85>, and <WTmean, 0.95>). However, M^{2}GP^f performs significantly better than M^{2}GP in only one scenario (i.e., <WTmean, 0.85>).

Figure 13.4 shows the distribution of the test performance of GP, MFGP, M^{2}GP, and M^{2}GP^f in the three homogeneous multitask scenarios. From the figure, we can observe that MFGP has (insignificantly) better test performance than GP. We can see that M^{2}GP obtains smaller objective values than GP and MFGP. This shows that the DFJSS tasks with the same objective but different utilisation levels are related and can be effectively solved together. Further, M^{2}GP^f tends to perform better than M^{2}GP in most scenarios. This demonstrates the efficacy of the presented origin-based offspring strategy.

Table 13.6 The mean (standard deviation) of the test performance of GP, MFGP, M²GP, and M²GP^f

Sce.	Task	GP	MFGP	M²GP	M²GP^f
1	<Fmean, 0.75>	337.57(1.80)	336.60(1.21)(≈)	335.86(0.91)(−)(−)	336.17(1.04)(−)(≈)(≈)
	<Fmean, 0.85>	388.79(4.30)	386.67(2.93)(≈)	385.14(1.94)(−)(−)	385.73(2.33)(−)(−)(≈)
	<Fmean, 0.95>	561.35(9.16)	556.55(5.83)(≈)	553.11(4.26)(−)(−)	552.74(4.75)(−)(−)(≈)
2	<Tmean, 0.75>	14.08(1.10)	13.60(0.25)(≈)	13.34(0.27)(−)(−)	13.33(0.25)(−)(−)(≈)
	<Tmean, 0.85>	41.61(2.73)	40.54(0.66)(≈)	39.75(0.87)(−)(−)	39.78(0.84)(−)(−)(≈)
	<Tmean, 0.95>	182.34(7.72)	180.39(4.46)(≈)	176.84(3.10)(−)(−)	176.65(4.17)(−)(−)(≈)
3	<WTmean, 0.75>	28.81(2.66)	27.26(0.61)(≈)	27.27(0.99)(−)(−)	26.92(0.72)(−)(−)(≈)
	<WTmean, 0.85>	81.23(7.63)	76.95(2.19)(≈)	76.43(3.19)(−)(−)	75.34(2.06)(−)(−)(≈)
	<WTmean, 0.95>	312.26(15.86)	303.05(8.84)(−)	297.72(10.38)(−)(−)	295.67(8.44)(−)(−)(≈)

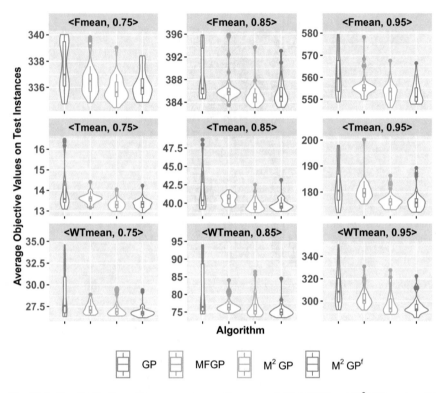

Fig. 13.4 The distribution of the average test performance of GP, MFGP, and M^2GP based on 30 independent runs in three *homogeneous multitask* scenarios (i.e., each row is a multitask scenario with three tasks)

13.5.2.2   Heterogeneous Multitask Scenarios

Table 13.7 shows the mean and standard deviation of the test performance of GP, MFGP, M^2GP, and M^2GP^f based on 30 independent runs in the three heterogeneous multitask scenarios. Different from the observations in the homogeneous multitask scenarios, all the three compared algorithms show similar performance, and their differences are not significant.

M^2GP^f performs significantly better than GP and MFGP in most scenarios, and outperforms M^2GP in <Tmean, 0.95>. It is likely to be because that the heterogeneous tasks are less related, and none of the algorithms can reach high-quality individuals for all the tasks. Moreover, it seems that the origin-based offspring reservation strategy helps to improve the GP performance for each task.

Figure 13.5 shows the distribution of the test performance of the compared algorithms in the three heterogeneous multitask scenarios. Obviously, M^2GP has a much better test performance than GP and MFGP. This suggests that the heterogeneous tasks, although less related, could still be effectively solved in a multitask scenario.

Table 13.7 The mean (standard deviation) of the test performance of GP, MFGP, M²GP, and M²GPᶠ over 30 independent runs in three **heterogeneous multitask** scenarios (i.e., each scenario contains two tasks)

Sce.	Task	GP	MFGP	M²GP	M²GPᶠ
1	<Fmax, 0.95>	2032.96(98.29)	2081.77(76.40)(+)	1991.15(88.56)(−)(−)	1981.28(37.19)(−)(−)(≈)
	<Tmax, 0.95>	1580.81(54.13)	1647.75(55.45)(+)	1576.03(54.94)(≈)(−)	1575.13(37.84)(≈)(−)(≈)
2	<Fmean, 0.95>	560.72(10.18)	556.10(6.10)(≈)	556.52(9.11)(≈)(≈)	553.79(7.43)(−)(−)(≈)
	<Tmean, 0.95>	180.81(6.83)	178.76(3.47)(≈)	180.20(6.51)(≈)(≈)	177.55(5.72)(−)(−)(−)
3	<WFmean, 0.95>	1136.33(25.65)	1121.67(12.74)(−)	1123.87(20.45)(−)(≈)	1121.05(22.20)(−)(−)(≈)
	<WTmean, 0.95>	311.20(16.82)	301.12(8.07)(−)	303.06(15.23)(−)(≈)	300.00(15.38)(−)(−)(≈)

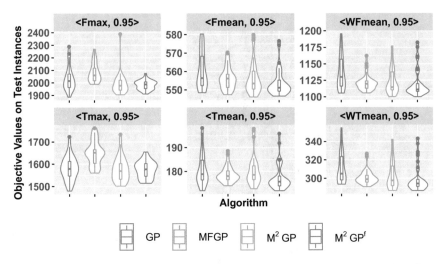

Fig. 13.5 The distribution of the average objective values on test instances of GP, MFGP, M^2GP, and M^2GP^f based on 30 independent runs in three **heterogeneous multitask** scenarios (i.e., each column is a multitask scenario with two tasks)

Furthermore, M^2GP^f seems to perform better than M^2GP. This again demonstrates the efficacy of the origin-based offspring reservation strategy.

13.5.2.3   Homogeneous Versus Heterogeneous Multitask Scenarios

In the previous experiments, <Fmean, 0.95>, <Tmean, 0.95>, and <WTmean, 0.95> are solved in both the homogeneous and heterogeneous multitask scenarios. To investigate which designed multitask scenarios can help learn better scheduling heuristics, we investigate how different multitask scenarios would affect how well each task is solved.

Figure 13.6 shows the distribution of the test performance of MFGP and M^2GP^f for the three common tasks mentioned above. Overall, for all the common tasks, M^2GP^f showed better performance (lower distribution) than MFGP in both the homogeneous and heterogeneous multitask scenarios. Another interesting observation is that both MFGP and M^2GP^f can obtain better test performance on the common tasks in the heterogeneous multitask scenarios rather than the homogeneous multitask scenarios.

In summary, M^2GP^f can learn effective scheduling heuristics for both the homogeneous and heterogeneous multitask scenarios. Furthermore, solving a task in a heterogeneous multitask scenario (different objectives but the same utilisation level) seems to help solve the task better.

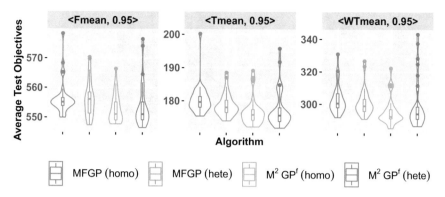

Fig. 13.6 The distribution of the test performance of MFGP and M^2GP^f in both homogeneous (denoted as homo) and heterogeneous (denoted as hete) multitask scenarios for their common tasks based on 30 independent runs

13.5.3 Insight of Learned High-Level Scheduling Heuristics

To understand how the different tasks interact with each other and help each other, we pick some routing and sequencing rules for the tasks in the second heterogeneous multitask scenario, which consists of two common tasks.

13.5.3.1 Routing Rules

Figures 13.7 and 13.8 show an example routing rules for the tasks <Fmean, 0.95> and <Tmean, 0.95>. From the figure, we can clearly see that these two scheduling heuristics have a major part of structure in common (highlighted in grey).

We further investigate the behaviour of the routing rule for minimising mean-tardiness, as shown in Fig. 13.8. The simplification of this rule is shown in Eq. (13.1).

$$
\begin{aligned}
R = \{ & WKR + PT - MWT - \frac{WKR}{PT} - Min\{ \\
& \frac{Max\{WKR, MWT\}}{PT}, \frac{WKR}{PT} - NOR\}\}/NIQ \\
& - \frac{Min\{WKR, TIS\}}{PT(NPT + W)} - \frac{WKR}{PT} \\
\approx \{ & PT - MWT - \frac{WKR}{PT} - Min\{ \\
& \frac{Max\{WKR, MWT\}}{PT}, \frac{WKR}{PT} - NOR\}\}/NIQ \\
& - \frac{Min\{WKR, TIS\}}{PT(NPT + W)} - \frac{WKR}{PT}
\end{aligned}
\tag{13.1}
$$

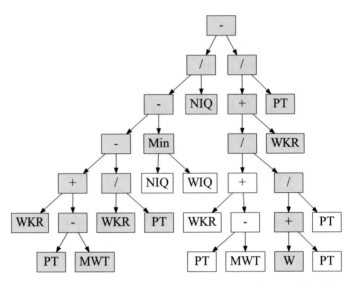

Fig. 13.7 An example routing rules for task 1 <Fmean, 0.95> in the scenario 2 of heterogeneous multitask

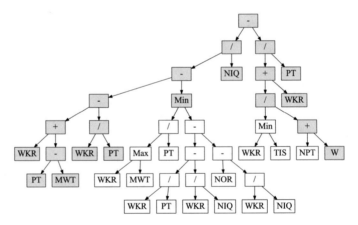

Fig. 13.8 An example routing rules for task 2 <Tmean, 0.95> in the scenario 2 of heterogeneous multitask

This rule uses WKR and PT very often. However, we know that WKR (i.e., the remaining work of the corresponding job of an operation for all the machines) is the same for all the machines, so can be treated as a constant when making a routing decision. Following the same logic, W, NOR, NPT, and TIS can also be treated as constants. After removing the redundant constant features, this rule can be simplified as shown in step 2 of Eq. (13.1).

From the simplified rule, we can see that this rule prefers the machine with a smaller processing time and a longer waiting time. This is consistent with our intu-

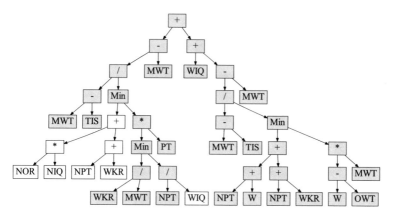

Fig. 13.9 An example learned sequencing rules for task 1 <Fmean, 0.95> in the scenario 2 of heterogeneous multitask

ition, since using efficient machines can reduce the production time. Additionally, this rule tends to select a machine with more operations in its queue. Noted that this does not indicate that the machine is very busy, since these operations can have short processing time.

13.5.3.2 Sequencing Rules

Figures 13.9 and 13.10 show the corresponding sequencing rules of the routing rules, as shown in Figs. 13.7 and 13.8, respectively. From the figure, it is observed that these two sequencing rules have a very large portion of common structures, and are different in only two minor parts. In the common grey part, the sequencing rules use MWT and NPT a lot. This indicates that MWT and NPT are both critical for operation sequencing decisions in minimising mean-flowtime and mean-tardiness.

The corresponding sequencing rule (Fig. 13.10) of the routing rule in Fig. 13.8 is shown in Eq. (13.2). From step 1 to step 2, "Min{W, PT * Min{$\frac{WKR}{MWT}$, $\frac{NPT}{WKR}$}}" can be converted to W, since W is often smaller than "PT * Min{$\frac{WKR}{MWT}$, $\frac{NPT}{WKR}$}". "(W − OWT) * MWT" tends to be smaller than zero, since W is often smaller than OWT. Thus, "Min{2NPT + W + WKR, (W − OWT) * MWT}" is replaced with "(W − OWT) * MWT". In the last step, W is removed, as it is much smaller than OWT.

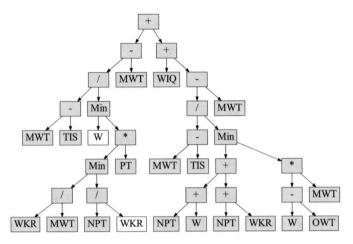

Fig. 13.10 An example learned sequencing rules for task 2 <Tmean, 0.95> in the scenario 2 of heterogeneous multitask

$$
\begin{aligned}
S &= \frac{MWT - TIS}{Min\{W, PT * Min\{\frac{WKR}{MWT}, \frac{NPT}{WKR}\}\}} - \\
&\quad \frac{MWT - TIS}{Min\{2NPT + W + WKR, (W - OWT) * MWT\}} \\
&\quad - 2MWT + WIQ \\
&\approx \frac{MWT - TIS}{W} + \frac{MWT - TIS}{(OWT - W) * MWT} \\
&\approx \frac{MWT - TIS}{W} + \frac{MWT - TIS}{OWT * MWT}
\end{aligned}
\tag{13.2}
$$

For operation sequencing, MWT is the same for all the operations, so can be treated as a constant. By looking at the simplified rule, we can see that this sequencing rule tends to select the operation that has stayed in the job shop floor for a long time (i.e., large TIS), and has waited in the queue of a machine for a long time (i.e., large OWT). It is consistent with our intuition that a long waiting time will delay the production, and the operation should be processed first. In addition, the machine prefers to choose an important operation with a large W. This is also consistent with our intuition that important jobs should be processed earlier to reduce the delay and improve customer satisfaction.

In summary, we can see that both the routing and sequencing rules for the tasks in heterogeneous multitask scenarios share substantial knowledge with each other. We observe the same pattern in the homogeneous multitask scenarios. We can conclude that the presented algorithm can solve the tasks in a mutually reinforcing way.

13.6 Chapter Summary

This chapter introduces an effective multitask GP hyper-heuristic algorithm to solve multiple DFJSS problems simultaneously. This chapter also presents an effective knowledge sharing mechanism with origin-based offspring reservation strategy for sharing knowledge between the tasks. The effectiveness of the presented algorithm is examined on both homogeneous and heterogeneous multitask DFJSS scenarios.

The results show that the presented M^2GP^f can achieve effective scheduling heuristics in both homogeneous and heterogeneous multitask DFJSS scenarios. In addition, M^2GP^f is robust in terms of the performance in both homogeneous and heterogeneous multitask scenarios. We also found that the task with a heterogeneous multitask scenario has more potential to be optimised well. The effectiveness of the presented multitask GP hyper-heuristic was examined by not only comparing the quality of learned scheduling heuristics, but also the structures and behaviours of the learned scheduling heuristics for all tasks in a multitask scenario. It has also been observed that the presented algorithm manages to solve the tasks in a mutually reinforcing way.

The presented multitask GP algorithm is the first time to use multitask GP for solving multiple DFJSS tasks simultaneously. It provides a basic framework for studying multitask GP for DFJSS. It expands the paradigm of evolutionary multitask to GP for complex dynamic combinatorial optimisation problems. In addition, it extends multitask learning to hyper-heuristic domain. In the next chapter, we will further enhance the effectiveness of multitask GP for DFJSS.

This chapter is a start point for using multitask GP for DFJSS, and provides a basic framework for studying multitask GPHH for DFJSS. In the next chapter, we will further enhance the effectiveness of multitask GP for DFJSS by selecting assisted tasks for positive knowledge sharing adaptively.

Chapter 14
Adaptive Multitask Genetic Programming for Dynamic Job Shop Scheduling

The success of MFEA depends on the positive knowledge sharing mechanism [84] between tasks. Existing studies found that transferring knowledge from more related tasks can lead to more positive transfer [44] to improve the performance of multitask learning. The knowledge sharing between unrelated tasks normally cannot help or worsen multitask learning, especially when there are more than two tasks. However, the relatedness between the tasks is not necessarily known in advance. How to measure the relatedness between tasks is an important but challenging issue in multitask learning, especially for combinatorial optimisation [169]. Task relatedness information is normally used to build multitask learning benchmarks for testing the multitask learning algorithms or to improve the quality of shared knowledge in multitask learning. Multitask test problems were proposed based on the characteristics of benchmark functions such as the fitness landscape and optimal solutions [54, 229]. The transfer ratio was adjusted to minimise the negative interactions between distinct tasks based on the relatedness between the benchmark function tasks in [14, 15]. However, these kinds of studies take the relatedness information between tasks as prior knowledge, which is not always available. This chapter will focus on measuring the relatedness between DFJSS tasks to improve the effectiveness of multitask GP in DFJSS.

14.1 Challenges and Motivations

Most of the existing studies [70, 80, 132, 133] on multitask learning mainly focus on optimising benchmark functions and the relatedness between tasks is normally based on the landscape of functions [54]. Specifically, a number of randomly sampled individuals were ranked based on their fitness for different tasks, and the correlation between the ranks was used to measure the relatedness between tasks in [54]. However, the extra individual evaluations for getting the ranks of individuals on all the tasks are time-consuming. The data generated in the evolutionary process was

F. Zhang et al., *Genetic Programming for Production Scheduling*, Machine Learning: Foundations, Methodologies, and Applications, https://doi.org/10.1007/978-981-16-4859-5_14

successfully used to measure the relatedness between tasks to guide the knowledge
sharing between tasks [14, 15, 44]. However, these studies only focus on the evolutionary algorithms with vector-based and fixed-length representation, which is not
directly applicable to GP with tree-based and variable-length representation.

DFJSS is a discrete minimisation problem, and unlike continuous numeric functions with known optimal solutions, the characteristics of DFJSS such as the fitness
landscape information are not known. The existing studies mainly work on vector-
based evolutionary algorithms with fixed-length solutions. The relatedness measure
based on the evolved individuals for different tasks are not directly applicable for GP
individuals with tree-based and variable-length representation. All the characteristics
of the investigated DFJSS problem and the used GP algorithm make it challenging
to measure the relatedness between tasks in this chapter. The goal of this chapter
is to develop an effective strategy to measure the relatedness of tasks in dynamic
scheduling with GP. Specially, we are interested in the following questions:

- How to extract information from individuals to represent the characteristics of
  tasks?
- How to measure the relatedness of tasks based on the extracted information?
- What is a good way to use the obtained relatedness information?
- What are the effects of the proposed algorithm, e.g., diversity of individuals for a
  task?

Detailed descriptions of the proposed algorithm are given in Sect. 14.2. The
experiment design is shown in Sect. 14.3. Results and discussions are presented
in Sect. 14.4. Further analyses are conducted in Sect. 14.5. Section 14.6 concludes
this chapter.

14.2   Algorithm Design and Details

This section will introduce the algorithm with a start of its framework. Then, how
to measure the relatedness between tasks is described in details. Finally, this section
describes how to select an assisted task for a specific task.

14.2.1   Framework of the Algorithm

Algorithm 14.1 shows the framework of the presented algorithm. The input is k tasks
to be solved. The output contains k best learned scheduling heuristics (h_1^*, h_2^*, ...,
h_k^*), each for a task. We use multiple subpopulations (i.e., $Subpop_1$, $Subpop_2$, ...,
$Subpop_k$) to solve the k tasks simultaneously and keep knowledge sharing between
them. $subpopsize_i$ indicates the number of individuals of subpopulation i. The fitness
of a heuristic h_i is denoted by $fitness_{h_i}$.

At the initialisation stage, we randomly initialise each subpopulation (line 1). *During the evaluation process*, the individuals in different subpopulations are evaluated, and the best individual for each task is recorded independently (from line 6 to line 16). Then, the phenotypic characterisation of each individual in each subpopulation is calculated with a set of decision situations along with the routing and sequencing reference rules (from line 19 to line 23). The behaviour matrices for subpopulations are built based on the phenotypic characterisations of individuals in the corresponding subpopulation (line 24). The relatedness between two tasks is calculated based on the built behaviour matrices (line 25). *During the evolution stage*, the crossover operator is applied to share knowledge between tasks. If the knowledge sharing condition is met ($rand \leq rmp$), the offspring of each subpopulation are generated for the next generation with the assisted task (from line 28 to line 32). Specifically, for each target task, one of the candidate tasks $Subpop_a$ is selected as the assisted task according to the presented assisted task selection strategy. When generating offspring, the individuals from the current subpopulation ($Subpop_i$) and the assisted subpopulation ($Subpop_a$) are used as the parents to realise knowledge transfer from the assisted task. If the knowledge sharing condition is not met ($rand > rmp$), the traditional GP crossover will be used to produce two offspring (from line 33 to line 35). Finally, the best learned scheduling heuristics for all tasks obtained at the previous generation are selected as the learned scheduling heuristics for tasks by the presented algorithm.

14.2.2 Task Relatedness Measure

Since the individuals in each subpopulation tend to be specific to the corresponding task in evolutionary algorithms over generations, the characteristics of a task can be represented by the behaviour of the individuals [14]. If the representative individuals for two tasks show similar behaviour, then the tasks are related to each other. The degree of relatedness between tasks depends on how similarly the representative individuals behave. In this chapter, we use all the individuals in its subpopulation as the representative individuals for each task.

14.2.2.1 Behaviour Matrix

We use the phenotypic characterisation to measure the behaviour of GP individuals, and compare individuals by their phenotype. This corresponds to line 21 of Algorithm 14.1, and the details of the way to calculate the phenotypic characterisation can be found in Chap. 8. For each task, we build a behaviour matrix to capture the characteristics of a task by using the phenotypic vectors of all the individuals for that task. This corresponds to line 24 of Algorithm 14.1. An example of the behaviour matrix for task T_i is shown as follows, where DS_i means the ith decision situation. In this example, the behaviour matrix is based on three individuals and four decision

Algorithm 14.1 The Framework of the Algorithm

Input: k tasks $T_1, T_2, ..., T_k$
Output: The best learned scheduling heuristics for each task $h_1^*, h_2^*, ..., h_k^*$
1: **Initialisation**: Randomly initialise the population with k subpopulations
2: set $h_1^*, h_2^*, ..., h_k^* \leftarrow null$
3: set $fitness_{h_1^*}, fitness_{h_2^*}, ..., fitness_{h_k^*} \leftarrow +\infty$
4: $gen \leftarrow 0$
5: **while** $gen < maxGen$ **do**
6:     // **Evaluation**: Evaluate the individuals in the population
7:     **for** i = 1 to k **do**
8:         **for** j = 1 to $subpopsize_i$ **do**
9:             Calculate $fitness_{h_j}$ based on fitness function of T_i
10:         **end for**
11:         **for** j = 1 to $subpopsize_i$ **do**
12:             **if** $fitness_{h_j} < fitness_{h_i^*}$ **then**
13:                 $h_i^* \leftarrow h_j$
14:             **end if**
15:         **end for**
16:     **end for**
17:     **if** $gen < maxGen - 1$ **then**
18:         // **Evolution**: Generate offspring for each subpopulation
19:         **for** i = 1 to k **do**
20:             **for** j = 1 to $subpopsize_i$ **do**
21:                 **Calculate phenotypic characterisation for individual h_j with a set of decision situations and the reference rules**
22:             **end for**
23:         **end for**
24:         **Build behaviour matrix for each subpopulation $Matrix_1, Matrix_2, ..., Matrix_k$**
25:         **Calculate the relatedness between each pair of tasks based on the built behaviour matrices**
26:         **for** i = 1 to k **do**
27:             **if** $newsubpopsize_i \leq subpopsize_i$ **then**
28:                 **if** $rand \leq rmp$ **then**
29:                     **Choose one task as assisted task T_a proportional to the relatedness**
30:                     **Choose the first parent from $Subpop_i$ with tournament selection**
31:                     **Choose the second parent from $Subpop_a$ with tournament selection**
32:                     Produce one offspring with origin-based offspring reservation strategy [231]
33:                 **else**
34:                     Choose two parents from $Subpop_i$, and produce two offspring with the traditional GP crossover
35:                 **end if**
36:             **end if**
37:         **end for**
38:     **end if**
39:     $gen \leftarrow gen + 1$
40: **end while**
41: **return** $h_1^*, h_2^*, ..., h_k^*$

situations, which is a 3 * 4 matrix. Each row indicates the phenotypic vector of an individual, while each column represents the behaviour of different individuals in the same decision situation.

$$Matrix_i = \begin{array}{c} \\ Ind_1 \\ Ind_2 \\ Ind_3 \end{array} \begin{array}{cccc} DS_1 & DS_2 & DS_3 & DS_4 \\ \left(\begin{array}{cccc} 3 & 2 & 3 & 1 \\ 2 & 3 & 1 & 2 \\ 1 & 1 & 2 & 2 \end{array}\right) \end{array}$$

14.2.2.2 Relatedness Calculation Between Tasks

For each task, the distribution of the behaviours of the representative individuals can reflect the characteristics of the task. Therefore, we define the relatedness between tasks based on the difference between the distributions of individual behaviours. This corresponds to line 25 of Algorithm 14.1.

Kullback–Leibler divergence, D_{KL} (also called relative entropy), is a measure of how one probability distribution is different from a second reference probability distribution. Due to the property of D_{KL}, it is a good candidate to measure the relatedness between tasks with the probability of the decisions in each decision situation in this chapter. It is noted that the behaviour matrices are discrete probability distributions. Assume there are two discrete probability distributions P and Q, the relatedness between them, which is represented as Kullback–Leibler value, can be defined as follows:

$$D_{KL}(P||Q) = \sum_{x \in X} P(x) \log(\frac{P(x)}{Q(x)}) \tag{14.1}$$

We use Eq. (14.1) to calculate the D_{KL} for each decision situation, and use the average value across all the decision situations as the relatedness value between two tasks. The D_{KL} value indicates how similar individuals' behaviour is for different tasks in the same decision situation.

An example of the calculation of the Kullback–Leibler divergence value of two tasks is shown as follows. There are two tasks (i.e., P and Q), and each task has three representative individuals. Assume the phenotypic vectors of the individuals for task P and Q in one (same) decision situation are $\overrightarrow{P} = (1, 2, 3)$ and $\overrightarrow{Q} = (1, 1, 1)$, respectively. Thus, the number of occurrences of 1, 2, 3 in $\overrightarrow{P}$ are $\overrightarrow{OP}(1) = 1$, $\overrightarrow{OP}(2) = 1$, and $\overrightarrow{OP}(3) = 1$. For $\overrightarrow{Q}$, we can have $\overrightarrow{OQ}(1) = 3$, $\overrightarrow{OQ}(2) = 0$, and $\overrightarrow{OQ}(3) = 0$. To avoid the division issue with 0 for Eq. (14.1), we add 1 to each number of the occurrence values for $\overrightarrow{P}$ and $\overrightarrow{Q}$ as suggested in [199], and thus $\overrightarrow{OP}(1) = 2$, $\overrightarrow{OP}(2) = 2$, $\overrightarrow{OP}(3) = 2$, $\overrightarrow{OQ}(1) = 4$, $\overrightarrow{OQ}(2) = 1$, and $\overrightarrow{OQ}(3) = 1$. The probabilities of all elements in $\overrightarrow{P}$ and $\overrightarrow{Q}$ are $\overrightarrow{P}(1) = \frac{1}{3}$, $\overrightarrow{P}(2) = \frac{1}{3}$, $\overrightarrow{P}(3) = \frac{1}{3}$, $\overrightarrow{Q}(1) = \frac{2}{3}$, $\overrightarrow{Q}(2) = \frac{1}{6}$, and $\overrightarrow{Q}(6) = \frac{1}{6}$. The relatedness between task P and Q in the same decision situation (i.e., $\overrightarrow{P}$ and $\overrightarrow{Q}$) is calculated as shown in Eq. (14.2).

$$D_{KL}(P||Q) = \vec{P}(1)\log(\frac{\vec{P}(1)}{\vec{Q}(1)}) + \vec{P}(2)\log(\frac{\vec{P}(2)}{\vec{Q}(2)})$$
$$+ \vec{P}(3)\log(\frac{\vec{P}(3)}{\vec{Q}(3)}) \tag{14.2}$$
$$= \frac{1}{3}\log(\frac{\frac{1}{3}}{\frac{2}{3}}) + \frac{1}{3}\log(\frac{\frac{1}{3}}{\frac{1}{6}}) + \frac{1}{3}\log(\frac{\frac{1}{3}}{\frac{1}{6}})$$
$$= 0.1$$

A smaller $D_{KL}(P||Q)$ means that the probability distribution (i.e., indicates the decision marking behaviour) in P is more similar with the probability distribution in Q, which leads to a more related task relationship. A larger relatedness value indicates a more related relationship between tasks. To make it easy to understand and implement, we transform the D_{KL} value to relatedness value. The calculation of the relatedness value of tasks P and Q (i.e., indicated by $R(P, Q)$) is shown in Eq. (14.3), where $D_{KL}(P||Q)$ is the Kullback–Leibler divergence value between tasks P and Q. Since $D_{KL}(P||Q) \geq 0$, the range of $R(P||Q)$ is (0, 1].

$$R(P, Q) = \frac{1}{1 + D_{KL}(P||Q)} \tag{14.3}$$

It is noted that a number of decision situations are used, and the average D_{KL} based on all the decision situations are used for calculating the relatedness between two tasks. The value of D_{KL} is just a tool to indicate the relatedness between tasks, the relative relatedness between all assisted tasks is more important for multitask learning.

14.2.3 Assisted Task Selection Strategy

The probabilities of candidate assisted tasks for a target task are designed proportionally to the relatedness between the candidate tasks and the target task. This corresponds to line 28 of Algorithm 14.1. $R(T_i, T_t)$ is the relatedness value of a candidate task T_i to a target task T_t. We give the more related tasks a higher chance to be selected as the assisted tasks. The probabilities of tasks to be selected (i.e., indicated by $Prob(T_i)$) as the assisted task are shown in Eq. (14.4).

$$Prob(T_i) = \frac{R(T_i, T_t)}{\sum_{i=1}^{|candidates|} R(T_i, T_t)} \tag{14.4}$$

First, we sum up the relatedness values of all the candidate assisted tasks (i.e., indicated by $|candidates|$) of a target task. Then, the probability of each candidate assisted task $Prob(T_i)$ is assigned proportionally to its relatedness to the target task.

14.3 Experiment Design

14.3.1 Comparison Design

Table 14.1 shows the details of the designed multitask scenarios represented by the optimised objective and utilisation level. The GP system with k subpopulations to solve k tasks independently, named GP, is used as the baseline algorithm. The second compared algorithm combines MFEA in [83] with GP, named MFGP. In addition, the state-of-the-art algorithm, named MTGP [231] that solves multiple DFJSS tasks with multiple subpopulations and an offspring reservation strategy is also compared. It is noted that MTGP selects the assisted task randomly for a target task. The presented multitask GP with adaptive assisted selection strategy is named ATMTGP, since it involves *assisted task* selection. To verify the effectiveness of the presented assisted task selection strategy for multitask learning, ATMTGP is compared with GP, MFGP, and MTGP based on their test performance on unseen instances.

14.3.2 Specialised Parameter Settings of Genetic Programming

The specialised parameter settings of GP are shown in Table 14.2. The transfer ratio indicates the frequency of a task to learn from other tasks, which is set to 0.3 as suggested in [83]. It is noted that the random parent selection for MFGP was suggested in [83]. For calculating the phenotypic characterisation, we follow the suggestions in [236] which is also the same as in Chap. 8. We use 20 routing decision situations with 7 candidate machines, and 20 sequencing decision situations with 7 candidate operations. Thus, the dimension of the phenotypic characterisation of an individual is 40. The decision situations are fixed to get the phenotypic characterisations of all the individuals, either learned for the same task or different tasks.

Table 14.1 The designed three multitask scenarios represented by the optimised objective and utilisation level

Scenario	Task 1	Task 2	Task 3
Scenario 1	<WTmean, 0.75>	<WFmean, 0.75>	<WTmax, 0.75>
Scenario 2	<WTmean, 0.85>	<WFmean, 0.85>	<WTmax, 0.85>
Scenario 3	<WTmean, 0.95>	<WFmean, 0.95>	<WTmax, 0.95>

Table 14.2 The specialised parameter settings of genetic programming

Parameter	Value
*Number of subpopulations	k
*Subpopulation size	1000
*The number of elites for each subpopulation	10
**Number of tasks	k
**Population size	$1000 * k$
**The number of elites for each task	10
**Parent selection	Random selection
The transfer ratio rmp	0.3

*: for GP, MTGP, and ATMTGP with multiple subpopulations only
**: for MFGP with one population only

14.4 Results and Discussions

14.4.1 Quality of Learned Scheduling Heuristics

Table 14.3 shows the mean and standard deviation of the objective values on the test
instances according to 30 independent runs of GP, MFGP, MTGP, and ATMTGP in
the three multitask scenarios. The results show that the presented ATMTGP performs
the best with the smallest average rank value according to Friedman's test. MFGP,
which incorporates the idea of MFEA into MFGP directly, performs even worse than
GP in most of the scenarios. One possible reason is that the tasks investigated in this
chapter are quite different since they have different objectives (i.e., an important
factor to lead the search direction), the traditional way that allocates individuals for
different tasks via skill factor is not applicable any more. We can also see that by using
the framework of multi-population with offspring reservation strategy [231], MTGP
achieves better performance than GP and MFGP. These findings are consistent with
the observations in [231]. It is noted that MTGP selects the assisted task randomly
for a target task.

ATMTGP performs better than MTGP for six (i.e., <WTmean, 0.75>, <WFmean,
0.75>, <WTmean, 0.85>, <WFmean, 0.85>, <WTmean, 0.95>, and <WFmean,
0.95>) out of the nine tasks. This verifies the effectiveness of the presented assisted
task selection strategy, since the only difference between MTGP and ATMTGP is
the assisted task selection strategy. We can also find that ATMTGP achieves similar performance as MTGP in only three tasks (i.e., <WTmax, 0.75>, <WTmax,
0.85>, and <WTmax, 0.95>). This is consistent with our expectations, since the
tasks with the objective of WTmean and the tasks with the objective of WFmean can
help each other (i.e., have a higher relatedness). However, the tasks with the objective
of WTmean, and the tasks with objective WFmean cannot help the tasks with the
objective of WTmax (i.e., have a lower relatedness). More details of the relatedness
analyses between them will be provided later.

Table 14.3 The mean (standard deviation) of the objective values on test instances of GP, MFGP, MTGP, and ATMTGP over 30 independent runs in three multitask scenarios (i.e., each scenario contains three tasks)

Sce.	Task	GP	MFGP	MTGP	ATMTGP
1	<WTmean, 0.75>	27.64(1.10)	29.73(3.26)(+)	27.16(0.84)(−)(−)	26.78(0.63)(−)(−)(−)
	<WFmean, 0.75>	737.71(4.75)	740.93(6.45)(+)	736.44(4.61)(≈)(−)	734.72(2.57)(−)(−)(−)
	<WTmax, 0.75>	2762.84(229.58)	3015.51(351.80)(+)	2535.32(105.08)(−)(−)	2571.53(142.73)(−)(−)(≈)*
2	<WTmean, 0.85>	77.95(3.95)	80.77(4.49)(+)	77.73(3.70)(≈)(−)	76.02(1.97)(−)(−)(−)
	<WFmean, 0.85>	836.73(8.74)	838.93(8.11)(≈)	834.90(8.60)(≈)(−)	832.36(5.38)(−)(−)(−)
	<WTmax, 0.85>	3838.35(254.07)	3888.66(305.84)(≈)	3607.01(242.96)(−)(−)	3598.40(220.16)(−)(−)(≈)
3	<WTmean, 0.95>	306.09(10.56)	306.85(10.85)(≈)	300.62(7.97)(−)(−)	296.62(5.99)(−)(−)(−)
	<WFmean, 0.95>	1121.95(11.62)	1133.35(15.21)(+)	1117.91(13.39)(≈)(−)	1110.47(6.52)(−)(−)(−)
	<WTmax, 0.95>	6218.45(708.01)	6165.27(556.55)(≈)	5930.66(594.25)(≈)(≈)	5669.19(597.66)(−)(−)(≈)

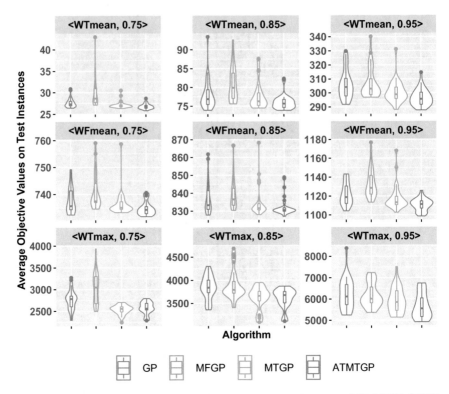

Fig. 14.1 The violin plot of the average objective values on test instances of GP, MFGP, MTGP, and ATMTGP based on 30 independent runs in three multitask scenarios (i.e., each column is a multitask scenario)

Figure 14.1 shows the violin plot of the average objective values on test instance of GP, MFGP, MTGP, and ATMTGP in the three multitask scenarios. It is clear that the presented algorithm ATMTGP obtains smaller objective values for the tasks minimising WTmean or WFmean among the involved four algorithms as shown in the first and the second rows in Fig. 14.1. More importantly, in general, the obtained objective values of ATMTGP on unseen instances are smaller than the values achieved by MTGP. This indicates that ATMTGP can detect the proper tasks as the assisted tasks for positive transfer, which verifies the effectiveness of the presented assisted task selection strategy from the perspective of accelerating positive knowledge transfer. It is noted that, for the tasks with the objective of WTmax in all scenarios, ATMTGP does not achieve significantly better performance than MTGP, since these tasks are less related to other tasks [44]. More details of the relatedness analyses between them will be provided later. However, we can still see that the obtained objective values of ATMTGP tend to be smaller than other algorithms as shown in tasks <WTmax, 0.85> and <WTmax, 0.95>.

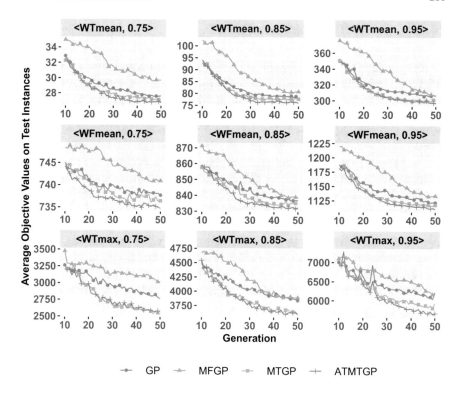

Fig. 14.2 The curves of the average objective values on test instances of GP, MFGP, MTGP, and ATMTGP based on 30 independent runs in three multitask scenarios (i.e., each column is a multitask scenario with three tasks)

14.4.1.1 Curves of the Average Objective Values on Test Instances

Figure 14.2 shows the curves of the average objective values on test instances of GP, MFGP, MTGP, and ATMTGP in the three multitask scenarios. The results show that ATMTGP can achieve better scheduling heuristics than other algorithms from an early stage for the desired tasks (i.e., <WTmean, 0.75>, <WFmean, 0.75>, <WTmean, 0.85>, <WFmean, 0.85>, <WTmean, 0.95>, and <WFmean, 0.95>), and keep this advantage during the whole evolutionary process. ATMTGP also shows its superiority in the undesired task scenario <WTmax, 0.95>. This further verifies the effectiveness of the presented assisted task selection strategy.

In summary, the presented ATMTGP can achieve better performance in the examined multitask scenarios. It can not only select proper tasks to realise positive knowledge sharing for a target task if there is a helpful task, but also reduce negative knowledge sharing for a target task if there are no useful assisted tasks.

14.4.2   Relatedness Between Tasks

To investigate why ATMTGP performs better than other algorithms, an important point is to analyse the relatedness between tasks. Figure 14.3 shows the curves of the calculated relatedness between tasks during the evolutionary process of ATMTGP in different multitask scenarios according to 30 independent runs. The relatedness between one task and itself is 1, and the main focus is to see the relatedness between a target task and other tasks. Overall, the results show that the relatedness values between the tasks are quite large at the beginning of the evolutionary process, and become smaller and smaller along with the generations. This is consistent with the intuition that the individuals at the early generations are not specific to tasks, and have a similar quality for all the tasks. However, in the later generations, the individuals are improved, and the individuals for each task become more specific for the corresponding task. Therefore, the measured relatedness value between tasks reduces along with the generations, since the tasks are different and the individuals tend to be good for the corresponding task only. For task 1 (i.e., WTmean) in different scenarios as shown the first column in Fig. 14.3, the results show that it is more related to task

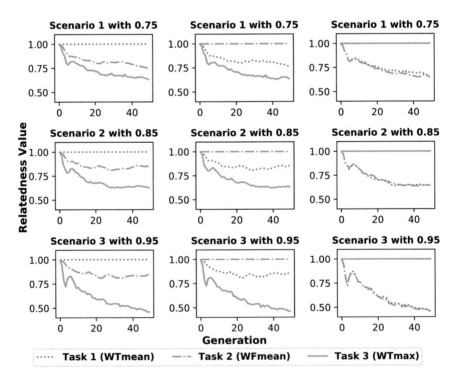

Fig. 14.3 The curves of the relatedness between tasks of ATMTGP based on 30 independent runs in three multitask scenarios (i.e., each row is a multitask scenario, the ith column represents the relatedness between task T_i and other tasks)

2 (i.e., WFmean). The task 2 (i.e., WFmean) in different scenarios as shown in the second column, which is more related to task 1 (i.e., WTmean). This is consistent with the findings for task 1, which shows that task 1 and task 2 are related. For task 3, the results show that task 1 and task 2 have the same relatedness values with task 3, which are lower than the relatedness between themselves. Taking the performance into consideration (i.e., task 1 and task 2 do not significantly help task 3), it concludes that both task 1 and task 2 are less related to task 3. For convenience, if the tasks cannot help each other and they have a relatively small relatedness value, we will say that they are not related below. Otherwise, they are related tasks. This points to another interesting question of the boundary values of relatedness between related and unrelated tasks in the investigated problems.

We investigate the relatedness values between tasks at all generations in all the scenarios, as shown in Fig. 14.3. The results show that the boundary relatedness values are different in different scenarios. For multitask scenario 1, as shown in the first row of Fig. 14.3, the relatedness values of the related tasks for a target task are around 0.75, while the relatedness values of the unrelated values for the target task are about 0.63. For multitask scenario 2, the relatedness values of related tasks are around 0.83, while the relatedness values of unrelated tasks are about 0.6. For multitask scenario 3, the relatedness values of the related tasks are around 0.85, while the relatedness values of unrelated tasks are about 0.48. Based on these observations, we can draw the conclusion that a complex task (i.e., a task with a larger utilisation level) requires a higher relatedness value than a simple task to identify two tasks that are related. In addition, a complex task requires a lower relatedness value than a simple task to identify two tasks that are unrelated.

Overall, we have the following findings based on the observations of the relatedness between tasks.

- The boundaries of the related and unrelated tasks in a more complex multitask scenario are larger than the boundaries in a simpler multitask scenario.
- When a problem becomes complex with a high utilisation level, the individuals for each task are more strict to the corresponding task and the tasks are not easy to relate to each other.
- The boundary relatedness value for distinguishing related and unrelated tasks for the investigated problems in this chapter depends on the utilisation level.

It is noted that the relatedness boundary may also depend on the way for calculating the relatedness values.

14.4.3 Selected Assisted Tasks

To investigate why the presented ATMTGP performs better than other algorithms, another important point is to investigate whether the assisted tasks are properly selected, i.e., more related tasks have higher chances to be selected as expected. We

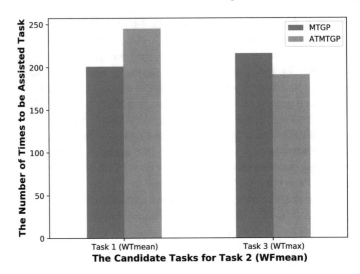

Fig. 14.4 The bar plot of the number of selected times of the assisted tasks (i.e., task 1 or task 3) for a target task (i.e., task 2) in scenario 2 with utilisation level 0.85 of ATMTGP based on 30 independent runs

take scenario 2 with utilisation level 0.85 as an example in this subsection. Figure 14.4 shows the bar plot of the number of selected times of candidate assisted tasks (i.e., task 1 and task 3) for the target task 2 in scenario 2 according to 30 independent runs. The results show that ATMTGP can successfully give a higher chance than MTGP to select the related task (i.e., task 1, WTmean) for task 2 (i.e., WFmean). In addition, ATMTGP gives a lower chance than MTGP to select an unrelated task (i.e., task 3, WTmax) for task 2 (i.e., WFmean). The same patterns can also be found in other scenarios.

14.4.4   Training Time

Table 14.4 shows the mean and standard deviation of the training time of GP, MFGP, MTGP, and ATMTGP based on 30 independent runs in the three multitask scenarios. The results show that there is no significant difference among the involved four algorithms. This means ATMTGP does not need extra computational cost to achieve better performance than other algorithms.

Table 14.4 The mean (standard deviation) of the training time (in minutes) of GP, MFGP, MTGP, and ATMTGP over 30 independent runs in three multitask scenarios

Scenario	GP	MFGP	MTGP	ATMTGP
1	163(20)	164(21)($\approx$)	165(20)($\approx$)($\approx$)	166(23)($\approx$)($\approx$)($\approx$)
2	184(17)	178(22)($\approx$)	176(24)($\approx$)($\approx$)	182(21)($\approx$)($\approx$)($\approx$)
3	193(21)	194(18)($\approx$)	193(26)($\approx$)($\approx$)	194(21)($\approx$)($\approx$)($\approx$)

14.5 Further Analyses

To deeply understand the proposed algorithm, the evolved scheduling heuristics obtained by ATMTGP, whether the performance improvement of ATMTGP benefits from population diversity or knowledge transfer, and the proposed assisted task selection strategy are further analysed.

14.5.1 Insights of Learned Scheduling Heuristics

The success of multitask learning relies on the knowledge transfer between different tasks. We choose the learned scheduling heuristics for sequencing the operations for different tasks in a multitask scenario as an example to investigate how the tasks help with each other. Figures 14.5, 14.6 and 14.7 show one of the learned sequencing rules for task 1 <WTmean, 0.75>, task 2 <WFmean, 0.75> and task 3 <WTmax, 0.75> in multitask scenario 1 from one run, respectively. It is obvious that the two sequencing rules for task 1 and task 2 share more knowledge between each other, since the major part of the rules is the same, which is highlighted in grey. However, the sequencing rule for task 3 is quite different from the sequencing rules for task 1 and task 2, according to the structure of the rule. One possible reason is that the tasks that minimise mean weighted objectives (i.e., WTmean and WFmean) have similar features, while they have different features from the tasks that minimise max weighted objectives (i.e., WTmax). This verifies the high relatedness between task 1 and task 2, and low relatedness of task 1 and task 2 with task 3. This finding is consistent with our previous observations in Sect. 14.4.1.

14.5.2 Population Diversity Versus Knowledge Transfer

In general, increasing the diversity of the population can improve the performance of an algorithm for handling a task to some extent. Learning from different tasks tends to improve the population diversity of multitask algorithms by getting genetic materials from other subpopulations. If the tasks are not related, is it possible to

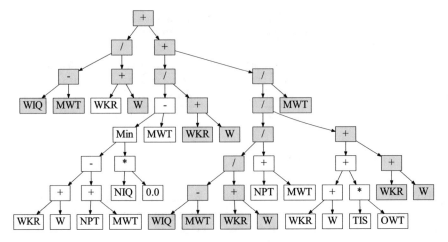

Fig. 14.5 One of the best learned sequencing rules for **task 1** <**WTmean, 0.75**> in multitask scenario 1

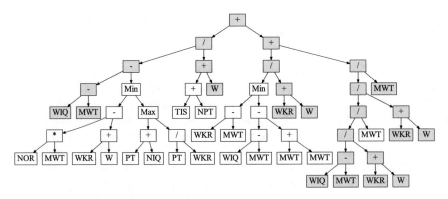

Fig. 14.6 One of the best learned sequencing rules for **task 2** <**WTmean, 0.75**> in multitask scenario 2

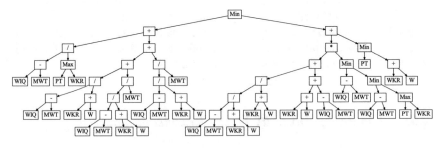

Fig. 14.7 One of the best learned sequencing rules for **task 3** <**WTmax, 0.75**> in multitask scenario 2

improve the performance of the algorithm by purely increasing diversity? Does the performance improvement of multitask algorithms mainly benefits from population diversity or transferred knowledge [82]?

To answer these two questions, we carry out experiments that use multitask algorithm MTGP to solve multitask scenarios with related and unrelated tasks, respectively. Each multitask scenario consists of two tasks. The multitask scenarios with unrelated tasks are designed to optimise WTmean (mean-weighted-tardiness) with job weights 1, 2, 4, and WTmax (max-weighted-tardiness) with job weights 1, 5, 10. The multitask scenarios with related tasks are designed to optimise WTmean (mean-weighted-tardiness) with job weights 1, 2, 4, and WFmean (mean-weighted-flowtime) with job weights 1, 2, 4. If the tasks in the multitask scenarios with related tasks can help each other, and the tasks in the multitask scenarios with unrelated tasks can not help each other, we can claim that the effectiveness of multitask learning mainly benefits from transferred knowledge. The experiment settings are the same as in Sect. 14.3.

Table 14.5 shows the mean and standard deviation of the objective values of GP and MTGP for multitask scenarios with two unrelated tasks (i.e., WTmean and WTmax) based on 30 independent runs. The results show that for the tasks with objective WTmax, MTGP performs better than GP in two (i.e., <WTmax, 0.75> and <WTmax, 0.85>) out of the three scenarios. However, the tasks with the objective of WTmax are not helpful for the tasks with the objective of WTmean, which is consistent with our findings in Sect. 14.4.1. According to Sect. 14.4.1, we know that the tasks with the objective of WTmax and WTmean are not related. For the tasks with objective WTmean, according to Table 14.5, they do not get benefit from either transferred knowledge or diversity. This shows that even transferring knowledge from unrelated tasks can help sometimes, but the effectiveness can not be guaranteed. Table 14.6 shows the mean and standard deviation of the objective values of GP and MTGP for the multitask scenarios with two related tasks (i.e., WTmean and WFmean) based on 30 independent runs. The results show that the performance of MTGP is significantly improved for all tasks in all scenarios.

Taking the results in Tables 14.5 and 14.6 into consideration, we can draw the conclusions that learning knowledge from other tasks may benefit a task by improving the diversity or sharing the useful knowledge. However, the key to improving the effectiveness of a multitask algorithm is to get useful knowledge from other tasks. This is consistent with the findings in [82]. The diversity effect investigated in this chapter cannot guarantee to make a significant improvement for handling the tasks. This further verifies the effectiveness of the presented adaptive assisted task selection strategy for finding useful tasks to learn knowledge from.

14.5.3 Is It Good to Always Choose the Most Related Task?

This chapter gives more related tasks higher chances to be selected as assisted tasks. It is interesting to know whether it is more effective to only learn from the most similar task. Intuitively, if two tasks are too similar, they might have no new genetic

Table 14.5 The mean (standard deviation) of the objective values on test instances of GP and MTGP over 30 independent runs in three multitask scenarios with two **unrelated tasks**

Scenario	Task	GP	MTGP
1	<WTmean, 0.75>	27.20(0.94)	27.15(0.93)($\approx$)
	<WTmax, 0.75>	3133.52(865.39)	2762.74(298.65)($-$)
2	<WTmean, 0.85>	75.66(2.09)	76.59(4.11)($\approx$)
	<WTmax, 0.85>	4097.84(1015.73)	3883.28(1358.91)($-$)
3	<WTmean, 0.95>	298.64(12.56)	293.91(6.35)($\approx$)
	<WTmax, 0.95>	5818.60(384.19)	5998.40(378.14)($\approx$)

* WTmean and WTmax are unrelated tasks

Table 14.6 The mean (standard deviation) of the objective values on test instances of GP and MTGP over 30 independent runs in three multitask scenarios with two **related tasks**

Scenario	Task	GP	MTGP
1	<WTmean, 0.75>	27.32(1.44)	26.78(0.82)($-$)
	<WFmean, 0.75>	736.17(3.62)	733.58(1.63)($-$)
2	<WTmean, 0.85>	75.80(2.98)	74.65(0.86)($-$)
	<WFmean, 0.85>	833.42(8.07)	828.47(1.73)($-$)
3	<WTmean, 0.95>	297.27(9.41)	290.91(4.37)($-$)
	<WFmean, 0.95>	1113.27(13.11)	1106.28(6.88)($-$)

* WTmean and WFmean are related tasks

materials to assist each other. Therefore, always learning from the most similar task might not be a good choice. We carry out another experiment with ATMTGP but higher chances the assisted task selection strategy from giving more related tasks higher chance to be selected to only select the most related tasks as assisted tasks. For convenience, we name ATMTGP that always selects the most related tasks as assisted tasks ATMTGP$_{most}$, while we name ATMTGP that gives more related tasks more chance to be selected as assisted tasks ATMTGP$_{more}$.

Table 14.7 shows the mean and standard deviation of the objective values on test instances of ATMTGP$_{more}$ and ATMTGP$_{most}$ according to 30 independent runs in three multitask scenarios as shown in Table 14.1. The results show that ATMTGP$_{most}$ performs worse than ATMTGP$_{more}$ in two out for the nine tasks (i.e., <WTmean, 0.75> and <WFmean, 0.95>). In addition, ATMTGP$_{most}$ does not show any superiority compared with ATMTGP$_{more}$ on other tasks. Taking the mean values into consideration, ATMTGP$_{most}$ performs even worse than ATMTGP$_{more}$ on task <WFmean, 0.75>, <WTmean, 0.85>, <WFmean, 0.85>, <WTmax, 0.85>, <WTmean, 0.95>, and <WTmax, 0.95>. This indicates that only learning from the most related task is not the most effective way in multitask learning. One possible reason is that if two tasks are too similar, there might have limited new genetic materials to learn from each other. This verifies the effectiveness of the presented assisted task selection strategy which is proportional to relatedness values of the assisted tasks. However, whether there is a more effective way to select assisted task is worth studying in the future work.

Table 14.7 The mean (standard deviation) of the objective values on test instances of ATMTGP$_{more}$ and ATMTGP$_{most}$ over 30 independent runs in three multitask scenarios

Scenario	Task	ATMTGP$_{more}$	ATMTGP$_{most}$
1	\<WTmean, 0.75\>	26.78(0.63)	27.67(1.45)(+)
	\<WFmean, 0.75\>	734.72(2.57)	736.91(7.92)($\approx$)
	\<WTmax, 0.75\>	2571.53(142.73)	2524.76(140.04)($\approx$)
2	\<WTmean, 0.85\>	76.02(1.97)	77.34(2.99)($\approx$)
	\<WFmean, 0.85\>	832.36(5.38)	832.61(3.53)($\approx$)
	\<WTmax, 0.85\>	3598.40(220.16)	3624.38(215.04)($\approx$)
3	\<WTmean, 0.95\>	296.62(5.99)	298.22(6.04)($\approx$)
	\<WFmean, 0.95\>	1110.47(6.52)	1114.76(7.68)(+)
	\<WTmax, 0.95\>	5669.19(597.66)	5916.75(515.25)($\approx$)

14.6 Chapter Summary

This chapter presents an effective adaptive assisted task selection strategy for choosing a proper task for multitask GP in DFJSS where the relatedness between tasks is unknown. The success of the presented algorithm depends on the presented effective way to measure the relatedness between tasks, and a properly designed algorithm to select the assisted tasks based on the relatedness information.

The results show that the presented ATMTGP can obtain significantly better scheduling heuristics for all the desired tasks in all the tested multitask scenarios. We also find that the boundary of relatedness level of tasks is problem dependent. For example, the relatedness with a value of 0.5 might indicate that a task is not related to another task in one scenario, while it may mean a middle level relatedness relationship between tasks in another scenario. Therefore, for developing assisted task selection strategy, the relative relatedness between tasks is more important than the absolute relatedness value. The effectiveness of the presented adaptive assisted task selection algorithm ATMTGP was examined by comparing not only the quality of learned scheduling heuristics, but also the relatedness analyses, the structures of the learned scheduling heuristics for all the tasks in a multitask scenario, and the discussions about diversity and knowledge transfer. It has been observed that the algorithm does manage to solve the tasks in a mutually reinforcing way.

The presented adaptive multitask GP algorithm can benefit the area of multitask learning in mainly three aspects. First, the presented algorithm broadens the study of multitask on measuring the relatedness between tasks according to individuals with tree-based and variable-length representation. Specifically, it extends the investigations of multitask with other popular but non-vector-based evolutionary algorithms. Second, a new assisted task selection strategy has been presented that the probabilities of tasks to be selected are proportional to the relatedness information. It can provide further guidance for designing the assisted task selection strategy rather than simply choosing the most related one, and can encourage people to dig more in this direction. Last, it is challenging to extract multitask problems when we do not know the characteristics of problems clearly. The way of extracting multitask scenarios based on optimised objectives in DFJSS can provide guidance for using the multitask approach to other problems, especially other combinatorial optimisation problems.

This chapter provides an effective multitask GP algorithm to learn knowledge from related tasks to realise positive knowledge sharing in DFJSS. The novelty is the proposed task relatedness measure for multitask DFJSS. In the next chapter, we will incorporate the surrogate technique into multitask GP algorithm to enhance the knowledge sharing between tasks in DFJSS.

Chapter 15
Surrogate-Assisted Multitask Genetic Programming for Learning Scheduling Heuristics

From the previous studies, we can see that the surrogate technique can make evolutionary algorithms more efficient (Chap. 7), and multitask learning can result in more effective GP for solving JSS problems (Chaps. 13 and 14). The existing studies mainly use surrogate models for single-task optimisation. To the best of our knowledge, only very limited studies were proposed to solve multitask problems by surrogates [61, 97, 136, 147]. The work [136] built a surrogate model based on Gaussian Process. In [97], surrogates are used to solve the problem with a smaller number of fitness evaluations. In [61, 147], multiple expensive optimisation problems are solved by multiple surrogate models, which acquire knowledge during the search and share with each other. Although these studies have shown good potential of the surrogate technique for multitask learning, they only use surrogates for each task independently. However, they do not consider the use of the surrogate for the knowledge sharing, which is the core mechanism of multitask learning.

15.1 Challenges and Motivations

The existing multitask approaches [61, 97, 136, 147] cannot be directly applied to evolve schedule rules for DFJSS. First, evaluating each individual on each task through a long simulation is too time consuming. Second, existing surrogate-assisted multitask methods are mainly developed for continuous optimisation benchmarks [248], but not discrete or combinatorial optimisation problems. The goal of this chapter is to explore how the surrogate technique can promote knowledge sharing in multitask learning rather than improving only the efficiency of algorithms. This has not been explored in multitask learning. Specially, we are interested in the following questions:

F. Zhang et al., *Genetic Programming for Production Scheduling*, Machine Learning: Foundations, Methodologies, and Applications,
https://doi.org/10.1007/978-981-16-4859-5_15

- How effective are the built surrogates for multiple tasks?
- How about the effectiveness of the constructed surrogate models?
- How to allocate individuals to different tasks properly with surrogate technique?

Detailed descriptions of the proposed algorithm are given in Sect. 15.2. The experiment design is shown in Sect. 15.3, followed by results and discussions in Sect. 15.4. Further analyses are conducted in Sect. 15.5. Finally, Sect. 15.6 concludes this chapter.

15.2   Algorithm Design

This section first gives the framework of the surrogate-assisted multitask GP algorithm. Then, it describes the way to build surrogates. Finally, it illustrates the knowledge sharing mechanism with an example.

15.2.1   Framework of the Algorithm

The overall framework of the proposed algorithm is shown in Algorithm 15.1. Given a set of k tasks, the algorithm learns a scheduling heuristics for each task. First, k subpopulations are randomly initialised, each for a task (line 1). Then, each individual is evaluated by the training simulations corresponding to the task it solves (from line 8 to line 18). For each subpopulation, a surrogate is built based on the phenotypic behaviour of the individuals and their fitness on the original simulation (from line 19 to line 21). Then, at each generation, each subpopulation generates a number of $n * subpopsize$ offspring (from line 24 to line 27). Then, the offspring with duplicated phenotypic characterisations are removed (line 28). Each remaining offspring is evaluated by the surrogates S_i ($i = 1, \ldots, k$) (line 31). For each subpopulation $Subpop_i$, the offspring are sorted based on their surrogate fitness given by S_i, and the best $subpopsize$ offspring are selected as the next $Subpop_i$ (from line 29 to line 34). We can evaluate the knowledge-carrying individuals efficiently using the surrogates.

15.2.2   Surrogate Model

In this work, we use the surrogate based on KNN and phenotypic characterisation [92], which has shown to be very effective in GP for dynamic scheduling. Specifically, to estimate the surrogate fitness of an individual, it is compared with all the individuals evaluated at the previous generation, and its surrogate fitness is set to the real fitness of the individual with the most similar phenotypic characterisation. The

---

Algorithm 15.1 Outline of the Algorithm

Input: Tasks $T_1, T_2, \ldots, T_k$
Output: The learned rules $h_1^*, h_2^*, \ldots, h_k^*$
1: **Initialisation**: Generate the individuals of k subpopulations randomly
2: set $h_1^*, h_2^*, \ldots, h_k^* \leftarrow null$
3: set $fitness(h_1^*), fitness(h_2^*), \ldots, fitness(h_k^*) \leftarrow +\infty$
4: $gen \leftarrow 0$
5: **while** $gen < maxGen$ **do**
6:     set $S \leftarrow null$
7:     set $newsubpop \leftarrow null$
8:     // **Evaluation** each individual (rule) in each subpopulation
9:     **for** i = 1 to k **do**
10:         **for** j = 1 to $subpopsize$ **do**
11:             Apply h_j to the training simulation of task T_i, and generate $Schedule_j$
12:             $fitness_{h_j} \leftarrow Obj(Schedule_j)$
13:         **end for**
14:         **for** j = 1 to $subpopsize$ **do**
15:             **if** $fitness_{h_j} < fitness_{h_i^*}$ **then**
16:                 $h_i^* \leftarrow h_j$
17:             **end if**
18:         **end for**
19:         Compute the phenotypic vector of the individuals in $Subpop_i$
20:         Update the surrogate model S_i by the phenotypic vectors and the individuals' fitness in $Subpop_i$
21:         $S \leftarrow S \cup S_i$
22:     **end for**
23:     **if** $gen < maxGen - 1$ **then**
24:         Select one or two individuals as parents
25:         // **Evolution**: generate the offspring for the next generation
26:         For each subpopulation, generate $n * subpopsize$ offspring using crossover/mutation/reproduction operators.
27:         Merge all the offspring together to form an offspring pool
28:         Remove the duplicates from the offspring pool
29:         // **Allocate offspring to tasks**
30:         **for** i = 1 to k **do**
31:             Predict the fitness of the offspring in the offspring pool by S_i
32:             Select the $subpopsize$ offspring with the best predicted fitness to form $newInds$
33:             $newsubpop \leftarrow newsubpop \cup newInds$
34:         **end for**
35:     **end if**
36:     $gen \leftarrow gen + 1$
37: **end while**
38: **return** $h_1^*, h_2^*, \ldots, h_k^*$

---

algorithm generates a large number of offspring, and uses the surrogates to pre-select the potentially good ones. It is similar to the previous studies of pre-selection [92, 162, 165], but focuses more on knowledge sharing in the context of multitask optimisation. As a result, the surrogates are used in a way that not only tackles a single task more efficiently, but also transfers promising building blocks among subpopulations effectively [238].

Algorithm 15.2 Duplicate removal from the offspring pool

Input: The generated offspring from all the subpopulations: Ind
Output: The set of offspring without phenotypic duplcate: $NonRepInd$
1: $PV \leftarrow \emptyset$
2: **for** i = 1 to $|Ind|$ **do**
3:    compute the phenotypic vector PV_i of Ind_i
4:    $PV \leftarrow PV \cup PV_i$
5: **end for**
6: **for** i = 1 to $|PV|$ **do**
7:    **for** j = i +1 to $|PV|$ **do**
8:        compute the distance $\delta(PV_i, PV_j)$ between PV_i and PV_j
9:        **if** $\delta(PV_i, PV_j) == 0$ **then**
10:            remove Ind_j from Ind
11:        **end if**
12:    **end for**
13: **end for**
14: **return** $NonRepInd \leftarrow Ind$

15.2.3 Knowledge Sharing with Surrogate

We describe the designed knowledge sharing mechanism with surrogate in the proposed algorithm with an example of three tasks, as shown in Fig. 15.1. The algorithm contains three subpopulation, each for solving a task. Each subpopulation consists of three individuals (for simplistic illustration). The individuals of different subpopulations are marked with different colours. For multitask learning, the key is to design an effective knowledge sharing mechanism for tasks to learn from each other. First, the three surrogate models (i.e., S_1, S_2, S_3) are built based on the phenotypic characterisations and the fitness of individuals in the subpopulations. Second, to get a number of individuals with knowledge from different tasks for effective knowledge sharing, $n * subpopsize$ (i.e., n is two in Fig. 15.1) are generated from each subpopulation independently. Then, all the offspring generated for all the tasks are merged into a single offspring pool, and the individuals with duplicated phenotypic charactersations are removed from the pool (as shown in Algorithm 15.2). In Fig. 15.1, the removed duplicated individuals are shown as dotted circles. Finally, for $Subpop_1$ ($Subpop_2$ or $Subpop_3$), the individuals in the offspring pool are evaluated by the surrogate S_1 (S_2 or S_3), and the best $subpopsize$ individuals are selected to the next generation. The newly generated subpopulations are denoted as $Subpop_1^*$, $Subpop_2^*$, and $Subpop_3^*$. Note that an individual can be allocated to multiple subpopulations, if its surrogate fitness is good enough for the corresponding tasks.

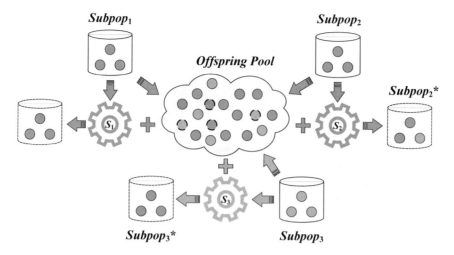

Fig. 15.1 An example of the proposed surrogate-assisted multitask genetic programming algorithm with three subpopulations for solving three tasks simultaneously

15.2.4 Algorithm Summary

Through building multiple surrogate models, the new algorithm is able to solve each single DFJSS task more efficiently, and also transfer promising building blocks of individuals for solving multiple DFJSS tasks more effectively. For each single DFJSS task, the KNN-based surrogate model can pre-select good offspring from many generated offspring. For knowledge sharing, the surrogate is used to estimate the offspring generated for different tasks, so that individuals from other subpopulation can be transferred if its surrogate fitness is good enough for this task.

15.3 Experiment Design

15.3.1 Comparison Design

The multitask scenarios used in the experiments of this chapter are shown in Table 15.1. The proposed algorithm in this chapter is denoted as SMT^{2}GP, since it includes surrogate, multi-tree, and multitask. For comparison, we select the baseline multi-tree GP (MTGP) [239] that solves each task independently. We also compare with the state-of-the-art multitask GP algorithm in DFJSS presented in Chap. 13 named M^{2}GP^f [234]. In addition, we compare with the SMTGP algorithm [92] that uses KNN-based surrogate model. It is a single-task algorithm, so we applied it to solve each task separately.

Table 15.1 The designed multitask scenarios with tasks represented by optimised objective and utilisation level

Scenario	Task 1	Task 2	Task 3
Scenario 1	<Fmean, 0.75>	<Fmean, 0.85>	<Fmean, 0.95>
Scenario 2	<Tmean, 0.75>	<Tmean, 0.85>	<Tmean, 0.95>
Scenario 3	<WTmean, 0.75>	<WTmean, 0.85>	<WTmean, 0.95>

Table 15.2 The specialised parameter settings of genetic programming

Parameter	Value
Number of subpopulations	3
Subpopulation size	400
The number of offspring for each task	400 * 12
Intermediate population size of SMTGP/SMT^{2}GP	400 * 12 * 3/400 * 4 * 3

The effectiveness of using surrogates in DFJSS can be verified by comparing MTGP with SMTGP. The effectiveness of the proposed multitask mechanism is verified by the comparison between SMTGP and SMT^{2}GP.

15.3.2   Specialised Parameter Settings of Genetic Programming

Table 15.2 shows the specialised parameter settings of GP in this chapter. Three subpopulations with a size of 400 individuals are used to solve the three tasks simultaneously. It is noted that increasing the number of individuals in the intermediate population can improve the performance of SMTGP. However, the improvement for SMTGP is marginal, especially when the number of individuals increases to a certain number. This is consistent with the finding drawn in [92]. Based on our preliminary results, 12 times of the subpopulation size (400 * 12) is a proper intermediate subpopulation size, and further increasing this number cannot lead to further performance improvement. As there are three tasks in the experiments, the intermediate subpopulation size is set to 400 * 4 for each subpopulatoin in SMT^{2}GP for fair comparison.

15.4 Results and Discussions

15.4.1 Quality of Learned Scheduling Heuristics

Figure 15.2 shows the distribution (violin plot) of the test performance of MTGP, M^2GP^f, SMTGP, and SMT^2GP in the three multitask scenarios in the experiments. The figure shows that M^2GP^f and SMTGP performed better than MTGP in all the scenarios. This shows that using surrogate or multitask optimisation is effective for solving DFJSS, which is consistent with [92, 234]. SMT^2GP outperformed SMTGP, M^2GP^f and MTGP. This demonstrates the effectiveness of SMT^2GP.

Figure 15.3 shows the average test curves of the compared algorithms. It can be seen that SMT^2GP converges faster than all the other algorithms for all the nine tasks. Since SMTGP and SMT^2GP are different only in the individual pre-selection, the advantage of SMT^2GP over SMTGP demonstrates the effectiveness of the proposed surrogate-based pre-selection with knowledge sharing.

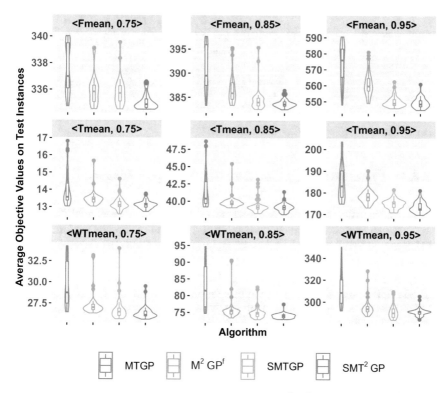

Fig. 15.2 The distribution of the test performance of MTGP, M^2GP^f, SMTGP, and SMT^2GP over 30 independent runs in the three multitask DFJSS scenarios, each with three tasks

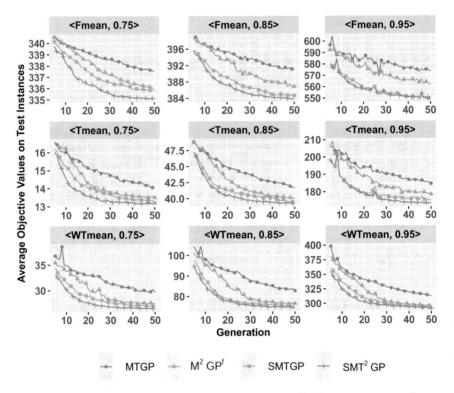

Fig. 15.3 The test curves over 30 independent runs of MTGP, M^2GP^f, SMTGP, and SMT^2GP in three multitask DFJSS scenarios, each with three tasks

15.4.2 Effectiveness of Constructed Surrogate Models

The surrogate is the foundation of the success for surrogate-assisted multitask learning. Table 15.3 shows the mean and standard deviation of the test performance of MTGP and SMTGP. From the table, it can be seen that SMTGP performed significantly better than MTGP on all the tasks.

Figure 15.4 shows the average test curves of over 30 independent runs of MTGP and SMTGP in the nine tasks of the three multitask scenarios. It can be seen that SMTGP converges much faster than MTGP for all tasks. Before generation five, MTGP and SMTGP usually showed similar convergence. This may be because that at early stage, the sample individuals in the KNN surrogate are not good enough, and thus the surrogate prediction is not accurate enough.

In summary, the results verify the effectiveness of the surrogate-assisted knowledge sharing mechanism for the multitask DFJSS scenarios.

Table 15.3 The mean (standard deviation) of the test performance of MTGP and SMTGP over 30 independent runs

Scenario	Task	MTGP	SMTGP
1	<Fmean, 0.75>	337.57(1.80)	335.60(1.20)(−)
	<Fmean, 0.85>	391.04(4.65)	384.35(1.86)(−)
	<Fmean, 0.95>	573.70(12.17)	548.77(4.60)(−)
2	<Tmean, 0.75>	14.03(1.01)	13.15(0.48)(−)
	<Tmean, 0.85>	41.61(2.73)	39.16(0.81)(−)
	<Tmean, 0.95>	184.60(8.04)	173.32(1.25)(−)
3	<WTmean, 0.75>	29.67(2.55)	26.53(0.59)(−)
	<WTmean, 0.85>	82.75(6.71)	74.84(2.63)(−)
	<WTmean, 0.95>	312.26(15.86)	289.48(6.73)(−)

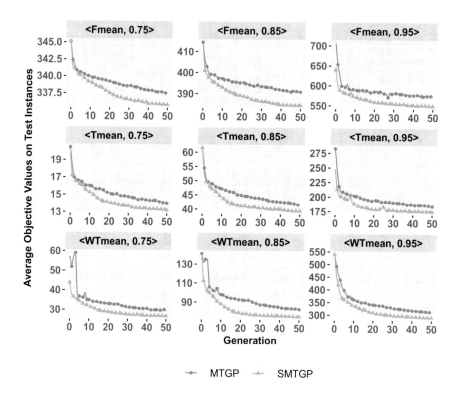

Fig. 15.4 The test curves of MTGP and SMTGP in the nine tasks

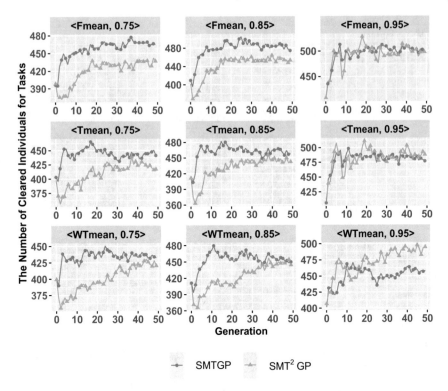

Fig. 15.5 The curves of the average number of cleared individuals over 30 independent runs of SMTGP and SMT^{2}GP in three multitask DFJSS scenarios, each with three tasks

15.4.3 Effectiveness of Diversity Preservation

To analyse the diversity of the offspring pool in the algorithm, we observe the number of cleared individuals. A smaller number of cleared offspring implies that the offspring pool has a higher diversity. Figure 15.5 shows the average curves of the number of cleared individuals over 30 independent runs of SMTGP and SMT^{2}GP in the nine tasks. From the figure, for the tasks with the utilisation level of 0.75 and 0.85, SMT^{2}GP has much fewer cleared individuals than SMTGP. This suggests that SMT^{2}GP has a better diversity of $Subpop_1$ for task 1 and $Subpop_2$ for task 2. This may be because that the final offspring in $Subpop_1$ and $Subpop_2$ are generated from all different subpopulations, and thus can be quite different due to the different parents selected (i.e., also can be produced by the individuals for different tasks). However, we cannot see this happening in task 3 with utilisation level of 0.95, especially in <WTmean, 0.95>.

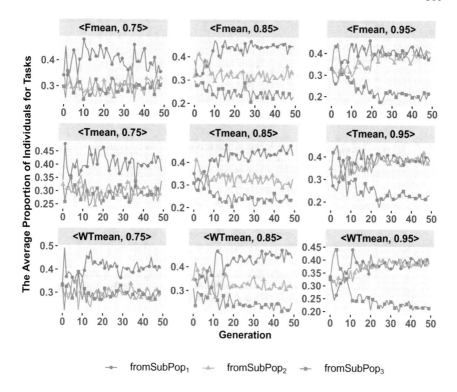

Fig. 15.6 The average number of individuals allocated to a specific task by SMT^{2}GP

15.4.4 Individual Allocation for Tasks

A key idea of multitask learning is to use proper individuals to solve appropriate tasks. To observe the task allocation, Fig. 15.6 shows the average number of allocated individuals over 30 independent runs of SMT^{2}GP in each of the nine tasks in the experiments. From the figure, one can see that the utilisation level is more important than the optimised objective.

For the subpopulation of task 1 (utilisation level of 0.75), the number of individuals originally generated for task 1 fluctuates around 40% during the evolutionary process. There are 30% individuals originally generated for task 2 and task 3, respectively. For task 2 (utilisation level of 0.85), there are 45% individuals from $Subpop_1$ for task 1, 30–35% from $Subpop_2$ for task 2, and less than 25% from $Subpop_3$ for task 3. For task 3 (utilisation level of 0.95), the numbers of individuals generated for task 1 and task 2 are quite similar, both between 35 and 40%. On the contrary, only a few individuals are from task 3. That is, the individuals from task 3 are not assigned properly. This somehow explains why $Subpop_3$ of SMT^{2}GP cannot show satisfactory diversity as expected.

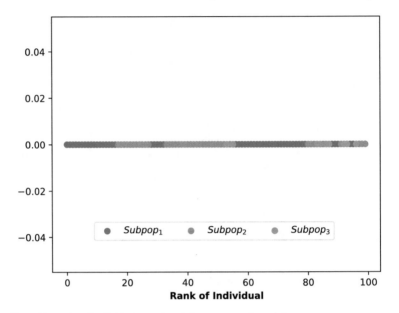

Fig. 15.7 The ranks of individual samples of the surrogate for task 3

In SMT^{2}GP, the new subpopulations contains individuals from different subpopulations due to the knowledge transfer. When using KNN, the individuals in the intermediate population will be selected if they are phenotypically similar with the top-ranked individuals in the surrogate pool. To investigate the effect of surrogate on choosing individuals, we set the *subpopsize* to 100 and take task 3 in the multitask scenario with mean-flowtime as an example. We collect the individual samples in the surrogate for task 3 at generation 20 (i.e., a steady-state), and evaluate the individuals on task 3 (i.e., mean-flowtime with utilisation level 0.95).

Figure 15.7 shows the ranks of individuals samples of the surrogate for task 3. For task 3, the results show that individuals from task 1 and task 2 have better ranks than those generated by task 3. The individuals which are close to the individuals originally from task 1 and task 2 based on phenotypic characteristics tend to be selected. This is the reason why the proportions of individuals for task 3 from $Subpop_1$, $Subpop_2$, and $Subpop_3$ are 0.4 (large), 0.4 (large), and 0.2 (small), respectively.

Individuals in the surrogate (named as *mapped individuals*) are close to the generated individuals for tasks that directly affect the choices of individuals in the intermediate population. Figure 15.8 shows the corresponding tasks of the mapped individuals in the surrogates for the individuals in the intermediate population for task 1, task 2, and task 3. The results show that the newly generated individuals for the current task T_i tend to be near the individual samples that are originally generated from the corresponding subpopulation $Subpop_i$ (i.e., $100 > 91 > 79$ corresponding to task 1, $66 > 63 > 50$ corresponding to task 2, and $6 \approx 7 > 1$ corresponding to task 3). It is consistent with our intuition that the newly generated individuals based

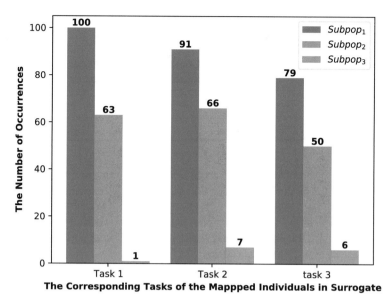

Fig. 15.8 The corresponding tasks of the mapped individuals in surrogate of the newly generated offspring for tasks

on the corresponding subpopulation tend to benefit the current task. Taking Fig. 15.7 into consideration, this is one reason for the bias to choose individuals from task 1 and task 2 for task 3, as shown in Fig. 15.6.

When we further look at the mapped individuals in task 3, most newly generated individuals for task 3 are close to the surrogate samples that are originally from task 1 and task 2. It is noted that although the surrogate samples for task 3 consist of large proportions of individuals that are originally for task 1 and task 2, the samples are good for task 3 (as shown in Fig. 15.7) and become the individuals for optimising task 3. The newly generated individuals for task 3, which have good quality, can still be kept into the next generation. Although the proportion of individuals samples in the surrogate S_3 is not high, this does not mean the individuals generated for task 3 are useless for task 3.

In summary, the individuals from task 1 and task 2 in the offspring pool are predicted to be better than those from task 3 by their surrogates. In addition, they are likely to behave more similarly with the individuals used by the KNN surrogate model for the same task. Therefore, the individuals generated for task 1 and 2 will be considered to be more promising than that from task 3. Figure 15.6 shows that task 1 has the least individuals removed, followed by task 2, and task 3 has the most individuals removed. Therefore, $Subpop_3$ always has the least number of individuals among the three subpopulations.

15.5   Further Analyses

To further understand the effect of the proposed algorithm, the sizes, structures, and behaviour of the learned scheduling heuristics are further analysed in this section.

15.5.1   Sizes of Learned Scheduling heuristics

Figure 15.9 shows the curves of the average sizes of the routing and sequencing rules obtained by MTGP, SMTGP, and SMT^{2}GP over 30 independent runs in each task of the three multitask scenarios. The results show that a more complex task requires a larger rule size for all the algorithms. For example, the rule sizes for the tasks with a utilisation level of 0.95 are larger than those for the tasks with a utilisation level of 0.75 or 0.85. The rule sizes of SMTGP and SMT^{2}GP are similar over generations, and the rule sizes of SMTGP and SMT^{2}GP are larger than MTGP in all the scenarios from the early stage. This implies that the use of the surrogate and multitask mechanisms

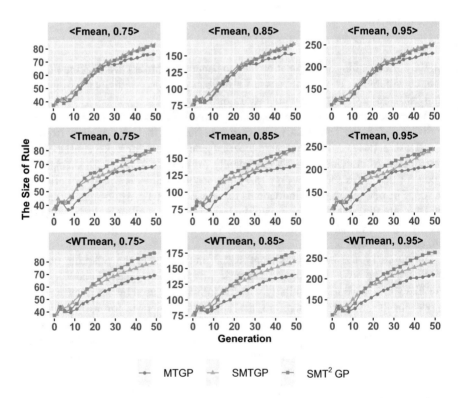

Fig. 15.9 The average routing plus sequencing rule size obtained by MTGP, SMTGP, and SMT^{2}GP

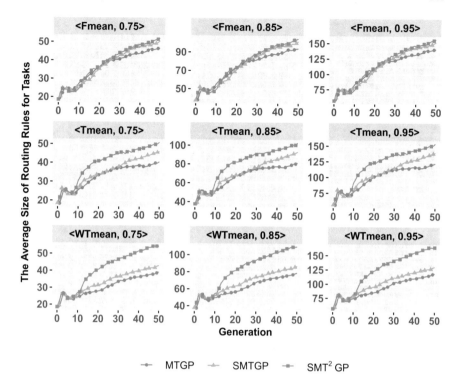

Fig. 15.10 The average **routing** rule sizes over 30 independent runs of the compared algorithms

may lead to larger scheduling heuristics than that of MTGP. A possible reason is that SMTGP tends to choose large and good scheduling heuristics in the intermediate population. For SMT²GP, another possible reason is that the parents may be from other subpopulations for complex tasks, and have large sizes. This might also be why SMTGP and SMT²GP can obtain good scheduling heuristics from the early stage.

SMTGP and SMT²GP showed similar rule sizes. To further study the effect of SMT²GP on the rule size, Figs. 15.10 and 15.11 show the curves of the average routing and sequencing rule sizes obtained by MTGP, SMTGP, and SMT²GP over 30 independent runs for each of the nine tasks, respectively. It can be seen that the sizes in each multitask scenario (i.e., the tasks with the same objective but with different utilisation level) show a similar trend but with different scales. It may be because the tasks in each multitask scenario are handled simultaneously by one population with three subpopulations, and the rule sizes of the subpopulations strongly interact with each other. We can see that the trend of individual allocation is highly dependent on the utilisation level, while the routing and sequencing rule size are highly related to the objective.

For the routing rule size, as shown in Fig. 15.10, SMT²GP tends to learn larger rules than SMTGP for most tasks. In addition, the routing rules obtained by SMTGP

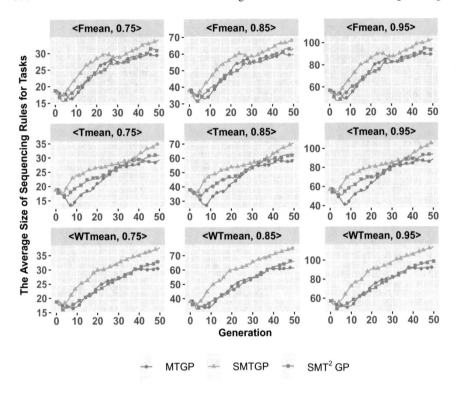

Fig. 15.11 The average **sequencing** rule sizes over 30 independent runs of the compared algorithms

have similar sizes to that of MTGP. For sequencing rule size, as shown in Fig. 15.11, SMTGP obtains larger rules than SMT^2GP for all the tasks. SMT^2GP and MTGP obtain similar sequencing rule sizes. This indicates that the rule size for tasks can be similar but with various routing and sequencing combinations. Furthermore, we can see that SMT^2GP improves the quality of the learned scheduling heuristics via routing rule, while SMTGP improves its performance through sequencing rule. Enhancing the quality of routing rule seems to be effective to improve the effectiveness of the final schedules in DFJSS.

15.5.2 Insight of Learned Scheduling Heuristics

As mentioned in the previous section, SMT^2GP has a significant impact on the routing rule. In this section, we choose three routing rules learned by SMT^2GP for each task in the multitask scenario 3 related to WTmean for further analysis. Figures 15.12, 15.13 and 15.14 show the best routing rules for <WTmean, 0.75>, <WTmean, 0.85>, and <WTmean, 0.95>, which are learned together in the same multitask scenario.

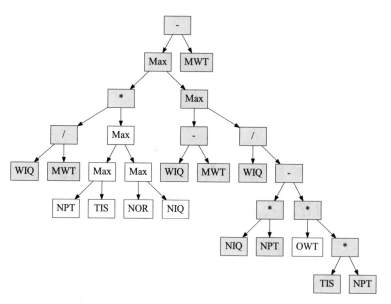

Fig. 15.12 An example learned routing rule for <WTmean, 0.75> in the multitask scenario 3

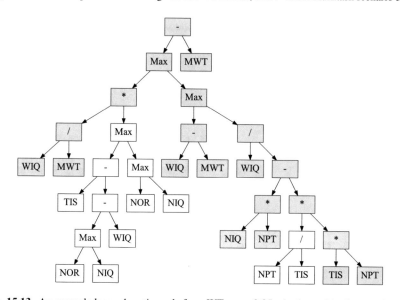

Fig. 15.13 An example learned routing rule for <WTmean, 0.85> in the multitask scenario 3

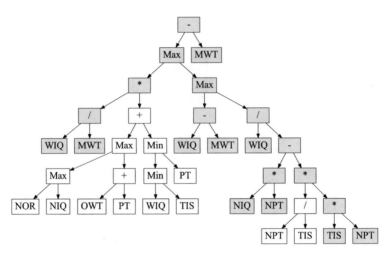

Fig. 15.14 An example learned routing rule for <WTmean, 0.95> in the multitask scenario 3

From the figures, we can see that the three routing rules show a major part of
structure (as shown in grey). A possible reason is that the three tasks have the same
objective, which require similar rule structures. For the learned building blocks,
except for the building blocks shown in grey, the routing rule for task 1 shares the
same component (i.e., Max{*, Max{NOR, NIQ}}) with the routing rule for task
2. Similarly, the routing rules for tasks 2 and 3 have common building blocks of
"Max{NOR, NIQ}" and "NPT/TIS", of which "Max{NOR, NIQ}" is also shared
by the routing rule of task 1. This shows that the rules of different tasks learn from
each other. The routing rules for task 1, task 2, and task 3 are of size 29, 35, and 37,
respectively. This is expected, as complex tasks often require larger rules.

For structural analysis, we simplify the routing rules as follows. Note that a
machine with a smaller priority value is considered as more prior. The routing rule
for task 1 in Fig. 15.12 can be simplified as Eq. (15.1).

$$
\begin{aligned}
R_1 = Max\{ &\frac{WIQ}{MWT} * Max\{NPT, TIS, NOR, NIQ\}, \\
&WIQ - MWT, \\
&\frac{WIQ}{NIQ * NPT - OWT * TIS * NPT}\} - MWT \\
\approx Max\{ &\frac{WIQ}{MWT} * TIS, WIQ - MWT, \\
&\frac{WIQ}{NPT(NIQ - OWT * TIS)}\} - MWT \\
= &\frac{WIQ}{MWT} * TIS - MWT
\end{aligned}
\tag{15.1}
$$

Due to the Max function in the root, only the component with the largest value decides the final value. "Max{NPT, TIS, NOR, NIQ}" can be replaced with TIS, since TIS is most likely to be bigger than NPT, NOR, and NIQ. The rule is converted to $\frac{WIQ}{MWT} * TIS - MWT$ using similar reasoning. After the simplification, this routing rule tends to select the machine with a small workload (WIQ) and a long idle time (MWT). Note that TIS is a constant here (the same for all the candidate machines of an operation) that will not significantly affect the final decision.

The simplification of the routing rule for task 2 in Fig. 15.13 is shown in Eq. (15.2).

$$
\begin{aligned}
R_2 &= Max\{\frac{WIQ}{MWT} * Max\{ \\
&\quad TIS - Max\{NIQ, NOR\} + WIQ, NOR, NIQ\}, \\
&\quad WIQ - MWT, \frac{WIQ}{NPT(NIQ - NPT)}\} - MWT \\
&\approx Max\{\frac{WIQ}{MWT} \\
&\quad * Max\{TIS - Max\{NIQ, NOR\} + WIQ\}, \\
&\quad WIQ - MWT, \frac{WIQ}{NPT(NIQ - NPT)}\} - MWT \\
&= \frac{WIQ}{MWT} * (TIS + WIQ - Max\{NIQ, NOR\}) \\
&\quad - MWT
\end{aligned}
\tag{15.2}
$$

First, "Max{TIS − Max{NIQ, NOR} + WIQ, NOR, NIQ}" is converted into "TIS − Max{NOR, NIQ} + WIQ", as "TIS − Max{NOR, NIQ} + WIQ" is usually larger than NOR and NIQ. We can see that this routing rule considers NIQ or NOR as well as WIQ and MWT. If NIQ is greater than NOR, it can be replaced with $\frac{WIQ}{MWT}$ * (TIS + WIQ − NIQ) − MWT. We can see that this rule considers NIQ, which was not in the rule for task 1. It tends to select the machine with many operations but a less workload in the queue. This means that the operations in the queue have short processing time. Therefore, the newly assigned operations are more likely to be processed earlier, since there will be more sequencing decision points. If NIQ is smaller than NOR, this rule can be transformed to $\frac{WIQ}{MWT}$ * (TIS + WIQ − NOR) − MWT. NOR is a constant for the candidate machine and should not greatly change the routing decisions.

The routing rule for task 3 in Fig. 15.14 can be simplified as Eq. (15.3). "Max{NOR, NIQ, OWT + PT}" can be converted to "OWT + PT", as "OWT + PT" is often greater than NOR and NIQ. "Min{WIQ, TIS, PT}" can be replaced with PT, which is usually much smaller than WIQ and TIS. In contrast with R_1 and R_2, R_3 tends to select the machine with high processing efficiency for a specific operation. It is consistent with our intuition that operations should be assigned to the most efficient machine for processing it. OWT can be considered as a constant as it is the same for all its candidate machines of an operation that are expected to be allocated.

$$
\begin{aligned}
R_3 &= Max\{\frac{WIQ}{MWT} * \{Max\{NOR, NIQ, OWT + PT\} \\
&\quad + Min\{WIQ, TIS, PT\}\}, WIQ - MWT, \\
&\quad \frac{WIQ}{NPT(NIQ - NPT)}\} - MWT \\
&\approx Max\{\frac{WIQ}{MWT} * (OWT + PT + PT), \\
&\quad WIQ - MWT, \frac{WIQ}{NPT(NIQ - NPT)}\} - MWT \\
&= \frac{WIQ}{MWT} * (OWT + 2PT) - MWT
\end{aligned}
\tag{15.3}
$$

In summary, a multitask scenario that optimises the same objective will obtain similar routing rules. All the learned rules are highly related to WIQ and MWT, and "WIQ/MWT" is a shared pattern between them. However, the obtained routing rules still differ from each other. For the task with medium complexity (task 2), the routing rule focuses more on the machine workload. For the most complex task (task 3), the routing rule considers the efficiency of the machine the most. This shows the efficacy of the newly developed surrogate-assisted multitask algorithm, since it not only shares the knowledge between different tasks but also keeps the unique characteristics of each task.

15.6   Chapter Summary

This chapter presents an effective multitask GP with surrogate to solve multiple DFJSS problems better. The presented algorithm builds surrogate models for all the tasks. The built surrogate models are used to improve the effectiveness of GP to both tackle each DFJSS problem alone, and also exchange promising building blocks of individuals among them.

The presented SMT^{2}GP has three main promising components compared with the existing multitask framework. First, it generates much more offspring than the existing methods, to gather many useful materials. Second, the generated offspring are approximatedly evaluated with surrogate rather than actual evaluations, so the efficiency can be greatly improved. Third, it allocates the offspring based on the surrogate fitness, which means that some offspring that are promising for multiple tasks can be allocated to multiple subpopulations. The results show that the presented SMT^{2}GP can learn highly competitive scheduling heuristics for DFJSS with high convergence speed for all the examined multitask scenarios. The effectiveness of SMT^{2}GP is examined by comparing the quality of the learned scheduling heuristics, the analyses of the diversity of individuals for tasks, structures and behaviour of the learned scheduling heuristics. It has also been observed that the individual allocations for the tasks are highly related to the utilisation level. This implies that the complexities significantly impact the knowledge sharing between the tasks in a multitask scenario. In addition, we find that the sizes of the learned rules over generations are highly related to the objective to be optimised in the task rather than the utilisation level.

In this book, Part I provides basic knowledge about production scheduling and evolutionary learning techniques especially genetic programming. Part II proposes effective GP algorithms for static JSS. Part III develops GP algorithms with different machine learning techniques for dynamic JSS. Part IV works on multi-objective GP for dynamic JSS. Part V proposes various multitask GP approaches for dynamic scheduling. The next chapter will summarise what we have achieved in this book and give a further discussion.

Part VI
Conclusions and Prospects

Chapter 16
Conclusions and Prospects

Scheduling plays an important role in production systems to ensure that production resources are well utilised, and customer orders are delivered on time. With recent technological advances in internet-of-things, artificial intelligence, and automation, modern production systems are digitalised and more flexible, and production environments can be monitored and diagnosed in real-time. Scheduling in such dynamic and complex environments is challenging, since scheduling needs to be more efficient and reactive, and scheduling decisions have to incorporate dynamic information and uncertainty. While conventional scheduling algorithms are still very useful in many cases, it is very hard and time-consuming due to trial-and-errors to adopt these algorithms to cope with new and real-world applications. Machine learning and hyper-heuristics have been actively investigated in the last decade as powerful solutions to the challenging production scheduling and other logistics problems. Instead of manually designing scheduling heuristics and algorithms for each problem, we can use machine learning and hyper-heuristics to automatically discover effective scheduling heuristics and algorithms from low-level heuristics, characteristics of scheduling heuristics, and dynamic information from the production environment. This automated design approach can significantly reduce the time required to develop solution methods by domain experts and increase the chance of discovering novel and effective scheduling heuristics and algorithms.

Genetic programming (GP) is a genetic-based and evolutionary learning and optimisation algorithm with many successful applications in regression, classification, computer vision, and software engineering. The flexible representation and search mechanisms make GP a suitable approach to designing production scheduling heuristics, since the structures and sizes of optimal heuristics are not known in advance. Another key practical advantage of GP for production scheduling is its interpretability. By applying appropriate representations, GP can produce user-friendly programs as priority functions commonly used in scheduling literature. There has been a growing interest in the GP approach to production scheduling in recent years not just from computer scientists and machine learning researchers but also those from traditional operations research and optimisation research communities. Because GP and pro-

F. Zhang et al., *Genetic Programming for Production Scheduling*, Machine Learning: Foundations, Methodologies, and Applications, https://doi.org/10.1007/978-981-16-4859-5_16

duction scheduling are both complex subjects with steep learning curves, researchers may find it challenging to approach this area or identify research opportunities. This book aims at meeting the growing demand in the field by providing a self-contained source of material that emphasises the theoretical aspects and application issues of GP for production scheduling.

This book is rooted in our studies in the last decade, which covers fundamental research topics ranging from representation, surrogate, genetic operators, fitness evaluations, to search strategies, and addresses practical and advanced topics and issues such as generalisation, computational efficiency, feature selection, multiple objectives, multitask, and interpretability. Throughout this book, empirical studies based on benchmark scheduling problems and simulated production environments are presented to demonstrate the effects of algorithm design decisions and their implications. In addition, this book covers the research on different types of job shop scheduling (JSS), from static to dynamic problems, and from non-flexible to flexible problems.

The rest of this chapter concludes the main research topics in this book. In addition, this chapter presents some potential research directions which are motivated by the studies in this book.

16.1   Main Conclusions

This section describes the main conclusions drawn from the 13 major contribution chapters, i.e., Chaps. 3–15.

Genetic Programming for Static Production Scheduling: Part II uses three chapters to investigate the learning and optimisation of schedule construction heuristics, iterative improvement heuristics, and augment operations research algorithms with static JSS problems.

Chapter 3 studies the advantages and disadvantages of three GP representations, i.e., decision-tree like representation, arithmetic representation and mixed representation, for schedule construction heuristics. The studies can guide the choice of GP representation for static JSS, and be expanded to cope with other production environments such as flexible JSS.

Chapter 4 presents a way to transform the construction heuristics introduced in Chap. 3 into improvement heuristics in static JSS. The main difference between schedule construction and schedule improvement heuristics is the heuristic method used to construct and refine schedules. The results show that the improved heuristics are significantly better than construction heuristics, and schedule improvement heuristics are promising to enhance the quality of obtained schedules.

Chapter 5 introduces a way to integrate GP and constraint programming solver, and the results show its advantages. This is an attempt of the applications of machine learning to scheduling and combinatorial optimisation, which is a growing hot topic nowadays.

Genetic Programming for Dynamic Production Scheduling: Part III involves four chapters to study GP for dynamic flexible JSS from four different aspects, i.e., representation, training efficiency improvement, search space reduction, and specialised genetic operators.

Chapter 6 introduces two ways, i.e., GP with multi-tree and GP with cooperative coevolution, to learn the routing rule and the sequencing rule simultaneously for dynamic flexible JSS. This chapter has discussed the advantages of both the representations with multi-tree and cooperative coevolution, and gives advice on which one can use in a particular GP algorithm.

Chapter 7 develops an effective GP with multi-fidelity surrogate models to improve the efficiency for learning scheduling heuristics in the dynamic flexible JSS problems. This book uses the degree of simplification of the problem to indicate the fidelity of the surrogate model. The surrogate models are designed by shortening the simulation of the original problem. The key idea of Chap. 7 is to solve the desired problem by utilising the advantages of surrogate models with different fidelities. The most important point is that an effective knowledge transfer mechanism is used to collaborate between the multiple surrogate models to learn from each other. The results verify the effectiveness and efficiency of the proposed algorithm. This work broadens the studies of the surrogate technique to JSS by investigating a new collaborative way to improve training efficiency with surrogates rather than only focusing on improving the accuracy of the surrogates. Similar ideas of multi-fidelity-based surrogate techniques are applicable to other problems.

Chapter 8 proposes a two-stage feature selection algorithm to select important features for the routing rule and the sequencing rule in dynamic flexible JSS, respectively. Chapter 8 further proposes novel individual adaptation strategies to help GP evolve scheduling heuristics only with the selected features without losing the qualities of the rules. The success of the proposed individual adaptation based on mimicking the behaviour of individuals lies in the conversion of variable-length representation to vector-based fixed representation for GP individuals with phenotypic characterisation. The results show that the proposed algorithm can detect important features effectively and efficiently. The individual adaptation strategies that generate individuals by mimicking the behaviours of individuals with only the selected features are also applicable to other problems, if one can design proper mechanisms to mimic individual behaviour with only the selected features based on the problem-specific characteristics.

Chapter 9 proposes two strategies to measure the importance of subtrees. The first strategy is based on the frequency of features, assuming that important subtrees tend to have more important features. The second strategy is based on the correlation between the behaviour of subtrees and the whole tree. If the behaviour of a subtree has a high (low) correlation with the behaviour of the whole tree, the subtree is important (unimportant). A recombinative guidance mechanism is then designed for crossover to replace unimportant subtrees of one parent with important subtrees from the other parent. The results show that the selected subtrees for crossover have a significant effect on the quality of the generated offspring. The learned rules by the proposed algorithm with correlation-based recombinative guidance have better performance than the compared algorithms due to its effectiveness for producing offspring. It is noted that the idea in this chapter is not limited to GP for dynamic flexible JSS but can benefit GP in general.

Genetic Programming for Multi-objective Production Scheduling: Part IV focuses on the study of multi-objective techniques on dynamic JSS with three chapters. Part IV introduces a multi-objective GP approach to learning scheduling heuristics for JSS gradually, from making a single scheduling decision to multiple scheduling decisions.

Chapter 10 proposes a multi-objective GP algorithm to learn scheduling heuristics for dynamic JSS. This chapter shows how GP can learn scheduling heuristics for handling multiple conflicting objectives. The results from extensive experiments have shown the effectiveness of the Pareto front of the learned scheduling heuristics on conflicting objectives. Chapter 10 has also provided valuable insights on how the decisions trade-offs are made.

Chapter 11 presents a multi-objective GP algorithm with a cooperative coevolution strategy for learning scheduling heuristics with multiple scheduling decisions, i.e., dispatching and due-date assignment. This work promotes the study of multi-objective GP with multiple scheduling decisions. The results show that cooperative coevolution is an effective way to handle multiple scheduling decisions. In addition, the proposed algorithm can achieve a better Pareto front than the compared algorithms.

Chapter 12 proposes a multi-objective GP algorithm with multi-tree representation to learn scheduling heuristics for dynamic flexible JSS with multiple scheduling decisions, i.e., machine assignment and operation sequencing. The results show that GP with multi-tree representation can be effectively used for multi-objective dynamic flexible JSS. In addition, the results show that the learned scheduling heuristics have good generalisation.

Multitask Genetic Programming for Production Scheduling: Part V focuses on the studies of multitask GP to learn scheduling heuristics for multiple JSS problems simultaneously, which consists of three chapters.

Chapter 13 firstly adapts the traditional multitask algorithm to GP with a number of modifications due to the specific characteristics of the problem investigated in this chapter. In addition, a novel origin-based offspring reservation strategy is proposed to enhance the knowledge sharing between tasks via crossover. The proposed algorithm is the first multitask GP approach for dynamic flexible JSS, which is a hyper-heuristic approach. The results show that the proposed algorithm is effective to learn scheduling heuristics for multiple dynamic flexible JSS tasks simultaneously. This is a starting point for the study in multitask GP for dynamic flexible JSS.

Chapter 14 proposes an adaptive multitask GP algorithm to positively learn knowledge from related tasks. An effective task relatedness measure has been studied. The relatedness of tasks in Chap. 13 is already known in advance, all the tasks are known to be related to each other in advance, and the relatedness of the tasks are assumed/treated equally important. However, the relatedness of tasks in this chapter is unknown, tasks can be related or unrelated, and the relatedness between tasks are different. To the best of our knowledge, this is the first attempt to reduce the negative transfer in the dynamic combinatorial optimisation problems by adaptively selecting assisted tasks based on relatedness. The results have verified the effectiveness of the proposed task relatedness measure in dynamic flexible JSS, and the developed adaptive multitask GP algorithm. This study broadens the study of measuring the relatedness between tasks in combinatorial optimisation problems, which is an important topic in multitask learning.

Chapter 15 proposes a surrogate-assisted GP algorithm for dynamic flexible JSS. The proposed algorithm uses the phenotypic characterisation and the fitness of fully-evaluated individuals to build the surrogates for all the tasks. The built surrogates are used to estimate the fitness of all the individuals for all the tasks, and allocates individuals to different tasks with the estimated fitness. The built surrogates are used to improve the effectiveness of solving a single task and share knowledge between tasks. The results show that the proposed surrogate-assisted GP can evolve highly competitive scheduling heuristics for dynamic flexible JSS with high convergence speed for all the examined multitask scenarios. It is a good case study to illustrate the effectiveness of surrogate-assisted evolutionary multitask for solving dynamic discrete and combinatorial optimisation problems.

16.2 Further Discussions and Future Prospects

The previous section gives a summary of the key findings in this book. This section further provides discussions on the potential directions to further enhance the proposed algorithms in this book.

Interpretability of Scheduling Heuristics: Learning interpretable models is a hot topic in machine learning [75], and this is particularly important for production scheduling as the human floor operators need to understand the rules before they can confidently use them in manufacturing. GP provides better interpretability due to its tree-based representation. However, the definition and measures of the interpretability of the learned scheduling heuristics by GP are still unclear. This book mainly uses the number of nodes, the number of unique features, and semantic analyses to investigate the interpretability of the learned scheduling heuristics. More promising measures for evaluating the interpretability of the obtained scheduling heuristics by GP are worth investigating. In addition, new algorithms to improve the interpretability of the learned scheduling heuristics need to be further studied.

Some potential directions of the interpretability of the learned scheduling heuristics obtained by GP are (1) propose effective criteria to measure the interpretability of the learned rules, for example, considering the types of features such as time-related feature, e.g., processing time, and the number related features, e.g., the number of operations left for a job, (2) develop grammar-based GP to restrict the search space to learn interpretable scheduling heuristics automatically during the evolutionary process.

Local Search: Local search [1] is a widely used technique in evolutionary algorithms to exploit good solutions by looking around their neighbourhoods. However, it is rarely applied in GP, since it is challenging to define neighbourhood relationship for GP individuals with the variable-length tree-based representation. A small change in the GP tree may lead to a big influence on the behaviour of a GP individual. In addition, the fitness evaluation of GP individuals in dynamic flexible JSS is time-consuming, since there are lots of priority calculations in the simulation. Thus, measuring the quality of the neighbours of an individual is time-consuming.

To handle these issues, some potential directions are (1) consider the phenotypic space of GP individuals to define the neighbourhood relationship. If two GP individuals have similar phenotypic characteristics, we can say they are neighbours in the local search domain, (2) utilise the surrogate technique to estimate the fitness of the neighbours of a good solution.

Multi/many-Objective Optimisation: Scheduling problems normally involve multiple objectives. A major challenge for evolving scheduling heuristics for dynamic flexible JSS is to balance and handle conflicting objective simultaneously such as minimising max-flowtime and mean-flowtime. This is particularly true in dynamic flexible JSS, since its parameters and variables can change over time [198]. To the best of our knowledge, handling multiple or many conflicting objectives for dynamic flexible JSS has not been tackled comprehensively by GP due to the difficulty of the problem.

Some potential directions that worth to be investigated are (1) develop a new dynamic multi/many-objective GP algorithm for dynamic flexible JSS based on the problem characteristics, (2) develop preference-based multi/many-objective GP algorithms for dynamic flexible JSS. The preference-based algorithms can be used to narrow down the Pareto-front search space of GP, and (3) develop new algorithms to analyse the behaviour of learned scheduling heuristics to identify how the trade-offs between different objectives are made.

Genotype Versus Phenotype: Genotype considers the structures of GP individuals, while phenotype focuses on the behaviour of GP individuals [175]. Although phenotypic characteristic has been successfully used in the literature (i.e., also in Chaps. 8, 9, 14, and 15) to measure the quality of GP individuals, the genotype can also play an important role for investigation of GP individuals. For example, the genotype can provide a better understanding of the genetic materials used in GP individuals. The variation of GP individuals occurs at the genotypic level but fitness is evaluated at the phenotypic level. Therefore, the mapping between the genotype and phenotype of GP individuals [16] in dynamic flexible JSS is an interesting topic.

Implicit Versus Explicit Knowledge Sharing: The knowledge sharing mechanisms in Chaps. 7, 13, and 14 are implemented implicitly via crossover. Explicit knowledge transfer represents another form to share knowledge, which can guarantee the effectiveness of knowledge sharing in different tasks [70]. As an explicit knowledge transfer, task adaptation aims to convert the search space of one task being good for the other task. A linearised domain adaptation was developed to transform the search space of a simple task to the search space similar to its constitutive complex task [13]. The transformed search space resembled a high correlation with its constitutive task and provided a platform for efficient knowledge transfer. All tasks were transformed into new space while maintaining the same geometric properties of solutions in [61]. The individuals were transformed to fit the other task by task space mapping strategy for generating higher-quality offspring in [131]. These kinds of studies [45] focus on adapting solutions of one task being good for other tasks, which is good to be also investigated in GP for JSS.

Surrogate: The surrogates in Chaps. 7, 8, and 15 (i.e., either based on the KNN model or simplified problem) have shown their effectiveness to estimate the fitness of GP individuals. There is a lot of information we can obtain from the simulation execution, such as the processing information in the early and middle stages. If one can utilise the information in an early stage to build surrogate models to estimate the performance of an individual in the later stage, the simulation can be dramatically shortened. Thus, training efficiency can be significantly improved. This is an interesting direction that uses historical data to predict the performance of an individual in the future.

Genetic Programming for Large Scale Problems: In real-world applications, the scale of the problems in production scheduling can be huge, e.g., with thousands of machines and tens of thousands of jobs. Learning scheduling heuristics for large scale scheduling problem with GP can be very time-consuming. Learning effective scheduling heuristics for large scale scheduling problem efficiently is an interesting direction. One possible way is to learn scheduling heuristics from small scale problems, and adapt to large scale scheduling problems somehow.

Genetic Programming for Other Combinatorial Optimisation Problems and Real-world Applications: The ideas of the proposed algorithms in this book can also be used for other combinatorial optimisation problems, such as learning routing policy for arc routing problems and learning scheduling policy for resource allocation. In addition, it is interesting to apply the proposed algorithms to real-world applications such as mobile manufacturing, tourism, transportation, and semiconductors.

Hybridising Genetic Programming with Exact Methods to Improve Efficiency of the Exact Methods for Production Scheduling: As mentioned in Chap. 1 of this book, exact methods are only suitable for small scale and static scheduling problems. However, the main advantage of exact methods is that they can find optimal solutions for the problems to be solved. The advantage of GP is that the learned scheduling heuristics can handle large scale and dynamic scheduling problems efficiently. Chapter 5 of this book has shown promising results of GP with the exact algorithm, i.e., constraint programming to learn effective scheduling heuristics. The research area of the hybrid algorithm of GP with exact methods for the scheduling problems is an interesting direction to dig more.

References

1. Aarts, E., Aarts, E.H., Lenstra, J.K.: Local Search in Combinatorial Optimization. Princeton University Press (2003)
2. Aha, D.W.: Case-based learning algorithms. In: Proceedings of the DARPA Case-Based Reasoning Workshop, vol. 1, pp. 147–158 (1991)
3. Amjad, M.K., Butt, S.I., Kousar, R., Ahmad, R., Agha, M.H., Faping, Z., Anjum, N., Asgher, U.: Recent research trends in genetic algorithm based flexible job shop scheduling problems. Math. Probl. Eng. **2018**, 1–32 (2018)
4. Applegate, D., Cook, W.: A computational study of the job-shop scheduling instance. ORSA J. Comput. **3**, 149–156 (1991)
5. Applegate, D., Cook, W.: A computational study of the job-shop scheduling problem. ORSA J. Comput. **3**(2), 149–156 (1991)
6. Ardeh, M.A., Mei, Y., Zhang, M.: Genetic programming hyper-heuristic with knowledge transfer for uncertain capacitated arc routing problem. In: Proceedings of the Genetic and Evolutionary Computation Conference, pp. 334–335 (2019)
7. Ardeh, M.A., Mei, Y., Zhang, M.: A novel genetic programming algorithm with knowledge transfer for uncertain capacitated arc routing problem. In: Pacific Rim International Conference on Artificial Intelligence, pp. 196–200. Springer (2019)
8. van Veldhuizen, D.A., Lamont, G.B.: Multiobjective evolutionary algorithm research: a history and analysis. Technical Report TR-98-03, Department of Electrical and Computer Engineering, Air Force Institute of Technology, Ohio (1998)
9. Ayodele, T.O.: Types of machine learning algorithms. New Adv. Mach. Learn. **3**, 19–48 (2010)
10. Bäck, T., Fogel, D.B., Michalewicz, Z.: Handbook of Evolutionary Computation. CRC Press (1997)
11. Baker, K.R.: Sequencing rules and due-date assignments in a job shop. Manag. Sci. **30**, 1093–1104 (1984)
12. Balas, E., Vazacopoulos, A.: Guided local search with shifting bottleneck for job shop scheduling. Manag. Sci. **44**, 262–275 (1998)
13. Bali, K.K., Gupta, A., Feng, L., Ong, Y.S., Siew, T.P.: Linearized domain adaptation in evolutionary multitasking. In: Proceedings of the IEEE Congress on Evolutionary Computation, pp. 1295–1302. IEEE (2017)
14. Bali, K.K., Gupta, A., Ong, Y.S., Tan, P.S.: Cognizant multitasking in multiobjective multifactorial evolution: MO-MFEA-II. IEEE Trans. Cybern. **51**(4), 1784–1796 (2020)

© The Editor(s) (if applicable) and The Author(s), under exclusive license to Springer Nature Singapore Pte Ltd. 2021 F. Zhang et al., *Genetic Programming for Production Scheduling*, Machine Learning: Foundations, Methodologies, and Applications, https://doi.org/10.1007/978-981-16-4859-5

15. Bali, K.K., Ong, Y.S., Gupta, A., Tan, P.S.: Multifactorial evolutionary algorithm with online transfer parameter estimation: MFEA-II. IEEE Trans. Evol. Comput. **24**(1), 69–83 (2019)

16. Banzhaf, W.: Genotype-phenotype-mapping and neutral variation—a case study in genetic programming. In: Proceedings of the International Conference on Parallel Problem Solving from Nature, pp. 322–332. Springer (1994)

17. Banzhaf, W., Nordin, P., Keller, R.E., Francone, F.D.: Genetic Programming: An Introduction, vol. 1. Morgan Kaufmann, San Francisco (1998)

18. Baykasoğlu, A., Göçken, M., Özbakir, L.: Genetic programming based data mining approach to dispatching rule selection in a simulated job shop. SIMULATION **86**(12), 715–728 (2010)

19. Baykasoglu, A., Gocken, M., Unutmaz, Z.D.: New approaches to due date assignment in job shops. Eur. J. Oper. Res. **187**, 31–45 (2008)

20. Beasley, J.: OR-Library: distributing test problems by electronic mail. J. Oper. Res. Soc. **41**, 1069–1072 (1990)

21. Bengio, Y., Lodi, A., Prouvost, A.: Machine learning for combinatorial optimization: a methodological tour d'horizon. Eur. J. Oper. Res. (2020)

22. Beyer, H.G., Schwefel, H.P.: Evolution strategies-A comprehensive introduction. Nat. Comput. **1**(1), 3–52 (2002)

23. Bhowan, U., Johnston, M., Zhang, M., Yao, X.: Evolving diverse ensembles using genetic programming for classification with unbalanced data. IEEE Trans. Evol. Comput. **17**(3), 368–386 (2012)

24. Blazewicz, J., Ecker, K., Pesch, E., Schmidt, G., Weglarz, J.: Handbook on Scheduling. Springer (2019)

25. Bokhorst, J.A., Nomden, G., Slomp, J.: Performance evaluation of family-based dispatching in small manufacturing cells. Int. J. Prod. Res. **46**(22), 6305–6321 (2008)

26. Boyle, E., Al-Akash, M., Gallagher, A.G., Traynor, O., Hill, A.D., Neary, P.C.: Optimising surgical training: use of feedback to reduce errors during a simulated surgical procedure. Postgrad. Med. J. **87**(1030), 524–528 (2011)

27. Branke, J., Nguyen, S., Pickardt, C.W., Zhang, M.: Automated design of production scheduling heuristics: a review. IEEE Trans. Evol. Comput. **20**(1), 110–124 (2016)

28. Branke, J., Hildebrandt, T., Scholz-Reiter, B.: Hyper-heuristic evolution of dispatching rules: a comparison of rule representations. Evol. Comput. **23**(2), 249–277 (2015)

29. Brucker, P., Schlie, R.: Job-shop scheduling with multi-purpose machines. Computing **45**(4), 369–375 (1990)

30. Burke, E., Kendall, G., Newall, J., Hart, E., Ross, P., Schulenburg, S.: Hyper-heuristics: an emerging direction in modern search technology. In: Handbook of Metaheuristics, pp. 457–474. Springer (2003)

31. Burke, E.K., Gendreau, M., Hyde, M., Kendall, G., Ochoa, G., Özcan, E., Qu, R.: Hyper-heuristics: a survey of the state of the art. J. Oper. Res. Soc. **64**(12), 1695–1724 (2013)

32. Burke, E.K., Hyde, M.R., Kendall, G., Ochoa, G., Ozcan, E., Woodward, J.R.: Exploring hyper-heuristic methodologies with genetic programming. In: Proceedings of the Computational Intelligence, pp. 177–201. Springer (2009)

33. Burke, E.K., Hyde, M., Kendall, G., Ochoa, G., Özcan, E., Woodward, J.R.: A classification of hyper-heuristic approaches. In: Handbook of Metaheuristics, pp. 449–468. Springer (2010)

34. Burke, E.K., Hyde, M., Kendall, G., Woodward, J.: A genetic programming hyper-heuristic approach for evolving 2-d strip packing heuristics. IEEE Trans. Evol. Comput. **14**(6), 942–958 (2010)

35. Camacho-Hernandez, G.A., Taylor, P.: Lessons from nature: structural studies and drug design driven by a homologous surrogate from invertebrates. AChBP. Neuropharmacology **179**, 108108 (2020)

36. Carlier, J., Pinson, É.: An algorithm for solving the job-shop problem. Manag. Sci. **35**(2), 164–176 (1989)

37. Carlier, J., Pinson, E.: Adjustment of heads and tails for the job-shop problem. Eur. J. Oper. Res. **78**(2), 146–161 (1994)

38. Chandola, V., Banerjee, A., Kumar, V.: Anomaly detection: a survey. ACM Comput. Surv. **41**(3), 1–58 (2009)
39. Chen, H., Chu, C., Proth, J.M.: An improvement of the Lagrangean relaxation approach for job shop scheduling: a dynamic programming method. IEEE Trans. Robot. Autom. **14**(5), 786–795 (1998)
40. Chen, K., Xue, B., Zhang, M., Zhou, F.: An evolutionary multitasking-based feature selection method for high-dimensional classification. IEEE Trans. Cybern. (2020). https://doi.org/10. 1109/TCYB20203042243
41. Chen, K., Zhou, F., Xue, B.: Particle swarm optimization for feature selection with adaptive mechanism and new updating strategy. In: Proceedings of the Australasian Joint Conference on Artificial Intelligence, pp. 419–431. Springer (2018)
42. Chen, Q., Zhang, M., Xue, B.: Structural risk minimization-driven genetic programming for enhancing generalization in symbolic regression. IEEE Trans. Evol. Comput. **23**(4), 703–717 (2019)
43. Chen, Q., Zhang, M., Xue, B.: Feature selection to improve generalization of genetic programming for high-dimensional symbolic regression. IEEE Trans. Evol. Comput. **21**(5), 792–806 (2017)
44. Chen, Y., Zhong, J., Feng, L., Zhang, J.: An adaptive archive-based evolutionary framework for many-task optimization. IEEE Trans. Emerg. Top. Comput. Intell. **4**(3), 369–384 (2019)
45. Chen, Z., Zhou, Y., He, X., Zhang, J.: Learning task relationships in evolutionary multitasking for multiobjective continuous optimization. IEEE Trans. Cybern. (2020). https://doi.org/10. 1109/TCYB20203029176
46. Cheng, R., Gen, M., Tsujimura, Y.: A tutorial survey of job-shop scheduling problems using genetic algorithms, Part II: hybrid genetic search strategies. Comput. Ind. Eng. **36**(2), 343–364 (1999)
47. Cheng, T.C.E.: Integration of priority dispatching and due-date assignment in a job shop. Int. J. Syst. Sci. **19**(9), 1813–1825 (1988)
48. Cheng, T.C.E., Jiang, J.: Job shop scheduling for missed due-date performance. Comput. Ind. Eng. **34**, 297–307 (1998)
49. Chiang, T.C., Shen, Y.S., Fu, L.C.: A new paradigm for rule-based scheduling in the wafer probe centre. Int. J. Prod. Res. **46**(15), 4111–4133 (2008)
50. Chiang, T.C., Fu, L.C.: Using dispatching rules for job shop scheduling with due date-based objectives. Int. J. Prod. Res. **45**(14), 3245–3262 (2007)
51. Chong, C.S., Sivakumar, A.I., Low, M.Y.H., Gay, K.L.: A bee colony optimization algorithm to job shop scheduling. In: Proceedings of the Conference on Winter Simulation, pp. 1954–1961. Winter Simulation Conference (2006)
52. Conroy, G.: Handbook of genetic algorithms. Knowl. Eng. Rev. **6**(4), 363–365 (1991)
53. Cowling, P., Kendall, G., Soubeiga, E.: A hyperheuristic approach to scheduling a sales summit. In: Proceedings of the International Conference on the Practice and Theory of Automated Timetabling, pp. 176–190. Springer (2000)
54. Da, B., Ong, Y.S., Feng, L., Qin, A.K., Gupta, A., Zhu, Z., Ting, C.K., Tang, K., Yao, X.: Evolutionary multitasking for single-objective continuous optimization: benchmark problems, performance metric, and baseline results. arXiv:1706.03470 (2017)
55. Daniel, W.W., et al.: Applied Nonparametric Statistics. Houghton Mifflin (1978)
56. Dash, M., Liu, H.: Feature selection for classification. Intell. Data Anal. **1**(3), 131–156 (1997)
57. Davis, J.P., Eisenhardt, K.M., Bingham, C.B.: Developing theory through simulation methods. Acad. Manag. Rev. **32**(2), 480–499 (2007)
58. Deb, K., Pratap, A., Agarwal, S., Meyarivan, T.: A fast and elitist multiobjective genetic algorithm: NSGA-II. IEEE Trans. Evol. Comput. **6**(2), 182–197 (2002)
59. Demirkol, E., Mehta, S., Uzsoy, R.: Benchmarks for shop scheduling problems. Eur. J. Oper. Res. **109**(1), 137–141 (1998)
60. Dick, G., Owen, C.A., Whigham, P.A.: Feature standardisation and coefficient optimisation for effective symbolic regression. In: Proceedings of the Genetic and Evolutionary Computation Conference, pp. 306–314 (2020)

61. Ding, J., Yang, C., Jin, Y., Chai, T.: Generalized multitasking for evolutionary optimization of expensive problems. IEEE Trans. Evol. Comput. **23**(1), 44–58 (2017)
62. Doerr, B., Kötzing, T., Lagodzinski, J.G., Lengler, J.: Bounding bloat in genetic programming. In: Proceedings of the Genetic and Evolutionary Computation Conference, pp. 921–928 (2017)
63. Dominic, P.D., Kaliyamoorthy, S., Kumar, M.S.: Efficient dispatching rules for dynamic job shop scheduling. Int. J. Adv. Manuf. Technol. **24**(1–2), 70–75 (2004)
64. Dorigo, M., Birattari, M., Stutzle, T.: Ant colony optimization. IEEE Comput. Intell. Mag. **1**(4), 28–39 (2006)
65. Douguet, D.: e-LEA3D: a computational-aided drug design web server. Nucleic Acids Res. **38**(Suppl 2), W615–W621 (2010)
66. Drake, J.H., Kheiri, A., Özcan, E., Burke, E.K.: Recent advances in selection hyper-heuristics. Eur. J. Oper. Res. **285**(2), 405–428 (2020)
67. Durasevic, M., Jakobovic, D.: A survey of dispatching rules for the dynamic unrelated machines environment. Expert Syst. Appl. **113**, 555–569 (2018)
68. Eiben, A.E., Smith, J.E.: What is an evolutionary algorithm? In: Introduction to Evolutionary Computing, pp. 25–48. Springer (2015)
69. Fahrmeir, L., Kneib, T., Lang, S., Marx, B.: Regression. Springer (2007)
70. Feng, L., Zhou, L., Zhong, J., Gupta, A., Ong, Y.S., Tan, K.C., Qin, A.: Evolutionary multitasking via explicit autoencoding. IEEE Trans. Cybern. **49**(9), 3457–3470 (2018)
71. Fisher, M., Raman, A.: Reducing the cost of demand uncertainty through accurate response to early sales. Oper. Res. **44**(1), 87–99 (1996)
72. Fogel, D.B.: An overview of evolutionary programming. In: Proceedings of the Evolutionary Algorithms, pp. 89–109. Springer (1999)
73. Gee, E.S., Smith, C.H.: Selecting allowance policies for improved job shop performance. Int. J. Prod. Res. **31**(8), 1839–1852 (1993)
74. Geiger, C.D., Uzsoy, R., Aytuğ, H.: Rapid modeling and discovery of priority dispatching rules: an autonomous learning approach. J. Sched. **9**(1), 7–34 (2006)
75. Gilpin, L.H., Bau, D., Yuan, B.Z., Bajwa, A., Specter, M., Kagal, L.: Explaining explanations: an overview of interpretability of machine learning. In: Proceedings of the IEEE International Conference on Data Science and Advanced Analytics, pp. 80–89. IEEE (2018)
76. Goh, C.K., Tan, K.C.: An investigation on noisy environments in evolutionary multiobjective optimization. IEEE Trans. Evol. Comput. **11**(3), 354–381 (2007)
77. Goh, C.K., Tan, K.C.: A competitive-cooperative coevolutionary paradigm for dynamic multiobjective optimization. IEEE Trans. Evol. Comput. **13**(1), 103–127 (2009)
78. Gomes, M.C., Barbosa-Póvoa, A.P., Novais, A.Q.: Reactive scheduling in a make-to-order flexible job shop with re-entrant process and assembly: a mathematical programming approach. Int. J. Prod. Res. **51**(17), 5120–5141 (2013)
79. Gong, G., Chiong, R., Deng, Q., Gong, X.: A hybrid artificial bee colony algorithm for flexible job shop scheduling with worker flexibility. Int. J. Prod. Res. **58**(14), 4406–4420 (2020)
80. Gong, M., Tang, Z., Li, H., Zhang, J.: Evolutionary multitasking with dynamic resource allocating strategy. IEEE Trans. Evol. Comput. **23**(5), 858–869 (2019)
81. Graves, S.C.: A review of production scheduling. Oper. Res. **29**(4), 646–675 (1981)
82. Gupta, A., Ong, Y.S.: Genetic transfer or population diversification? Deciphering the secret ingredients of evolutionary multitask optimization. In: Proceedings of the IEEE Symposium Series on Computational Intelligence, pp. 1–7. IEEE (2016)
83. Gupta, A., Ong, Y.S., Feng, L.: Multifactorial evolution: toward evolutionary multitasking. IEEE Trans. Evol. Comput. **20**(3), 343–357 (2015)
84. Gupta, A., Ong, Y.S., Feng, L.: Insights on transfer optimization: because experience is the best teacher. IEEE Trans. Emerg. Top. Comput. Intell. **2**(1), 51–64 (2017)
85. Gupta, A., Ong, Y., Feng, L., Tan, K.C.: Multiobjective multifactorial optimization in evolutionary multitasking. IEEE Trans. Cybern. **47**(7), 1652–1665 (2017)
86. Guyon, I., Elisseeff, A.: An introduction to variable and feature selection. J. Mach. Learn. Res. **3**, 1157–1182 (2003)

87. Guzek, M., Bouvry, P., Talbi, E.G.: A survey of evolutionary computation for resource management of processing in cloud computing. IEEE Comput. Intell. Mag. **10**(2), 53–67 (2015)
88. Han, L., Kendall, G., Cowling, P.: An adaptive length chromosome hyper-heuristic genetic algorithm for a trainer scheduling problem. In: Recent Advances in Simulated Evolution and Learning, pp. 506–525. World Scientific (2004)
89. Hansen, P., Mladenovic, N.: Variable neighborhood search: principles and applications. Eur. J. Oper. Res. **130**(3), 449–467 (2001)
90. Hao, X., Qu, R., Liu, J.: A unified framework of graph-based evolutionary multitasking hyper-heuristic. IEEE Trans. Evol. Comput. (2020). https://doi.org/10.1109/TEVC20202991717
91. Haupt, R.: A survey of priority rule-based scheduling. Oper. Res. Spektrum **11**(1), 3–16 (1989)
92. Hildebrandt, T., Branke, J.: On using surrogates with genetic programming. Evol. Comput. **23**(3), 343–367 (2014)
93. Hildebrandt, T., Heger, J., Scholz-Reiter, B.: Towards improved dispatching rules for complex shop floor scenarios: a genetic programming approach. In: Proceedings of the Genetic and Evolutionary Computation, pp. 257–264 (2010)
94. Hoffmann, P.: A dynamic limit order market with fast and slow traders. J. Financ. Econ. **113**(1), 156–169 (2014)
95. Holland, J.H., Holyoak, K.J., Nisbett, R.E., Thagard, P.R.: Induction: Processes of Inference, Learning, and Discovery. MIT Press (1989)
96. Holthaus, O., Rajendran, C.: Efficient jobshop dispatching rules: further developments. Prod. Plan. Control **11**(2), 171–178 (2000)
97. Huang, S., Zhong, J., Yu, W.: Surrogate-assisted evolutionary framework with adaptive knowledge transfer for multi-task optimization. IEEE Trans. Emerg. Top, Comput (2019)
98. Iba, H., de Garis, H.: Extending genetic programming with recombinative guidance. Adv. Genet. Grogram. **2**, 69–88 (1996)
99. Iwasaki, Y., Suzuki, I., Yamamoto, M., Furukawa, M.: Job-shop scheduling approach to order-picking problem. Trans. Inst. Syst., Control Inf. Eng. **26**(3), 103–109 (2013)
100. Jakobović, D., Budin, L.: Dynamic scheduling with genetic programming. In: European Conference on Genetic Programming, pp. 73–84. Springer (2006)
101. Jayamohan, M., Rajendran, C.: New dispatching rules for shop scheduling: a step forward. Int. J. Prod. Res. **38**(3), 563–586 (2000)
102. Jayamohan, M., Rajendran, C.: Development and analysis of cost-based dispatching rules for job shop scheduling. Eur. J. Oper. Res. **157**(2), 307–321 (2004)
103. Jin, Y.: Surrogate-assisted evolutionary computation: recent advances and future challenges. Swarm Evol. Comput. **1**(2), 61–70 (2011)
104. John, D.J.: Co-evolution with the Bierwirth-Mattfeld hybrid scheduler. In: Proceedings of the Genetic and Evolutionary Computation Conference, p. 259 (2002)
105. Jurado, S., Nebot, À., Mugica, F., Avellana, N.: Hybrid methodologies for electricity load forecasting: entropy-based feature selection with machine learning and soft computing techniques. Energy **86**, 276–291 (2015)
106. Kanet, J.J., Li, X.: A weighted modified due date rule for sequencing to minimize weighted tardiness. J. Sched. **7**(4), 261–276 (2004)
107. Karunakaran, D., Mei, Y., Zhang, M.: Multitasking genetic programming for stochastic team orienteering problem with time windows. In: Proceedings of the IEEE Symposium Series on Computational Intelligence, pp. 1598–1605. IEEE (2019)
108. Kennedy, J.: Particle swarm optimization. In: Encyclopedia of Machine Learning, pp. 760–766 (2010)
109. Kennedy, J., Eberhart, R.: Particle swarm optimization. In: Proceedings of the International Conference on Neural Networks, vol. 4, pp. 1942–1948. IEEE (1995)
110. Kim, B., Koyejo, O., Khanna, R., et al.: Examples are not enough, learn to criticize! criticism for interpretability. In: Proceedings of the Advances in Neural Information Processing System, pp. 2280–2288 (2016)
111. Kotsiantis, S.B., Zaharakis, I., Pintelas, P.: Supervised machine learning: a review of classification techniques. Emerg. Artif. Intell. Appl. Comput. Eng. **160**(1), 3–24 (2007)

112. Koulamas, C.: A new constructive heuristic for the flowshop scheduling problem. Eur. J. Oper. Res. **105**(1), 66–71 (1998)

113. Koulinas, G., Kotsikas, L., Anagnostopoulos, K.: A particle swarm optimization based hyper-heuristic algorithm for the classic resource constrained project scheduling problem. Inf. Sci. **277**, 680–693 (2014)

114. Koza, J.R.: Genetic Programming: A Paradigm for Genetically Breeding Populations of Computer Programs to Solve Problems, vol. 34. Stanford University, Department of Computer Science Stanford, CA (1990)

115. Koza, J.R.: Genetic Programming: On the Programming of Computers by Means of Natural Selection. MIT Press (1992)

116. Koza, J.R.: Genetic programming as a means for programming computers by natural selection. Stat. Comput. **4**(2), 87–112 (1994)

117. Koza, J.R., Keane, M.A., Streeter, M.J., Mydlowec, W., Yu, J., Lanza, G.: Genetic Programming IV: Routine Human-Competitive Machine Intelligence, vol. 5. Springer Science & Business Media (2006)

118. Koza, J.R., Poli, R.: Genetic programming. In: Search Methodologies, pp. 127–164. Springer (2005)

119. Kreipl, S.: A large step random walk for minimizing total weighted tardiness in a job shop. J. Sched. **3**, 125–138 (2000)

120. Kumari, A.C., Srinivas, K., Gupta, M.: Software module clustering using a hyper-heuristic based multi-objective genetic algorithm. In: Proceedings of the IEEE International Advance Computing Conference, pp. 813–818. IEEE (2013)

121. van Laarhoven, P.J.M., Aarts, E.H.L., Lenstra, J.K.: Job shop scheduling by simulated annealing. Oper. Res. **40**(1), 113–125 (1992)

122. Lamé, G., Dixon-Woods, M.: Using clinical simulation to study how to improve quality and safety in healthcare. BMJ Simul. Technol. Enhanc, Learn (2018)

123. Law, A.: Simulation Modeling and Analysis. McGraw-Hill Higher Education (2015)

124. Lawler, E.L., Wood, D.E.: Branch-and-bound methods: a survey. Oper. Res. **14**(4), 699–719 (1966)

125. Lawrence, S.: Resource Constrained Project Scheduling: An Experimental Investigation of Heuristic Scheduling Techniques. Ph.D. thesis, Graduate School of Industrial Administration, Carnegie-Mellon University, Pittsburgh, Pennsylvania (1984)

126. Lensen, A., Xue, B., Zhang, M.: Particle swarm optimisation representations for simultaneous clustering and feature selection. In: Proceedings of the IEEE Symposium Series on Computational Intelligence, pp. 1–8. IEEE (2016)

127. Li, G., Lin, Q., Gao, W.: Multifactorial optimization via explicit multipopulation evolutionary framework. Inf. Sci. **512**, 1555–1570 (2020)

128. Li, H., Ong, Y.S., Gong, M., Wang, Z.: Evolutionary multitasking sparse reconstruction: framework and case study. IEEE Trans. Evol. Comput. **23**(5), 733–747 (2018)

129. Li, L., Zhang, F., Liu, C., Niu, B.: A hybrid artificial bee colony algorithm with bacterial foraging optimization. In: Proceedings of the IEEE International Conference on Cyber Technology in Automation, Control, and Intelligent Systems, pp. 127–132. IEEE (2015)

130. Lian, Z., Jiao, B., Gu, X.: A similar particle swarm optimization algorithm for job-shop scheduling to minimize makespan. Appl. Math. Comput. **183**(2), 1008–1017 (2006)

131. Liang, Z., Zhang, J., Feng, L., Zhu, Z.: A hybrid of genetic transform and hyper-rectangle search strategies for evolutionary multi-tasking. Expert Syst. Appl. **138**, 112798 (2019)

132. Lin, J., Liu, H.L., Tan, K.C., Gu, F.: An effective knowledge transfer approach for multi-objective multitasking optimization. IEEE Trans. Cybern. (2020). https://doi.org/10.1109/TCYB20202969025

133. Lin, J., Liu, H.L., Xue, B., Zhang, M., Gu, F.: Multi-objective multi-tasking optimization based on incremental learning. IEEE Trans. Evol. Comput. **24**(5), 824–838 (2020)

134. Lipowski, A., Lipowska, D.: Roulette-wheel selection via stochastic acceptance. Physica A **391**(6), 2193–2196 (2012)

135. Liu, D., Jiang, T., Wang, Y., Miao, R., Shan, F., Li, Z.: Clearness of operating field: a surrogate for surgical skills on in vivo clinical data. Int. J. Comput. Assist. Radiol. Surg. **15**(11), 1817–1824 (2020)

136. Liu, D., Huang, S., Zhong, J.: Surrogate-assisted multi-tasking memetic algorithm. In: Proceedings of the IEEE Congress on Evolutionary Computation, pp. 1–8 (2018)

137. Liu, Y., Wang, L., Wang, X.V., Xu, X., Zhang, L.: Scheduling in cloud manufacturing: state-of-the-art and research challenges. Int. J. Prod. Res. **57**(15–16), 4854–4879 (2019)

138. Lourenco, H.R.: Job-shop scheduling: computational study of local search and large-step optimization methods. Eur. J. Oper. Res. **83**(2), 347–364 (1995)

139. Luke, S.: Essentials of Metaheuristics. Lulu (2009). http://cs.gmu.edu/~sean/book/metaheuristics/

140. Luke, S., Panait, L., Balan, G., Paus, S., Skolicki, Z., Bassett, J., Hubley, R., Chircop, A.: ECJ: A java-based evolutionary computation research system. **880** (2006). http://cs.gmu.edu/eclab/projects/ecj

141. Manne, A.S.: On the job-shop scheduling problem. Oper. Res. **8**(2), 219–223 (1960)

142. Mei, Y., Nguyen, S., Xue, B., Zhang, M.: An efficient feature selection algorithm for evolving job shop scheduling rules with genetic programming. IEEE Trans. Emerg. Top. Comput. Intell. **1**(5), 339–353 (2017)

143. Mei, Y., Zhang, M., Nyugen, S.: Feature selection in evolving job shop dispatching rules with genetic programming. In: Proceedings of the Genetic and Evolutionary Computation Conference, pp. 365–372. ACM (2016)

144. Michalski, R.S.: A theory and methodology of inductive learning. In: Machine Learning, pp. 83–134. Elsevier (1983)

145. Miller, J.F., Harding, S.L.: Cartesian genetic programming. In: Proceedings of the Genetic and Evolutionary Computation Conference, pp. 2701–2726 (2008)

146. Miller, T.: Explanation in artificial intelligence: insights from the social sciences. Artif. Intell. **267**, 1–38 (2019)

147. Min, A.T.W., Ong, Y.S., Gupta, A., Goh, C.K.: Multiproblem surrogates: transfer evolutionary multiobjective optimization of computationally expensive problems. IEEE Trans. Evol. Comput. **23**(1), 15–28 (2017)

148. Minton, S.: An analytic learning system for specializing heuristics. In: IJCAI, vol. 93, pp. 922–929. Citeseer (1993)

149. Mitchell, T.M., et al.: Machine Learning. McGraw-hill, New York (1997)

150. Miyazaki, S.: Combined scheduling system for reducing job tardiness in a job shop. Int. J. Prod. Res. **19**(2), 201–211 (1981)

151. Mizrak, P., Bayhan, G.M.: Comparative study of dispatching rules in a real-life job shop environment. Appl. Artif. Intell. **20**, 585–607 (2006)

152. Mohan, J., Lanka, K., Rao, A.N.: A review of dynamic job shop scheduling techniques. Procedia Manuf. **30**, 34–39 (2019)

153. Montgomery, D.C.: Design and Analysis of Experiments. Wiley (2001)

154. Nag, K., Pal, N.R.: A multiobjective genetic programming-based ensemble for simultaneous feature selection and classification. IEEE Trans. Cybern. **46**(2), 499–510 (2016)

155. Neshatian, K., Zhang, M.: Unsupervised elimination of redundant features using genetic programming. In: Proceedings of the Australasian Joint Conference on Artificial Intelligence, pp. 432–442. Springer (2009)

156. Nguyen, S., Zhang, M., Johnston, M., Tan, K.C.: A computational study of representations in genetic programming to evolve dispatching rules for the job shop scheduling problem. IEEE Trans. Evol. Comput. **17**(5), 621–639 (2013)

157. Nguyen, S.B.S., Zhang, M.: A hybrid discrete particle swarm optimisation method for grid computation scheduling. In: Proceedings of the IEEE Congress on Evolutionary Computation, pp. 483–490. IEEE (2014)

158. Nguyen, S., Mei, Y., Xue, B., Zhang, M.: A hybrid genetic programming algorithm for automated design of dispatching rules. Evol. Comput. **27**(3), 467–496 (2019)

159. Nguyen, S., Mei, Y., Zhang, M.: Genetic programming for production scheduling: a survey with a unified framework. Complex Intell. Syst. **3**(1), 41–66 (2017)
160. Nguyen, S., Zhang, M., Alahakoon, D., Tan, K.C.: Visualizing the evolution of computer programs for genetic programming. IEEE Comput. Intell. Mag. **13**(4), 77–94 (2018)
161. Nguyen, S., Zhang, M., Alahakoon, D., Tan, K.C.: People-centric evolutionary system for dynamic production scheduling. IEEE Trans. Cybern. **51**(3), 1403–1416 (2019)
162. Nguyen, S., Zhang, M., Johnston, M., Tan, K.C.: Selection schemes in surrogate-assisted genetic programming for job shop scheduling. In: Proceedings of the Asia-Pacific Conference on Simulated Evolution and Learning, pp. 656–667. Springer (2014)
163. Nguyen, S., Zhang, M., Johnston, M., Tan, K.C.: Automatic programming via iterated local search for dynamic job shop scheduling. IEEE Trans. Cybern. **45**(1), 1–14 (2015)
164. Nguyen, S., Zhang, M., Tan, K.C.: Enhancing genetic programming based hyper-heuristics for dynamic multi-objective job shop scheduling problems. In: IEEE Congress on Evolutionary Computation, pp. 2781–2788 (2015)
165. Nguyen, S., Zhang, M., Tan, K.C.: Surrogate-assisted genetic programming with simplified models for automated design of dispatching rules. IEEE Trans. Cybern. **47**(9), 2951–2965 (2017)
166. Nowicki, E., Smutnicki, C.: A fast taboo search algorithm for the job shop problem. Manag. Sci. **42**(6), 797–813 (1996)
167. Núnez, M.: The use of background knowledge in decision tree induction. Mach. Learn. **6**(3), 231–250 (1991)
168. Oltean, M., Grosan, C.: A comparison of several linear genetic programming techniques. Complex Syst. **14**(4), 285–314 (2003)
169. Ong, Y.S., Gupta, A.: Evolutionary multitasking: a computer science view of cognitive multitasking. Cogn. Comput. **8**(2), 125–142 (2016)
170. Ong, Y.S., Zhou, Z., Lim, D.: Curse and blessing of uncertainty in evolutionary algorithm using approximation. In: Proceedings of the IEEE International Conference on Evolutionary Computation, pp. 2928–2935. IEEE (2006)
171. Panwalkar, S.S., Iskander, W.: A survey of scheduling rules. Oper. Res. **25**, 45–61 (1977)
172. Pardalos, P., Shylo, O.: An algorithm for the job shop scheduling problem based on global equilibrium search techniques. CMS **3**, 331–348 (2006)
173. Park, J., Mei, Y., Nguyen, S., Chen, G., Zhang, M.: Evolutionary multitask optimisation for dynamic job shop scheduling using niched genetic programming. In: Proceedings of the Australasian Joint Conference on Artificial Intelligence, pp. 739–751. Springer (2018)
174. Park, J., Mei, Y., Nguyen, S., Chen, G., Zhang, M.: Investigating a machine breakdown genetic programming approach for dynamic job shop scheduling. In: Proceedings of the European Conference on Genetic Programming, pp. 253–270. Springer (2018)
175. Paterson, N.R., Livesey, M.: Distinguishing genotype and phenotype in genetic programming. In: Late Breaking Papers at the Genetic Programming, pp. 141–150 (1996)
176. Pawlak, T.P., Krawiec, K.: Synthesis of constraints for mathematical programming with one-class genetic programming. IEEE Trans. Evol. Comput. **23**(1), 117–129 (2018)
177. Pearl, J.: Heuristics: Intelligent Search Strategies for Computer Problem Solving. Addison-Wesley (1984)
178. Peng, C., Wu, G., Liao, T.W., Wang, H.: Research on multi-agent genetic algorithm based on Tabu search for the job shop scheduling problem. PloS ONE **14**(9), e0223182 (2019)
179. Peterson, L.E.: K-nearest neighbor. Scholarpedia **4**(2), 1883 (2009)
180. Petrovic, S., Castro, E.: A genetic algorithm for radiotherapy pre-treatment scheduling. In: European Conference on the Applications of Evolutionary Computation, pp. 454–463. Springer (2011)
181. Pezzella, F., Morganti, G., Ciaschetti, G.: A genetic algorithm for the flexible job-shop scheduling problem. Comput. Oper. Res. **35**(10), 3202–3212 (2008)
182. Pickardt, C.W., Hildebrandt, T., Branke, J., Heger, J., Scholz-Reiter, B.: Evolutionary generation of dispatching rule sets for complex dynamic scheduling problems. Int. J. Prod. Econ. **145**(1), 67–77 (2013)

183. Pillay, N., Qu, R.: Hyper-Heuristics: Theory and Applications. Springer (2018)
184. Pinedo, M., Singer, M.: A shifting bottleneck heuristic for minimizing the total weighted tardiness in a job shop. Nav. Res. Logist. **46**(1), 1–17 (1999)
185. Pinedo, M.L.: Scheduling: Theory, Algorithms, and Systems, 3rd edn. Springer, New York (2008)
186. Pinedo, M.: Planning and Scheduling in Manufacturing and Services. Springer (2005)
187. Pinedo, M.: Scheduling, vol. 29. Springer (2012)
188. Poersch, J.M.: A new paradigm for learning language: connectionist artificial intelligence. Ling. Ensino **8**(1), 161–183 (2005)
189. Poli, R.: Evolution of graph-like programs with parallel distributed genetic programming. In: Proceedings of the International Conference on Genetic Algorithms, pp. 346–353 (1997)
190. Poli, R., Langdon, W.B., McPhee, N.F., Koza, J.R.: A Field Guide to Genetic Programming. Lulu.com (2008)
191. Poli, R., McPhee, N.F.: General schema theory for genetic programming with subtree-swapping crossover: Part I. Evol. Comput. **11**(1), 53–66 (2003)
192. Poli, R., McPhee, N.F.: General schema theory for genetic programming with subtree-swapping crossover: Part II. Evol. Comput. **11**(2), 169–206 (2003)
193. Potter, M.A., de Jong, K.A.: Cooperative coevolution: an architecture for evolving coadapted subcomponents. Evol. Comput. **8**(1), 1–29 (2000)
194. Rafter, J.A., Abell, M.L., Braselton, J.P.: Multiple comparison methods for means. SIAM Rev. **44**(2), 259–278 (2002)
195. Ragatz, G.L., Mabert, V.A.: A simulation analysis of due date assignment rules. J. Oper. Manag. **5**(1), 27–39 (1984)
196. Raghavjee, R., Pillay, N.: A genetic algorithm selection perturbative hyper-heuristic for solving the school timetabling problem. ORiON **31**(1), 39–60 (2015)
197. Rajendran, C., Holthaus, O.: A comparative study of dispatching rules in dynamic flowshops and jobshops. Eur. J. Oper. Res. **116**(1), 156–170 (1999)
198. Raquel, C., Yao, X.: Dynamic multi-objective optimization: a survey of the state-of-the-art. In: Evolutionary Computation for Dynamic Optimization Problems, pp. 85–106. Springer (2013)
199. Russell, S.J., Norvig, P.: Artificial Intelligence: A Modern Approach. Pearson Education (2010). http://vig.pearsoned.com/store/product/1,1207,store-12521_isbn-0136042597,00.html
200. Saidi, R., Bouaguel, W., Essoussi, N.: Hybrid feature selection method based on the genetic algorithm and Pearson correlation coefficient. In: Machine Learning Paradigms: Theory and Application, pp. 3–24. Springer (2019)
201. Sels, V., Gheysen, N., Vanhoucke, M.: A comparison of priority rules for the job shop scheduling problem under different flow time-and tardiness-related objective functions. Int. J. Prod. Res. **50**(15), 4255–4270 (2012)
202. Simon, F.Y.P., et al.: Integer linear programming neural networks for job-shop scheduling. In: Proceedings of the IEEE International Conference on Neural Networks, pp. 341–348. IEEE (1988)
203. Sotskov, Y.N., Shakhlevich, N.V.: NP-hardness of shop-scheduling problems with three jobs. Discret. Appl. Math. **59**(3), 237–266 (1995)
204. Stone, P., Veloso, M.: Learning to solve complex planning problems: finding useful auxiliary problems. In: Proceedings of the AAAI Fall Symposium on Planning and Learning (1994)
205. Sun, X., Gong, D., Jin, Y., Chen, S.: A new surrogate-assisted interactive genetic algorithm with weighted semisupervised learning. IEEE Trans. Cybern. **43**(2), 685–698 (2013)
206. Sun, Y., Yen, G.G., Mao, H., Yi, Z.: Manifold dimension reduction based clustering for multi-objective evolutionary algorithm. In: IEEE Congress on Evolutionary Computation, pp. 3785–3792. IEEE (2016)
207. Sun, Y., Yen, G.G., Yi, Z.: Evolving unsupervised deep neural networks for learning meaningful representations. IEEE Trans. Evol. Comput. **23**(1), 89–103 (2018)
208. Sutton, R.S., Barto, A.G.: Reinforcement Learning: An Introduction. MIT Press (2018)

209. Taillard, E.: Benchmarks for basic scheduling problems. Eur. J. Oper. Res. **64**(2), 278–285 (1993)
210. Tan, B., Ma, H., Mei, Y.: A hybrid genetic programming hyper-heuristic approach for online two-level resource allocation in container-based clouds. In: Proceedings of the Congress on Evolutionary Computation, pp. 2681–2688. IEEE (2019)
211. Tan, K.C., Yang, Y.J., Goh, C.K.: A distributed cooperative coevolutionary algorithm for multiobjective optimization. IEEE Trans. Evol. Comput. **10**(5), 527–549 (2006)
212. Tay, J.C., Ho, N.B.: Evolving dispatching rules using genetic programming for solving multiobjective flexible job-shop problems. Comput. Ind. Eng. **54**(3), 453–473 (2008)
213. Tay, J.C., Ho, N.B.: Evolving dispatching rules using genetic programming for solving multiobjective flexible job-shop problems. Comput. Ind. Eng. **54**, 453–473 (2008)
214. Teramoto, K., Morinaga, E., Wakamatsu, H., Arai, E.: A neighborhood limitation method for job-shop scheduling based on simulated annealing. Trans. Inst. Syst. Control Inf. Eng. **33**(6), 171–181 (2020)
215. Uy, N.Q., Hoai, N.X., O'Neill, M., McKay, R.I., López, E.G.: Semantically-based crossover in genetic programming: application to real-valued symbolic regression. Genet. Program Evolvable Mach. **12**(2), 91–119 (2011)
216. Uysal, A.K.: An improved global feature selection scheme for text classification. Expert Syst. Appl. **43**, 82–92 (2016)
217. Van Breedam, A.: Improvement heuristics for the vehicle routing problem based on simulated annealing. Eur. J. Oper. Res. **86**(3), 480–490 (1995)
218. Van Laarhoven, P.J., Aarts, E.H.: Simulated annealing. In: Simulated Annealing: Theory and Applications, pp. 7–15. Springer (1987)
219. van Veldhuizen, D.A., Lamont, G.B.: Multiobjective evolutionary algorithm test suites. In: Proceedings of the ACM Symposium on Applied Computing, pp. 351–357 (1999)
220. Vepsalainen, A.P., Morton, T.E.: Priority rules for job shops with weighted tardiness costs. Manag. Sci. **33**(8), 1035–1047 (1987)
221. Wang, S., Mei, Y., Zhang, M.: Novel ensemble genetic programming hyper-heuristics for uncertain capacitated arc routing problem. In: Proceedings of the Genetic and Evolutionary Computation Conference, pp. 1093–1101 (2019)
222. Wang, S.C.: Artificial neural network. In: Interdisciplinary Computing in Java Programming, pp. 81–100. Springer (2003)
223. Wang, Z., Tang, K., Yao, X.: Multi-objective approaches to optimal testing resource allocation in modular software systems. IEEE Trans. Reliab. **59**(3), 563–575 (2010)
224. Xiong, J., Xing, L.N., Chen, Y.W.: Robust scheduling for multi-objective flexible job-shop problems with random machine breakdowns. Int. J. Prod. Econ. **141**(1), 112–126 (2013)
225. Xu, B., Mei, Y., Wang, Y., Ji, Z., Zhang, M.: Genetic programming with delayed routing for multiobjective dynamic flexible job shop scheduling. Evol. Comput. **29**(1), 75–105 (2021)
226. Xu, R., Wunsch, D.: Clustering, vol. 10. Wiley (2008)
227. Xue, B., Zhang, M., Browne, W.N., Yao, X.: A survey on evolutionary computation approaches to feature selection. IEEE Trans. Evol. Comput. **20**(4), 606–626 (2016)
228. Yska, D., Mei, Y., Zhang, M.: Genetic programming hyper-heuristic with cooperative coevolution for dynamic flexible job shop scheduling. In: European Conference on Genetic Programming, pp. 306–321. Springer (2018)
229. Yuan, Y., Ong, Y.S., Feng, L., Qin, A.K., Gupta, A., Da, B., Zhang, Q., Tan, K.C., Jin, Y., Ishibuchi, H.: Evolutionary multitasking for multiobjective continuous optimization: benchmark problems, performance metrics and baseline results. arXiv:1706.02766 (2017)
230. Yuan, Y., Ong, Y.S., Gupta, A., Tan, P.S., Xu, H.: Evolutionary multitasking in permutation-based combinatorial optimization problems: realization with TSP, QAP, LOP, and JSP. In: Proceedings of the IEEE Region 10 Conference, pp. 3157–3164. IEEE (2016)
231. Zhang, F., Mei, Y., Nguyen, S., Tan, K.C., Zhang, M.: Multitask genetic programming based generative hyper-heuristics: a case study in dynamic scheduling. IEEE Trans. Cybern. (2021). https://doi.org/10.1109/TCYB20213065340

232. Zhang, F., Mei, Y., Nguyen, S., Zhang, M.: Genetic programming with adaptive search based on the frequency of features for dynamic flexible job shop scheduling. In: Proceedings of the European Conference on Evolutionary Computation in Combinatorial Optimization, pp. 214–230. Springer (2020)

233. Zhang, F., Mei, Y., Nguyen, S., Zhang, M.: Guided subtree selection for genetic operators in genetic programming for dynamic flexible job shop scheduling. In: Proceedings of the European Conference on Genetic Programming, pp. 262–278. Springer (2020)

234. Zhang, F., Mei, Y., Nguyen, S., Zhang, M.: A preliminary approach to evolutionary multi-tasking for dynamic flexible job shop scheduling via genetic programming. In: Proceedings of the Genetic and Evolutionary Computation Conference, pp. 107–108. ACM (2020)

235. Zhang, F., Mei, Y., Nguyen, S., Zhang, M.: Correlation coefficient based recombinative guidance for genetic programming hyper-heuristics in dynamic flexible job shop scheduling. IEEE Trans. Evol. Comput. **25**(3), 552–566 (2021)

236. Zhang, F., Mei, Y., Nguyen, S., Zhang, M.: Evolving scheduling heuristics via genetic programming with feature selection in dynamic flexible job shop scheduling. IEEE Trans. Cybern. **51**(4), 1797–1811 (2021)

237. Zhang, F., Mei, Y., Nguyen, S., Zhang, M.: Collaborative multi-fidelity based surrogate models for genetic programming in dynamic flexible job shop scheduling. IEEE Trans. Cybern. (2021). https://doi.org/10.1109/TCYB20213050141

238. Zhang, F., Mei, Y., Nguyen, S., Zhang, M., Tan, K.C.: Surrogate-assisted evolutionary multitask genetic programming for dynamic flexible job shop scheduling. IEEE Trans. Evol. Comput. (2021). https://doi.org/10.1109/TEVC2021306570

239. Zhang, F., Mei, Y., Zhang, M.: Genetic programming with multi-tree representation for dynamic flexible job shop scheduling. In: Proceedings of the Australasian Joint Conference on Artificial Intelligence, pp. 472–484. Springer (2018)

240. Zhang, F., Mei, Y., Zhang, M.: Surrogate-assisted genetic programming for dynamic flexible job shop scheduling. In: Proceedings of the Australasian Joint Conference on Artificial Intelligence, pp. 766–772. Springer (2018)

241. Zhang, F., Mei, Y., Zhang, M.: Can stochastic dispatching rules evolved by genetic programming hyper-heuristics help in dynamic flexible job shop scheduling? In: Proceedings of the IEEE Congress on Evolutionary Computation, pp. 41–48. IEEE (2019)

242. Zhang, F., Mei, Y., Zhang, M.: Evolving dispatching rules for multi-objective dynamic flexible job shop scheduling via genetic programming hyper-heuristics. In: Proceedings of the IEEE Congress on Evolutionary Computation, pp. 1366–1373. IEEE (2019)

243. Zhang, F., Mei, Y., Zhang, M.: A new representation in genetic programming for evolving dispatching rules for dynamic flexible job shop scheduling. In: Proceedings of the European Conference on Evolutionary Computation in Combinatorial Optimization, pp. 33–49. Springer (2019)

244. Zhang, F., Mei, Y., Zhang, M.: A two-stage genetic programming hyper-heuristic approach with feature selection for dynamic flexible job shop scheduling. In: Proceedings of the Genetic and Evolutionary Computation Conference, pp. 347–355. ACM (2019)

245. Zhang, M., Gao, X., Lou, W.: A new crossover operator in genetic programming for object classification. IEEE Trans. Syst., Man, Cybern., Part B **37**(5), 1332–1343 (2007)

246. Zhang, Q., Cao, M., Zhang, F., Liu, J., Li, X.: Effects of corporate social responsibility on customer satisfaction and organizational attractiveness: a signaling perspective. Bus. Ethics Eur. Rev. **29**(1), 20–34 (2020)

247. Zhang, X., Tian, Y., Jin, Y.: A knee point-driven evolutionary algorithm for many-objective optimization. IEEE Trans. Evol. Comput. **19**(6), 761–776 (2015)

248. Zhong, J., Feng, L., Cai, W., Ong, Y.S.: Multifactorial genetic programming for symbolic regression problems. IEEE Trans. Syst., Man, Cybern. Syst. (2018)

249. Zhou, H., Cheung, W., Leung, L.C.: Minimizing weighted tardiness of job-shop scheduling using a hybrid genetic algorithm. Eur. J. Oper. Res. **194**(3), 637–649 (2009)

250. Zhou, L., Feng, L., Tan, K.C., Zhong, J., Zhu, Z., Liu, K., Chen, C.: Toward adaptive knowledge transfer in multifactorial evolutionary computation. IEEE Trans. Cybern. (2020). https://doi.org/10.1109/TCYB20202974100

251. Zhou, Q., Jiang, P., Shao, X., Hu, J., Cao, L., Wan, L.: A variable fidelity information fusion method based on radial basis function. Adv. Eng. Inform. **32**, 26–39 (2017)
252. Zhou, Z.H., Yu, Y., Qian, C.: Evolutionary Learning: Advances in Theories and Algorithms. Springer (2019)
253. Zhu, X., Goldberg, A.B.: Introduction to Semi-Supervised Learning. Synthesis Lectures on Artificial Intelligence and Machine Learning **3**(1), 1–130 (2009)
254. Zitzler, E., Laumanns, M., Thiele, L.: SPEA2: Improving the strength Pareto evolutionary algorithm for multiobjective optimization. In: Evolutionary Methods for Design, Optimisation and Control with Application to Industrial Problems, pp. 95–100 (2002)
255. Zitzler, E., Thiele, L.: Multiobjective evolutionary algorithms: a comparative case study and the strength Pareto approach. IEEE Trans. Evol. Comput. **3**(4), 257–271 (1999)

Index

© The Editor(s) (if applicable) and The Author(s), under exclusive license to Springer
Nature Singapore Pte Ltd. 2021 F. Zhang et al., *Genetic Programming for Production
Scheduling*, Machine Learning: Foundations, Methodologies, and Applications,
https://doi.org/10.1007/978-981-16-4859-5

Printed in the United States
by Baker & Taylor Publisher Services